ELECTRONIC COMPUTERS Made Simple

The Made Simple series
has been created
especially for self-education
but can equally well
be used as
an aid to group study.
However complex the subject,
the reader is taken
step by step,
clearly and methodically
through the course. Each volume
has been prepared by
experts,
taking account of
modern educational requirements,
to ensure the most
effective way of
acquiring knowledge.

In the same series

- Accounting
- Acting and Stagecraft
- Additional Mathematics
- Administration in Business
- Advertising
- Anthropology
- Applied Economics
- Applied Mathematics
- Applied Mechanics
- Art Appreciation
- Art of Speaking
- Art of Writing
- Biology
- Book-keeping
- British Constitution
- Business and Administrative Organisation
- Business Economics
- Business Statistics and Accounting
- Calculus
- Chemistry
- Childcare
- Commerce
- Company Law
- Computer Programming
- Computers and Microprocessors
- Cookery
- Cost and Management Accounting
- Data Processing
- Dressmaking
- Economic History
- Economic and Social Geography
- Economics
- Effective Communication
- Electricity
- Electronic Computers
- Electronics
- English
- English Literature
- Export
- Financial Management
- French
- Geology
- German
- Housing, Tenancy and Planning Law
- Human Anatomy
- Human Biology
- Italian
- Journalism
- Latin
- Law
- Management
- Marketing
- Mathematics
- Metalwork
- Modern Biology
- Modern Electronics
- Modern European History
- Modern Mathematics
- Money and Banking
- Music
- New Mathematics
- Office Practice
- Organic Chemistry
- Personnel Management
- Philosophy
- Photography
- Physical Geography
- Physics
- Practical Typewriting
- Psychiatry
- Psychology
- Public Relations
- Rapid Reading
- Retailing
- Russian
- Salesmanship
- Secretarial Practice
- Social Services
- Sociology
- Spanish
- Statistics
- Teeline Shorthand
- Transport and Distribution
- Twentieth-Century British History
- Typing
- Woodwork

ELECTRONIC COMPUTERS
Made Simple

Henry Jacobowitz

Advisory Editor
Leslie Basford, BSc

Made Simple Books
HEINEMANN : London

© 1982 edition by William Heinemann Ltd

Made and printed in Great Britain
by Richard Clay (The Chaucer Press) Ltd, Bungay, Suffolk
for the publishers William Heinemann Ltd,
10 Upper Grosvenor Street, London W1X 9PA

First edition, April 1967
Reprinted, December 1968
Reprinted, February 1970
Reprinted, April 1971
Reprinted, July 1974
Reprinted, December 1976
Reprinted, July 1978
Reprinted, November 1980
Reprinted, June 1982

This book is sold subject to the
condition that it shall not, by
way of trade or otherwise, be lent,
re-sold, hired out, or otherwise
circulated without the publisher's
prior consent in any form of binding
or cover other than that in which it is
published and without a similar condition
including this condition being imposed
on the subsequent purchaser

SBN 434 98532 5

Foreword

It is just 40 years since ENIAC, the first of the electronic computers, reached full development. In this relatively short period, we have seen the computer evolve from a sophisticated laboratory apparatus to an essential tool in industry and commerce.

Simultaneously, the number of people whose lives are affected directly or indirectly by computers has grown from a very few to practically the whole population. Our electricity supply is beginning to be computer controlled; our bank accounts are computed electronically—and our cheques carry magnetic symbols for machines to read; and, within the foreseeable future, texts such as this may have their type set up by computer.

This book is dedicated to those people who are more than casually curious about this latest outgrowth of scientific culture, and to those who are planning to qualify themselves for some aspect of computer work. It has been thoroughly recommended for those interested in a non-rigorous treatment of the subject and who need an easily comprehensible book before a more specialized text is attempted. It is also particularly appropriate as a companion reader for students taking evening courses, studying at technical colleges, or working for recognized examinations. It is the primary recommendation for the City and Guilds course in Digital Computers—Advanced Telecommunications (No. 300/16).

It offers a comprehensive treatment of what has often been regarded as a complicated and 'advanced' subject. The reader who has a basic grasp of electricity and electronics, together with a genuine interest, should find here a solidly based insight into the 'hardware' of electronic computers. He will be equipped with a comprehensive reference, to which he may usefully return again and again. And if this volume whets rather than satiates his curiosity, he will be well prepared for going on to tackle more-specialized texts.

This book is unusual in that it aims to cover the basic aspects of both analogue and digital computers: perhaps a word of explanation is required. While it may appear over-ambitious to include two such widely differing techniques within the compass of a single volume, the developments of the past few years have confirmed the wisdom of the choice. Once austerely separate, analogue and digital techniques are beginning to intermingle freely and complement each other. The 'ideal' computer of the future will undoubtedly be a blend of the best of both techniques. Consequently, both types must be understood.

L. B.

C. G. C.

Table of Contents

	FOREWORD	v
1	COMPUTERS—WHAT THEY ARE AND WHAT THEY DO	1
	Brief History of Computing Machines	5
	Analogue and Digital Computing Principles Compared	11
	Review and Summary	14
2	INTRODUCTION TO THE ANALOGUE COMPUTER	16
	Classification of Analogue Computers	17
	A Few Mathematical Concepts and Analogies	20
	Review and Summary	34
3	BUILDING BLOCKS OF ANALOGUE COMPUTERS—I: MECHANICAL AND ELECTRICAL DEVICES	36
	Mechanical Computing Devices	36
	Electrical Computing Devices—Passive Networks	45
	Review and Summary	52
4	BUILDING BLOCKS OF ANALOGUE COMPUTERS—II: OPERATIONAL AMPLIFIERS	54
	Feedback in Amplifiers	55
	Computing with Operational Amplifiers	60
	Review and Summary	69
5	BUILDING BLOCKS OF ANALOGUE COMPUTERS—III: SERVOMECHANISMS AND FUNCTION GENERATORS	72
	Closed-loop (Servo) Control System	72
	Servo Function Generators	87
	Review and Summary	98
6	OPERATION OF COMPLETE ANALOGUE COMPUTERS	100
	Machine Equations and Scale Factors	100
	The Computer Block Diagram	106
	A Typical Analogue Computer Application: Aircraft Flight Simulation	110
	Review and Summary	115
7	INTRODUCTION TO THE DIGITAL COMPUTER	117
	Early Digital Computers	117
	Classification of Digital Computers	118
8	SURVEY OF NUMBER SYSTEMS	122
	The Unitary System	122
	The Decimal System	122
	The Binary Number System	123
	The Biquinary or 'Two–Five' System	132
	Octal Number Notation	133

	Hexadecimal Notation	136
	Binary-coded Decimal Notation	137
	Review and Summary	138
9	BUILDING BLOCKS OF DIGITAL COMPUTERS—I: COMPUTER LOGIC (BOOLEAN ALGEBRA)	140
	Fundamentals of Boolean Algebra	143
	Functions of Binary Variables	144
	Functions of Three or More Variables	159
	Laws of Rearrangement	164
	Review and Summary	171
10	BUILDING BLOCKS OF DIGITAL COMPUTERS—II: ELECTRONIC DEVICES	173
	Electronic Devices for Performing Arithmetic and Logic Operations	174
	Logic Chains for Arithmetic	197
	Review and Summary	208
11	BUILDING BLOCKS OF DIGITAL COMPUTERS—III: MAGNETIC AND OTHER DEVICES	211
	Review of Magnetism	211
	Magnetic Core Logic	218
	Delay-Line Storage	225
	Cryotrons	228
	Review and Summary	230
12	THE COMPLETE DIGITAL COMPUTER—I: OPERATION OF COMPUTER MEMORY AND ARITHMETIC UNIT	233
	System Operation	233
	The Storage of Information in the Computer Memory	236
	Address Selection	241
	Operation of the Arithmetic Unit	245
	Algebraic Addition and Subtraction	251
13	THE COMPLETE DIGITAL COMPUTER—II: COMPUTER PROGRAMMING, CONTROL, AND COMMUNICATION	260
	Program Selection and Computer Co-ordination by Control Unit	260
	Selection of Instructions	263
	Techniques of Programming	272
	Communication with the Computer through the Input–Output Units	276
14	PRESENT TRENDS AND FUTURE PROSPECTS	288
	Computer Process Control	288
	Computers and Man	299
	New Techniques and Components	308
	APPENDIX: GLOSSARY OF COMPUTER TERMS	314
	INDEX	326

CHAPTER 1

COMPUTERS—WHAT THEY ARE AND WHAT THEY DO

Formerly, when a mathematician or philosopher wanted to define a 'thing' he took into account all its possible and actual properties and then formulated his definition of the thing in terms of those properties and attributes. The trouble with such absolute definitions was that other philosophers and scientists would often disagree as to exactly what the defining properties of the thing really were, and often it was found later that no such thing existed anywhere. Scientific method has become much more rigorous. Now, when a scientist defines, he considers the operations the thing performs, or he must cause it to perform, and the sum total of these operations then constitutes an 'operational definition'.

What Computers Are and What They Are Not

We can play this game of definitions with the amazing instrument called a 'computer'. We could define modern automatic computers by listing the ideal properties they should have. Thus, we could say that computers are an extension of man's thinking, so-called 'giant brains' that will make man 'obsolete'; or we could say, more humbly, that computers are merely tools or instruments which carry out the instructions given them by men, that they can calculate faster than any man, but can do essentially nothing qualitatively different than any mathematically trained person can do—much slower. We could consider the argument as to whether computers can think or not, and whether they have other human attributes such as capacity for learning, feeling, reproducing themselves, and so on. This would undoubtedly involve us in secondary definitions, as to just what constitutes thinking, learning, feeling, etc., and we might get hopelessly ensnarled, in the manner of some philosophers.

To avoid this fruitless endeavour, let us turn to an 'operational' definition: computers are what they can do. Thus, we must look at some of the things computers can do, and have been doing for quite a while. From this list of things we can distil the attributes and qualities that make up modern computers, and, hence, what they are. We might also explore some of the historical sources of modern electronic computers, and classify the various types available at the present time. Then, for the greater part of this book, we shall study the elements and building blocks that make up the various types of computers. Much later on we shall explore the overall operation of the 'systems' of elements that constitute automatic electronic computers. Finally, after understanding the essence of present-day computer development, we will be able to look into the future, at possible combinations of computer classes and extensions of operations now being developed, as well as at fascinating new components and processes. We may then be able to redefine the computer in terms of future potentialities, and, in the process, possibly answer the question as to whether computers can think.

What Computers Do

Here are some of the operations present-day automatic computers perform:

Solve complicated mathematical equations.

Compute the behaviour of nuclear reactors.
Predict the performance of an aircraft or missile still on the drawing board.
Prepare electricity bills and make out pay slips.
Control the flight of rockets and guided missiles.
Predict the weather.
Simulate the flight behaviour of giant bombers—on the ground.
Calculate the most efficient assignment of troops.
Control the processing of petroleum and chemicals.
Attempt to forecast economic conditions and predict election results.
Maintain control of inventories.
Automatically operate machine tools in a predetermined way.
Control the smooth flow of supplies with minimum waste.
Solve problems of physics and engineering design.
Determine which products can be marketed most profitably.
Explore the relations of biological and physiological processes.
Navigate submarines under ice, and satellites into their orbits.
Compute and print a table of mathematical functions in a few minutes.
List and store the status of seat reservations throughout an airline.
Run entire factories automatically.

Notes Towards a Definition

Since computers are what they do, let us analyse this list. Even from this brief listing it is evident that computer applications encompass the entire range of reasonable human activities. We put the stress on 'reasonable', since all the applications listed have one thing in common—they are eminently reasonable and logical in nature; or stated in another way, they are free of human emotions and value judgements. Every one of the listed tasks can be and is being carried out by human beings, though evidently much more slowly and inefficiently. Some of these, such as the solution of complicated differential equations, are so difficult and laborious that they are rarely attempted by mathematicians, thus slowing scientific progress. Others, like weather and economic forecasting, involve such voluminous and rapidly changing data that even batteries of highly trained statisticians are too slow to complete the task successfully in the brief time available.

In addition to being reasonable or 'calculable', all computer applications involve meaningful symbols and data; that is, the information they process conveys the same meaning to everyone. This may appear to be a small point, but note that it immediately eliminates all areas of opinion and value judgements, such as the comparative beauty of a painting or the moral value of an action. Thus, computers perform logical operations on meaningful data and information; they do not choose between values and opinions. The term 'meaningful' is not synonymous with 'important', 'true', or 'significant' in this context, but must be interpreted quite narrowly as something that means the same thing and is understood in the same way by everyone. Thus, the logical operations performed by computers do not question the factual truth or significance of the initial data, but explore only their implications. For example, consider the statements (premises):

1. Nicotine in cigarettes causes lung cancer.
2. Cigarette filters eliminate nicotine.

The implication in these two statements is obvious, and you do not need a computer to arrive at the conclusion:

3. Smoking filtered cigarettes does not cause lung cancer.

This conclusion is reasonable and computable (logically valid), though it may be that neither of the two starting premises is factually true or even significant. The answer or conclusion simply follows from the starting premises by the rules of logic. As another example, consider the following statements from physics and geometry:

1. Light travels in straight lines through empty space.
2. Two parallel, straight lines will meet at infinity.

Hence:

3. Two parallel light rays travelling through empty space will meet at infinity.

The conclusion (3) follows logically from the physical and geometrical premises (1 and 2, respectively), and it is factually correct *if* these premises are correct. Given these premises *and* the rules of logic, either a computer or a reasonable human being must arrive at the same conclusion. As a matter of fact, both premises have been challenged in recent times, and their denials form the base of the general theory of relativity and non-Euclidian geometry, respectively.

To compute the answer to a problem, a computer must be given both the data of the problem and built-in rules of logic or mathematics for solving the problem. These are exactly the same conditions as those required for a human computer. Consider, for example, a simple arithmetic problem, multiplying 432 by 23:

$$\begin{array}{r} 432 \\ \times 23 \\ \hline 1296 \\ 864 \\ \hline \end{array}$$

Carry (1)

Ans: 9936

To solve this problem, you first computed the partial product, $3 \times 432 = 1296$, by 'remembering' the multiplication table; you then obtained the second partial product, $2 \times 432 = 864$, in the same way and placed it underneath the first product, 'shifted' one place to the left. Finally, you added the two partial products together, remembering to 'carry 1', when the sum was greater than '9'. A computer solves the problem in essentially the same manner. Given the two numbers and the instruction to 'multiply', the computer will perform the same operations you have carried out on paper: using its built-in rules for multiplication, it will first obtain the partial products by 'remembering' the multiplication table contained in its 'memory store'. It will then add the partial products, using the rule for shifting and 'remembering' to 'carry 1' whenever necessary. The answer is printed, punched on tape, or read on an indicator.

Definition of an Automatic Computer

We have found out so far that computers perform logical (mathematical) operations on given information without questioning the factual truth of the starting information or the validity of the operations. This is true for *all* computers, whether mechanical, electrical, or electronic—even for the abacus or strings of beads used thousands of years ago. The one important quality that puts the modern computer in a class by itself is its automatic operation. Since a computer is able to operate by itself, it can perform a long sequence of related operations with the original information without the need for a human operator to transfer partial results, give additional instructions, and collect the answers.

In an automatic computer all instructions necessary to solve a complicated problem or a sequence of problems are originally written out as a 'program' and translated into machine language for the continuing guidance and control of the computer. The operator needs only to press the 'start' button. Using its own source of energy, the computer then 'reads off' the data of the problem and the instructions and automatically proceeds to solve the problem, drawing upon its built-in rules of logic and mathematics and storing any partial results in its 'memory' until they are needed. The final answers are then typed or punched on paper, displayed on an indicator, plotted on graph paper in the form of curves, or 'read out' in a number of alternative ways. Moreover, if so instructed by the 'program', the computer can adjust itself to changing circumstances during the course of the problem. It is this flexibility of control and the completely automatic operation that make modern computers so completely different from anything invented in the past, and apparently capable of almost human behaviour.

We now have the final ingredient towards the definition of modern computers: automatic operation in accordance with a predetermined 'program' of instructions. The 'operational definition' we have arrived at inductively, therefore, should read:

A computer automatically performs logical (mathematical) operations on input information and puts out answers, according to a predetermined 'program' of instructions.

Do Computers 'Think'?

From our list of applications and the basic definition above, it certainly appears that automatic computers display much of the characteristic behaviour associated with human thinking. Computers perform logical operations, such as comparing and choosing between alternatives, matching up equals, selecting the next instruction to be carried out, etc. Computers also perform mathematical operations, such as counting, adding, subtracting, multiplying, dividing, computing powers and roots, logarithms, trigonometric functions, integrals, derivatives, and what have you. Finally, computers exhibit such specifically 'human' attributes as remembering, making logical decisions, and adjusting themselves to changed circumstances, and some of the latest machines even appear to profit from past experience and seem to be capable of 'learning'. But before equating this complex behaviour with human thinking, let us recall that computers, up to now, lack all critical judgement and capacity for 'creative' thinking. They blindly follow the program of instructions and their built-in rules of logic. As we have seen, computers never question the truth or significance of the intial data (problem) or the validity of the instructions and built-in rules of operations; they simply compute the implications of the given data and instructions according to fixed rules. These data, rules, and instructions are always conceived of by human beings. Even the most fundamental axioms and rules of logic and mathematics are not 'self-evident', but rather are adopted by human consensus or convention. The most significant developments in science and mathematics have usually been the result of questioning these very axioms and rules, previously thought to be self-evident. No computer can do this, nor can it use past accumulated knowledge to arrive creatively at new possibilities and inventions. Perhaps, then, as long as computers cannot duplicate these highest forms of human thinking, they will remain the robots we have always thought them to be.

BRIEF HISTORY OF COMPUTING MACHINES

The earliest computing device undoubtedly consisted of the five fingers of each hand, and this is still the preferred device of every child who learns to count. Since there are ten discrete fingers (digits) available for counting, both digital computation and the decimal system have enjoyed a huge popularity throughout history. However, improvements were made to replace the digits of the hand by a more reliable 'count-10' device.

From Pebbles and Beads to the Abacus

It probably did not take more than a few million years of human evolution before someone had the idea that pebbles could be used just as well as fingers to count things. Thus, ten pebbles or ten of anything were kept in a handy container to represent the numbers 1 to 10, instead of the ten fingers. The form the pebble container should take for handy calculations kept many of the best minds of the Stone Age busy for centuries. It was not till about five thousand years ago in the Tigris–Euphrates Valley (and as late as 460 B.C. in Egypt) that

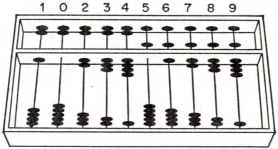

Fig. 1. The Japanese abacus (soroban), showing number set-up.

there arose the idea of arranging a clay board with a number of grooves into which the pebbles were placed. By sliding the pebbles along the grooves from one side of the board to the other, counting became almost 'semi-automatic'; even to the point of allowing one hand to be kept free for other performances.

The grooved pebble container was too big a thing to be kept secret for long, and the processes of cultural diffusion (e.g. deported slaves) saw to it that it became known in China, Japan, and Rome. When the diversity of these races were confronted with this leap into the future, a flowering of ingenuity—a sort of minor renaissance—resulted, which swept the pebble computer to a high plateau of development. One group came up with the idea of drilling holes in the pebbles and stringing the resulting beads in groups of ten on a frame of wire; another used reeds instead. In either case, the beads could be moved easily and rapidly along the wire or reeds, and a tremendous speed-up in calculations resulted. This device, in somewhat more sophisticated form, became known as the abacus in China and the soroban in Japan (Fig. 1). Simple as it looks, the abacus is an amazingly versatile device for performing all arithmetic operations, and in the hands of a skilled operator it can be as fast as desk-type computers. The abacus or soroban is used to this day by Japanese tradesmen and Chinese laundrymen throughout the world.

From 'Napier's Bones' to the Slide Rule

After reaching this first milestone, the development of computing devices seems to have stagnated for the next two thousand years, there having been, apparently, few scientific and business calculating needs during the Middle Ages that required more than ten fingers or the abacus. One man who tried to change this state was Gerbert, who learned about Arabic numerals and a calculating machine based upon them from the Moors. For years he tried unsuccessfully to introduce both in medieval Europe.

The real beginning of modern computers goes back to the seventeenth century, from which we date our 'modern era' in just about every field of endeavour. Having divorced themselves from all past speculations and authorities, such intellectual giants as Descartes, Pascal, Leibniz, and Napier made a new beginning in philosophy, science, and mathematics, which was to revolutionize the ancient view of the world. In mathematics, particularly, such tremendous progress was made, and the attendant calculations became so laborious, that the need for more sophisticated computing machines became urgent.

The development of logarithms by John Napier in 1614, and their conversion to the base 10 by Henry Briggs in 1615, stimulated the invention of various devices that substituted the addition of logarithms for multiplication. One such device, invented by John Napier in 1617, was a mechanical arrangement of numbering rods, which could do multiplication. These became known later, fittingly, as 'Napier's bones'. A slide rule without moving parts, based upon Napier's logarithms, was invented in 1620 by Edmund Gunter. This was improved upon by the introduction of a sliding scale by William Oughtred in 1632. He gave it the name 'astrolabe' because of its astronomical uses. The astrolabe was the true forerunner of the modern slide rule and nomogram. After this auspicious beginning the pace of progress became increasingly rapid, and we can present only a few highlights of the evolution that led to the development of mechanical computing machines.

Development of the Desk Calculator

Perhaps most significant in the evolution of the mechanical calculators was the introduction, in 1642, of the 'toothed wheels' (gears) by Blaise Pascal, the famous French philosopher and mathematician. Although limited to addition and subtraction, the toothed counting wheel is still used in adding machines, and it may be of interest to look for a moment at its simple operating principle. As shown in Fig. 2, several wheels with teeth numbered from 0 to 9

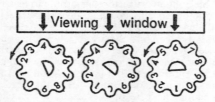

Fig. 2. The number '456' represented by three toothed wheels.

are arranged in a row representing ones, tens, hundreds, thousands, etc. The number that is to be operated upon is represented by the tooth that faces the index above each wheel (a viewing window in an actual machine). For example, the number 456 is represented by the position of the three toothed wheels in Fig. 2. If you wanted to add 111 to this number, you would simply turn each wheel by one tooth (or notch), so that the teeth indicating 5, 6, and 7 would face the index or viewing windows. The result of the addition of 456 and 111, or 567, then appears at the viewing window. One of Pascal's early desk calculators, using the toothed-wheel principle, is illustrated in Fig. 3.

It was not long before scientists realized that Pascal's toothed wheels could also perform multiplication by repeated addition of a number. The German philosopher and mathematician, Baron von Leibniz, added this improvement

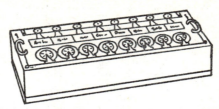

Fig. 3. Early desk calculator invented in 1642 by Blaise Pascal.

to the Pascal machine in 1671, but did not complete his first calculating machine until 1694. The Leibniz 'reckoning machine' (Fig. 4) was the first two-motion calculator designed to multiply by repeated addition, but mechanical flaws prevented it from becoming popular.

As with Leibniz's machine, so it was with many others over the next hundred years: the ideas were good, but the execution was not. In the age of one-millionth-inch tolerances it is perhaps hard to conceive that the mechanics of

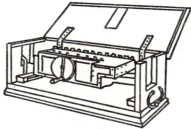

Fig. 4. Reckoning machine, invented by Baron von Leibniz in 1694, used repeated addition by tooth wheels to accomplish multiplication.

that day were unable to produce simple mechanisms using gears and cranks that operated reliably. It was not until 1820 that Thomas de Colmar improved Pascal's calculator sufficiently to make it practicable for multiplication. Over the next sixty years Thomas made some 1500 six-place multiplying machines. The machine is still made today in Paris.

Babbage's Folly

The Newton of the computer field was Charles Babbage, a professor of mathematics at Cambridge University, who set out to build an automatic computer long before a practical adding machine was available. It was in 1812 that Babbage first conceived the idea of building a machine that could solve differential equations and print out the answers. He worked on his 'difference engine' with the help of the British Government for some twenty years, but finally gave up in 1842. The difference engine was the type of computer that would nowadays be called an 'accumulator mechanism'.

Although equally unsuccessful in practice, the ideas behind Babbage's next project—the 'analytical engine'—proved to be the seeds for the development of large-scale modern digital computers. In 1833 Babbage conceived a computer that would be largely automatic and, for his day, would operate extremely rapidly. To perform arithmetic, the machine still used Pascal's toothed wheels, but the operation was to be sufficiently rapid to complete one addition per second (compared to approximately a million additions per second in the most advanced modern machines). To achieve this speed, Babbage had to overcome a main stumbling block, namely, the slow entry of the data by a human operator laboriously transferring numbers from a 'work sheet'. Somehow the work sheet as well as the human operator had to become part of the machine, so that it could operate automatically. This idea, as conceived by Babbage, is the basis of all automatic computing.

Babbage formulated this grand idea as follows: he would divide his machine into three parts: the store, the mill, and the control. The store would hold all the data that was needed during the computation of a problem. Known as 'memory storage' in present-day digital computers, this part would take the place of the inefficient and frequently unreadable work sheet. The data was to be stored in the form of holes punched into cards, an invention that was finally realized some sixty years later by Herman Hollerith. The mill was to be the calculating part of his analytical engine, i.e. the decimal tooth wheels that would operate on the data made available by the punched cards. Finally, the human operator would be replaced by an automatic operator, the control. Babbage even conceived the idea of having a separate store (memory) for the instructions (program) that would tell the machine when to add or subtract, multiply or divide. Babbage's analytical engine never worked; it was too far ahead of its time.

Further Developments

Though Babbage's ideas did not come into their own until a century later, other workers in the field improved the existing mechanical calculators considerably, so that many of the basic nineteenth-century designs are still in use

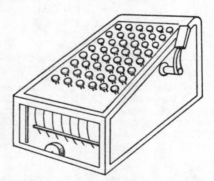

Fig. 5. Adding machine introduced by Burroughs in 1885.

today. Dorr Felt patented a key-driven adding machine in 1850 and developed a practical machine in 1886. In 1887 a patent was issued for an improved machine, which is the forerunner of present-day comptometers, but which was dubbed a 'macaroni box' in Felt's own day. At about the same time William Seward Burroughs produced one of the first commercial adding machines, a rough illustration of which is shown in Fig. 5. Many improvements were made upon these early designs in succeeding years, the printing feature (in 1889) being perhaps the most important. Figs. 6 and 7 show early models of the

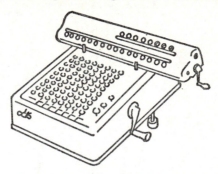

Fig. 6. Early model of Monroe Calculating Machine.

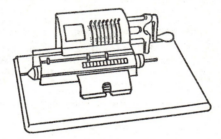

Fig. 7. Early model of Marchant Calculating Machine.

Monroe and Marchant calculating machines, which were introduced in 1911. Electric motors were incorporated into calculating machines by about 1920.

Something Different: The Analogue Principle

All the mechanical calculators we have considered up to this point are essentially extensions of the ten-finger (digital) counting system. Working with discrete numbers, which represent digits, they are known as digital machines. Although we will have much more to say about digital machines later on, we must now recognize another principle of computing: computing by measuring or analogy. The differences between the digital and analogue principles will become much clearer in the next section, but perhaps some glimmer of the essential idea will come through in the following description of some early analogue devices.

It is a simple mathematical problem to compute the area bounded by a regular plane figure or curve. If the curve is irregular and not described by a simple mathematical function, the problem becomes more difficult and is subject to numerical approximations. This problem was first tackled, in 1814, by a Bavarian engineer, J. A. Hermann, who invented a mechanical device for measuring the area under a curve. Now called a planimeter, the device measured the area under a curve by passing a tracer along the boundary line of the curve. Many such devices appeared in succeeding years, with the forerunner of the modern polar planimeter (Fig. 8) first being brought out by Amsler in

1854. Although a planimeter only measures the area of a plane figure, it is a true computing device. This becomes evident when it is realized that the area bounded by a curve is the value of the definite integral of the mathematical function represented by the curve.

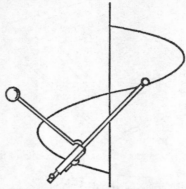

Fig. 8. Polar planimeter for measuring the area of a plane figure.

Another analogue device for integrating mathematical functions was invented by James Thomson, the brother of Lord Kelvin. As the illustration (Fig. 9) shows, this is the forerunner of the modern ball-and-disc integrator, which we shall study in detail in a later chapter. Again, a simple mechanical motion that is not based upon counting, but is analogous to the function to be computed, is utilized. In 1876 Lord Kelvin conceived the idea of connecting a

Fig. 9. Sketch of Thomson's 'ball-and-disc' integrator.

number of these mechanical integrators together to solve differential equations. Based upon this principle, Lord Kelvin, in 1878, built a 'harmonic synthesizer' designed to predict tides. Note how the analogue or measuring principle, while still using simple mechanical devices, permits computations of a very high level of mathematical complexity.

In the early part of the twentieth century both the analogue and (a little later) the digital computing devices received tremendous impetus, and many new machines—automatic to varying degrees—appeared. Before we can discuss these more recent developments, which will take up the remainder of the book, we must clarify the fundamental differences between the analogue and digital computing principles.

ANALOGUE AND DIGITAL COMPUTING PRINCIPLES COMPARED

Suppose you wanted to check the claim that there are 'forty-three beans in a cup' of a certain coffee. You could satisfy your sceptical curiosity in two basically different ways. The simplest way would be to count the number of beans in a cup filled with beans of that coffee. (We shall disregard for the moment that the advertiser may have had something else in mind.) Since you would undoubtedly use your fingers for counting out the coffee beans, it should be evident that this is the 'digital' or counting method of computation. The second way would be to determine the average weight of a coffee bean (by weighing a few, of course); then weigh a cup filled with beans, subtract the weight of the cup to obtain the net weight of a cup of beans, and finally, divide the net weight by the average weight of a coffee bean. The result will be the number of beans in a cup of coffee. This second method is called the 'analogue' principle of computation, since it makes use of a measuring (weighing) process that varies analogously with the number of coffee beans; the assumption is made here that the weight of the coffee beans increases proportionally to their number. The essential difference between digital and analogue computation is therefore one between counting discrete objects (numbers) and measuring continuous data.

Here is another simple problem: how to determine the number of toy blocks packed tightly in a box. Given this problem, almost any child would use the digital method; that is, he would pour the blocks on the floor and count them piece by piece. Say he arrives at the figure 1728 as the number of blocks in the box. Had he given this problem to his college-trained father, the latter would probably have used the analogue method of computation, which is considerably faster in this case. He would have used a ruler or tape measure and determined the dimensions of the box and those of a single block. Say, the box turns out to be 1 ft wide, 2 ft long, and 6 in high, while a single block measures out as a 1-in cube. Having completed the analogue computation, the father would then complete the problem on paper by multiplying the dimensions to determine the number of cubic inches in the box: i.e. 12 in (wide) × 24 in (long) × 6 in (high) = 1728 in^3, and since each block is 1 in^3 in volume, there are 1728 blocks in the box. He might even have solved the problem mentally in cubic feet by remembering that there are 1728 in^3 in a cubic foot (that is, 1 ft × 2 ft × $\frac{1}{2}$ ft = 1 ft^3 or 1728 in^3).

In this particular problem we have used the cubic dimensions as the physical quantity that varies analogously to the number of blocks. We could, of course, also have used the weight of the box and that of a block to determine the analogue computation, as shown in the previous example. A somewhat more interesting example of analogue and digital computation is the motor-car speedometer. This device is coupled by gears and a flexible shaft to the car's main drive shaft. The speed of the car, indicated by the speedometer pointer, is measured by an analogous physical quantity, namely the rate of rotation of the main drive shaft. Since speed is the mathematical 'derivative' of the distance travelled (with respect to time), the analogous measuring process consists of differentiation, in this case. Clearly, the speedometer is an analogue computer. To complicate things, however, the speedometer also contains a mileage counter that indicates the distance the car has travelled. The mileage counter is also operated by the flexible shaft coupled to the main drive shaft, and hence the distance travelled by the car is measured again by an analogue device. To assure that the $\frac{1}{10}$-mile dial of the mileage indicator travels only a fraction of an inch when the car traverses $\frac{1}{10}$ mile, reducing gears are inserted between the input shaft rotation and the mileage dial. The gear reduction is a form of analogue division. Whenever the $\frac{1}{10}$-mile dial completes a full rotation the 1-mile dial moves by one position or digit. In effect, the 1-mile dial counts or adds the number of turns (or miles) completed by the $\frac{1}{10}$-mile dial. Similarly, whenever the 1-mile dial completes one rotation (indicating that the car has travelled 10 miles) the 10-mile dial moves by one digit, thus counting the number of turns of 10 miles each completed by the 1-mile dial. The indication of the 100-mile, 1000-mile, and 10,000 mile dials is accomplished in the same manner. Since each dial computes the distance travelled by counting or adding the number of turns completed by the $\frac{1}{10}$-mile dial, and always indicates precisely a single, discrete digit, the mileage indicator output, evidently, is a form of digital computer.

We now have a simple principle for distinguishing between digital and analogue computing devices, depending upon whether the computation is performed by a numerical counting of discrete data (digital) or by continuous measurement of a physical quantity analogous to the numbers in the problem under consideration (analogue). Thus, we can immediately identify the abacus and the various adding machines and desk calculators as counting or digital devices, while the slide rule, planimeter, ball-and-disc integrator, etc., are clearly measuring or analogue devices. In addition to this fundamental differentiating principle, there are other important distinguishing marks between the two types of computation. Some of these follow directly from the fundamental distinction, while others will be pointed out later on.

Digital Characteristics

Since digital computation essentially consists of the counting or adding of discrete items (such as gear teeth, holes punched in paper, or electrical pulses), all mathematical operations of even the most complex problem must be broken down into counting or adding. For example, subtraction is accomplished in a digital computer by adding the complement of the number (1000 minus the number). Multiplication is usually carried out by repeated addition, division by repeated subtraction. Even powers and roots, integration (a form of addition) and differentiation, etc., are all carried out by conversion to arithmetic, specifically, addition. Digital computation, therefore, is controlled arithmetic. No matter how difficult the problem, it must be analysed and broken down into arithmetical steps, and all simple steps must then be completed in sequence until the problem is solved (a process known as sequential or serial operation). Evidently, the simple arithmetic operations, and the devices that perform them, have little resemblance to the physical problem under consideration, but the method is essentially the same as if you were to carry out all calculations on paper in the simplest arithmetical terms. Naturally, if you are going to solve a difficult problem, say in calculus, in terms of addition and subtraction, you will have to perform a staggering amount of arithmetic that, easily, might take years to complete. The digital computer must carry out the same amount of labour, but it has a vast number of basically simple devices for performing the arithmetic operations and a 'memory' for storing numbers and partial results. Moreover, in present-day computers all these arithmetic devices are electronic, which means that they can carry out an arithmetic operation in about a millionth of a second. Thus, time is no object.

The digital nature of computation and the great number of devices needed has additional consequences, both good and bad. Obviously, you can always add another (identical) arithmetic device, such as an adder, to retain another digit. Hence, the accuracy of a digital computer is essentially unlimited or absolute, just as counting by your fingers. On the other hand, the use of a fantastic number of elementary devices to handle complex problems and achieve the required accuracy raises the cost of a digital computer to a formidable figure. Despite this inherently high cost, the easy and rapid manner in which a digital computer can handle and store vast masses of data makes it ideally suited to complex numerical and statistical problems in business and science, where very high accuracy is required. We shall devote the entire second part of the book to its operation.

Analogue Characteristics

An analogue computer sets up a mathematical analogy to the problem in question. The computing device may be mechanical (such as a planimeter or ball-and-disc integrator), electrical (a resistor network), electromechanical (a motor), or electronic (an amplifier), but regardless of its form it must represent a quantity of the problem continuously and in a mathematically analogous manner. If the problem is simple, a single analogue device may represent the entire problem. Thus, the varying rate of rotation of a shaft may continuously represent the speed of a car, as we have seen, while the total amount (angle) the shaft has rotated can represent the total distance (miles) travelled. A more complex problem can be represented by a more sophisticated analogous device. Evidently, even a very difficult problem, such as simulating the behaviour of an aircraft, can be represented by relatively few highly sophisticated analogue devices, which solve various portions of the problem (such as air speed, elevation, angle of attack, etc.) at the same time. (This is known as parallel operation.) Fewer devices, even if relatively complicated, means lower cost and less trouble in preparing a problem (programming) for the computer. On the other hand, the fact that each device measures some physical or mathematical quantity creates a physical limitation to the accuracy of measurement, which is usually no more than one part in ten thousand in an analogue computer. A considerable advantage of the analogue computer is that it actually represents a physical problem or system, and hence is capable of giving the designer genuine insight into the behaviour of that system under varying conditions. Analogue computers are, therefore, best suited to serve as models for and simulate some physical system having varying stimuli.

Analogue *versus* Digital Comparison Chart

Now that we have become familiar with some basic analogue and digital characteristics, we can set up the following comparison chart for future reference:

Analogue Computer	Digital Computer
Sets up analogy of problem.	Breaks down problem into arithmetic.
Represents physical variable by continuous measurement of analogous quantity (shaft rotation, voltage).	Represents numbers by discrete, coded pattern (digital data), such as perforations in card or presence of pulses.
Basic operation performed by relatively few 'single-purpose' devices (integrators, multipliers, summers, resolvers, etc.).	Operations performed by relatively many interchangeable arithmetic devices (adders, registers, accumulators etc.).
Relatively few devices needed; hence comparatively low cost and ease of programming.	Many devices needed; hence, high cost and difficult programming.
Distinct elements used for each operation (parallel channels).	Identical elements used in sequence (primarily series operation).

Analogue Computer	Digital Computer
Accuracy limited to about 1 part in 10^4.	Unlimited accuracy (to 1 part in 10^{12} or more).
Data storage (memory) dispersed in various non-interchangeable devices.	Data storage concentrated in space, interchangeable and unlimited in duration.
Analogue computer serves as model and 'mirrors' relations of actual system; operations usually carried out in actual (real) time of physical system.	Digital computer compounds arithmetic data, unrelated to system it represents. Time of operations usually does not correspond to 'real' time.
Represents physical or mathematical quantities.	Can represent numbers, as well as letters and other symbols.
Best suited to represent measurable quantities and simulate response of physical systems by mathematical analogies.	Best suited to handle discrete random processes, statistical data, and numerical problems of business and scientific nature.

Hybrid Computers

There is another category of computers, which should be mentioned in passing. These are the hybrid computers, which make use of both analogue and digital components and techniques. The car speedometer may be considered a hybrid, although it does not employ true digital techniques. The digital differential analyser is a true hybrid, since it uses digital circuits and techniques in a machine that is organized like, and fulfils the purpose of an analogue computer. The most recent analogue computers use digital memory storage for intermediate results or repetitive problems; other hybrids exist. To use both types of computing techniques for solving a problem, 'analogue-to-digital' and 'digital-to-analogue' converters are required, which will make analogue data palatable to a digital computer, and vice versa. Since there has been a considerable swing towards using the best of both techniques in the most recent computers, we shall study hybrid types and converters in the last section of the book.

REVIEW AND SUMMARY

Computers are neither 'giant brains' nor merely tools, but they must be operationally defined in terms of what they can do.

Any computer performs reasonable (logical, mathematical) operations on meaningful information. A modern (automatic) computer automatically performs logical (mathematical) operations on input information and puts out answers, according to a predetermined program of instructions.

Computers do not question the factual truth or significance of the starting (input) data, but explore only their implications. Computers are not (yet) capable of inductive or creative thinking.

There are two main classes of computers: digital and analogue. Digital computers operate by numerical counting (adding) of discrete data, using the method of controlled arithmetic. Analogue computers set up an analogy of the problem; they operate by continuously measuring a physical quantity analogous to the numbers in the problem under consideration.

Analogue computers serve as models for and simulate the relations of an actual physical system, usually operating in real time, while digital computers compound primarily arithmetic data unrelated to the system represented.

Analogue computers employ relatively few distinct, single-purpose devices in parallel-channel operation; hence, cost is relatively low and programming is easy; digital computers employ many interchangeable arithmetic devices in usually sequential operation; hence, cost is relatively high and programming is complex.

The accuracy of analogue computers is limited to the accuracy of measurement; accuracy of digital computers is essentially unlimited.

Analogue computers are best suited for simulating the response of physical systems, while digital computers are best suited for handling numerical problems, statistical data, and discrete random processes. Digital computers can be set up, however, to serve as mathematical models of physical systems.

Hybrid computers make use of both analogue and digital techniques. They employ 'analogue-to-digital' and 'digital-to-analogue' converters for transforming the data into suitable form for either type of computation. The digital differential analyser is one type of hybrid, using digital components in an analogue organization.

CHAPTER 2

INTRODUCTION TO THE ANALOGUE COMPUTER

We have already met some of the early analogue computing devices, such as the slide rule, the planimeter, and the ball-and-disc integrator. You will also recall the early use (in 1878) of a complete mechanical analogue computer, employing Thomson's integrators, the 'harmonic synthesizer' for predicting tides. The first appearance of electrical analogue computers was in the 1920s, when the General Electric Company and Westinghouse both invented analogue machines for simulating the behaviour of power networks. These were known as d.c. network analysers, and they made use of a multiplicity of resistors hooked up in a fashion analogous to the actual power networks. This permitted the exploration of a large-scale power flow and loading on the small scale of the computer. A much more versatile machine, the a.c. network analyser, was introduced in 1929. This computer realistically simulated the performance of alternating-current power networks, showing both the magnitude and phase of the currents. The machine occupies a large-sized room.

The Differential Analyser

Meanwhile continual improvements were being made on the mechanical analogue computing devices, such as the ball-and-disc integrator. The torque output of the integrator was sufficiently increased by Hannibal Ford to permit its use in a naval gunfire computer during the First World War. These mechanical improvements culminated in the development at Massachusetts Institute of Technology of a large-scale completely mechanical 'differential analyser' by Dr. Vannevar Bush in 1931. Dr. Bush, following in the footsteps of Lord Kelvin's brilliant example, connected together a variety of mechanical devices that could add, multiply, integrate, differentiate, etc. The entire assembly was capable of solving complex differential equations arising out of physical problems. We have already seen how the rotation of a shaft can represent distance travelled and velocity, and how a planimeter or ball-and-disc device can integrate a function. Dr. Bush added many more devices, such as gear-boxes (for multiplication and division), differential gears (for addition and subtraction), curve tracers, and devices that could generate mathematical functions.

In the mid-1930s Hartree and Porter at Manchester University devised a model mechanical differential analyser. This was the basis of the 'Metadyne', a full-size analyser built by Metropolitan-Vickers (now part of Associated Electrical Industries).

Although these differential analysers worked very well in solving the differential equations of a particular problem, it was troublesome to change to another problem. All the mechanical interconnexions between the various analogue devices had to be changed to represent the mathematical relations of the new problem. To avoid this difficulty, the M.I.T. staff developed a more advanced differential analyser, in which all interconnexions could be made by electrical means. This machine, first brought out in 1942 amid great secrecy, solved important military problems during the Second World War and is still being used for advanced work. Although in principle it is still a mechanical analogue computer, the second M.I.T. differential analyser uses about 200 miles of wire, 3000 relays, 150 electric motors, and 2000 electronic valves. For the mathematical work itself, the machine utilizes about 130 rotating shafts, 28 gearboxes, 16 adders, 18 integrators, and three 'function tables'.

The 'Operational Amplifier'

In addition to the versatile mechanical differential analyser, more specialized mechanical analogue computers appeared in the 1930s, chiefly for equation solving and harmonic analysis. Among these were a mechanical device for solving simultaneous algebraic equations, known as 'Wilbur's mechanism'. Several mechanical harmonic analysers also made their debut. It was, however, an electronic development that really put the large-scale analogue computer on the map. The work of George A. Philbrick in the late 1930s and the independent research by Lovell at the Bell Telephone Laboratories led to the introduction of the 'operational amplifier' during the Second World War. We shall study the operational amplifier in great detail in a later chapter, but it was this device more than anything else that made possible accurate and stable analogue computations of any kind by purely electronic means. The present-day large-scale, general-purpose analogue computers are directly based upon it. But before delving more deeply into these developments we must clarify the various classes of analogue computers.

CLASSIFICATION OF ANALOGUE COMPUTERS

The chart overleaf (Fig. 10) illustrates the variety of devices embraced by the general category of analogue computers. As a matter of interest, the same chart can be drawn for digital computers.

Analogue computers may be divided into general-purpose and fixed-purpose machines, depending upon whether they provide generally valid mathematical solutions useful for multiple purposes or solve the problem inherent in a specific situation only. General-purpose and fixed-purpose analogue computers are also categorized as 'indirect' and 'direct' types, respectively; the latter terms, however, are not in exact correspondence with the former. Indirect analogue computers solve algebraic and differential equations of the linear and non-linear type. These equations may represent physical systems of various kinds, but the main point is that a general-purpose indirect computer may be set up to meet any type of problem situation that can be expressed in mathematical terms. In contrast, the direct type of analogue computer sets up a direct analogy to the behaviour, form, and parameters of the problem, such as is represented by a miniature aeroplane, for example.

General-purpose (Indirect) Analogue Computers

Let us consider for a moment the general-purpose (indirect) analogue computers. As shown in the chart, these may be broken down into primarily mechanical, fluid, or electrical types, though rarer categories (the pneumatic, for example) also exist. We are already somewhat familiar with the mechanical types. These range all the way from the simple slide rule, where lengths on a stick are analogous to the logarithms of numbers, to the sophisticated mechanical differential analyser, which can take complex systems of differential equations in its stride. Various types of linkages and nomograms also belong in this category.

The fluid type of general-purpose computer is rather rare. The electrolytic tank is the main representative of this group. In this device two or more electrodes are inserted into a tank filled with a conductive liquid (electrolyte) and electricity is applied to the electrodes. Depending on the configuration of the tank and electrodes, certain potential fields are created which 'simulate' known types of differential equations. By moving 'probes' through the tank the potential field can be explored and, hence, a solution of the equation represented by

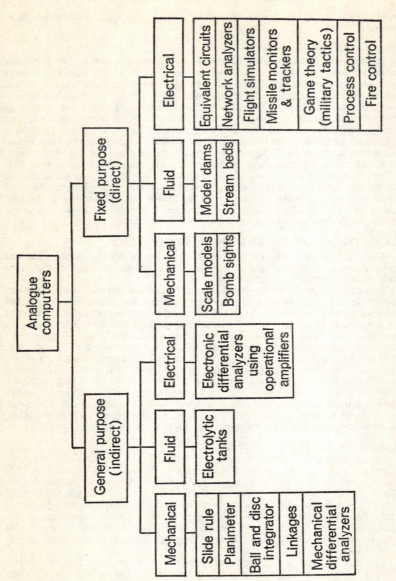

Fig. 10. Classification of analogue computers.

it can be obtained. The electrolytic tank also can be used as a fixed-purpose direct analogue device, to simulate the potential field inside an electron tube, for example, where the exact equation representing the field is unknown or not easily soluble.

The electrical type of general-purpose analogue computer would perhaps better be called 'electronic', since the electronic differential analyser using operational amplifiers is its primary representative. This is the most flexible and useful kind of general-purpose analogue computer, since its electronic and electrical components can easily be interconnected to perform almost any type of mathematical operation. A major portion of this part of the book is devoted to its study, since most probably it will be the only surviving type of general-purpose analogue computer. Figs. 11 and 12 illustrate two table-top

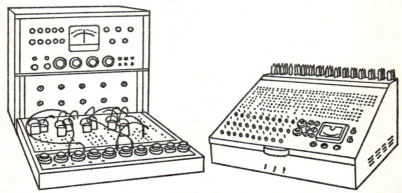

Fig. 11. Sketch of Donner 3400 Desk-top General Purpose Analogue Computer (from photograph courtesy of Donner Scientific Co.).

Fig. 12. Sketch of Heathkit General-purpose Analogue Computer (from photograph courtesy of Heath Co.).

general-purpose electronic analogue computers, each costing a few hundred pounds. The one shown in Fig. 12 is a do-it-yourself kit using fifteen operational amplifiers and various other components, which might come in handy to demonstrate the analogue portions of this book. Available large-scale analogue installations are practically unlimited in complexity, size, and cost.

Fixed-purpose (Direct) Analogue Computers

Like the general-purpose (indirect) computers, the fixed-purpose (direct) analogue computers may also be broken down into mechanical, fluid, and electrical categories. The mechanical types are, perhaps, the easiest to comprehend, since they are represented by a great variety of scale models. The behaviour of a tiny experimental aeroplane in a wind tunnel tells the designer all he needs to know about the aerodynamic performance of its full-size counterpart. Not every model will serve as a scale model, however. A true scale model preserves the physical form and structure of its prototype, while its physical dimensions are scaled down to convenient size. Fairly elaborate dimensional analysis is required in properly scaling down a machine or device. Even an electrical-transmission system can be reduced to miniature size.

Mechanical bombsights fall into the category of fixed-purpose analogue computers, though they are not generally of the direct type. The famous

Norden bombsight used during the Second World War, for example, automatically solves the equations involved in causing a bomb to hit its target and does not set up a direct analogy. It is, therefore, an indirect type of analogue computer.

Fluid fixed-purpose computers are primarily scale models, such as table-top model dams and stream beds. The same considerations as described above for the mechanical scale models apply also to the fluid types. As previously mentioned, we could also include the electrolytic field tanks when they are used to represent an analogy to a specific problem.

The number of electrical and electronic fixed-purpose analogue computers is unlimited. A great number of small electronic companies manufacture specialized electrical analogue computers. We have already mentioned the d.c. and a.c. network analysers, which simulate electrical-power-distribution systems by simple resistance and impedance networks. A little later we shall deal with equivalent circuits, which directly represent the mathematical relations prevailing in a mechanical, hydraulic, pneumatic, or acoustic device by a simple electric circuit. By measuring the current flowing, for a given impressed voltage, through various parts of an equivalent circuit, the equivalent parameters in the problem device can be determined.

Most fixed-purpose analogue computers are of the electronic type, using operational amplifiers, but in contrast to electronic differential analysers, they are tailored to do a specific job. Some Second World War veterans are familiar with the fire-control analogue computers which helped to defeat the German V-1 missiles. (They were not effective against the faster V-2 missiles, however.) Present-day special-purpose computers are miniaturized by the use of light-weight 'solid-state' devices, such as transistors. The Polaris missile, for example, has a solid-state analogue 'think' device which monitors its flight performance. If the missile does not achieve sufficient velocity in the initial portion of its flight the analogue computer 'aborts' the flight. The missile gets the 'green light' only if it performs as programmed.

In addition to guiding flight, fixed-purpose analogue computers can simulate the flight of even the most complicated jet bomber. Every characteristic of the bomber, such as lift and drag, flutter and stall, etc., is exactly duplicated on the ground by a giant analogue computer. Analogue 'missile trackers' can use radar information to provide a continuous plot of the flight of a missile. Other fixed-purpose analogue computers can utilize principles of game theory to solve problems in military tactics and strategy, do operations research, assist in the proper blending of petrol, and control a variety of industrial and chemical processes.

A FEW MATHEMATICAL CONCEPTS AND ANALOGIES

In the next few pages we shall indulge in some higher mathematics and physics. If you wish, you may skip this portion for the time being and not lose too much in the long run. If you decide to struggle through it, however, you will gain genuine insight into the manner in which physical problems are formulated mathematically through differential equations and how apparently different problems are mathematically analogous. You will then appreciate how an analogue computer can solve a variety of problems in different fields by setting up a single mathematical analogy.

Distance, Velocity, and Acceleration

First, we shall review some elementary ideas. You will recall from the discussion of the car speedometer that speed is the mathematical 'derivative' of the distance travelled with respect to time. This murky statement was offered

Introduction to the Analogue Computer   21

as an explanation of how the same rotating shaft could represent both speed and distance travelled. Let us see how this comes about.

As every schoolboy knows, speed is the distance covered in a certain time, usually expressed as miles per hour or feet per second. We can express this familiar relationship by the simple mathematical equation:

$$\text{Speed} = \frac{\text{Distance covered}}{\text{Time required}} = \frac{\text{Miles}}{\text{Hours}} = \frac{\text{Feet}}{\text{Seconds}}$$

Using the symbols v for velocity (a more scientific term than speed), s for the distance covered, and t for the time required, we obtain speed or velocity, $v = \frac{s}{t}$, which is the familiar classroom formula. You may have thought this to be the last word on the subject, and in some respects it really is. For instance, if a 'body' always covers equal distances in equal intervals of time it is said to be moving with uniform velocity, and this velocity is given by the ratio s/t. Thus, if a satellite always covers 5000 miles of its orbit in 15 minutes its speed is $\frac{5000 \text{ miles}}{\frac{1}{4} \text{ hour}} = 20{,}000$ miles per hour.

This is fine as far as it goes. If you travelled to your week-end cottage 120 miles away in 2 hours you would report this feat by saying that you travelled at an average speed of 60 miles per hour. You're very careful to add the term 'average', since you're aware that you had to do 70 miles per hour (as shown on the speedometer) most of the time to compensate for unavoidable slow-downs and stops. In other words, even though you obviously travelled with non-uniform velocity the formula $v = s/t$ is still valid, provided you divide the total distance covered by the total time elapsed to obtain the average velocity. If you wanted to know your velocity at any time, rather than the average over a period of two hours, you could still use the same formula, provided you shortened the time interval sufficiently. For example, if you wanted to determine the speed of your car between 3.45 and 3.46 p.m. you could throw out a marker from your car at 3.45 and one at 3.46, then come back later and measure the distance covered in that minute. If you covered $1\frac{1}{2}$ miles in that minute your speed was $1\cdot 5/1 = 1\cdot 5$ mile/min or 90 mile/h.

You may suspect that your speed wasn't uniform even in that single minute, since you recall that halfway through this brief span, at 3.45.30 p.m. to be exact, a mosquito bit you in the neck, whereupon you wildly accelerated for about 10 seconds and then almost stopped to nurse your wounds. Thus, you decide to shorten the time interval to 1 second. Now, however, you run into considerable experimental difficulties and discover that you need elaborate equipment to determine the exact moment to throw out the markers and measure the distance correctly. Moreover, it occurs to you that even 1 second is a pretty long time for the speed to remain constant, and what you're really after is the speed at the instant when you're passing a particular point. This instantaneous speed appears to be what the car's speedometer is indicating at any time.

We could modify the previous formula to express the speed at any instant. Thus, the Instantaneous speed $= \frac{\text{Infinitesimal distance covered}}{\text{Infinitesimal time required}}$ where by 'infinitesimal' we mean smaller than anything that can be measured. Again using the symbol v to represent instantaneous speed (or velocity) and ds and dt to represent infinitesimal distance and time, respectively, we can write in mathematical shorthand

$$\text{Instantaneous speed, } v = \frac{ds}{dt}$$

The expression ds/dt may be read 'the (instantaneous) rate of change of distance with respect to time', or 'the derivative of distance with respect to time'.

In order to give some significance to the rather abstract concept of 'instantaneous rate of change' or 'derivative' and find one way of computing it, let us look at a graphic plot of a memorable week-end trip (Fig. 13). The dramatic events happening between 3 and 5 p.m. on Friday afternoon are all recorded there. We see at once that the average speed for the entire trip was indeed 60 mile/h, since a total distance of 120 miles was covered in 2 h. It also appears that the trip took place in three main 'laps'. The first half-hour from 3.00 to 3.30 p.m. was relatively slow owing to city traffic. This is indicated by the shallow slope of the graph. The next hour's driving, from 3.30 to 4.30 p.m., took place on a motorway at considerably increased speed, as indicated by the steeper average slope of the graph. A number of untoward events took place during this portion, which we shall explore in more detail later on. The last lap, from 4.30 to 5.00 p.m., was over a country road to the cottage, and, apparently, at somewhat slower average speed.

Let us see what the average speed was during each of the three laps. We will use our old standby, $v = s/t$, but since only portions of the total distance and time are involved, we shall write

$$\text{Average speed } v = \frac{\Delta s}{\Delta t}$$

where Δs stands for the 'change in distance' and Δt for the 'change in time'. (Note that these symbols are not the same as ds and dt, which stand for infinitesimal changes in distance and time, respectively.) Applying this relation to the average speed during city driving in the first half hour, we measure time along the horizontal axis to the point where the motorway starts. The length along the horizontal axis (abscissa) is, of course, Δt, which equals 30 min ($\frac{1}{2}$ h) in this case. The distance along the vertical axis (ordinate) is Δs, which equals 20 miles. Hence, for the first lap,

$$v = \frac{\Delta s}{\Delta t} = \frac{20 \text{ mile}}{30 \text{ min}} = \frac{20 \text{ mile}}{\frac{1}{2} \text{ h}} = 40 \text{ mile/h}.$$

Since we have divided the vertical distance (Δs) by the horizontal length (Δt), it is evident that $\frac{\Delta s}{\Delta t}$ ($= 40$ mile/h) also measures the average rise or slope of the graph from 3.00 to 3.30 p.m.; that is, a straight line drawn from '0' (zero time and distance) to the point where the motorway starts (at 3.30 p.m., 20 miles out) would have a slope of 40 (in the scale of mile/h). We see, therefore, that the average speed during any particular time interval is measured by the average slope of the time–distance graph.

Using this same reasoning for the next lap on the motorway, we see from the graph that 75 miles were covered during the hour between 3.30 and 4.30 p.m., and hence,

$$\text{Average Speed} = \frac{\Delta s}{\Delta t} = \frac{75}{1} = 75 \text{ mile/h}.$$

This, then, is also numerically equal to the average slope of the graph during that period. During the final lap on the country road the change in time, $\Delta t = 30$ min or $\frac{1}{2}$ h, and change in distance, $\Delta s = 25$ miles. Hence, during this interval,

$$\text{Average speed} = \frac{\Delta s}{\Delta t} = \frac{25 \text{ mile}}{0\cdot 5 \text{ h}} = 50 \text{ mile/h}$$

which determines also the average slope of the graph.

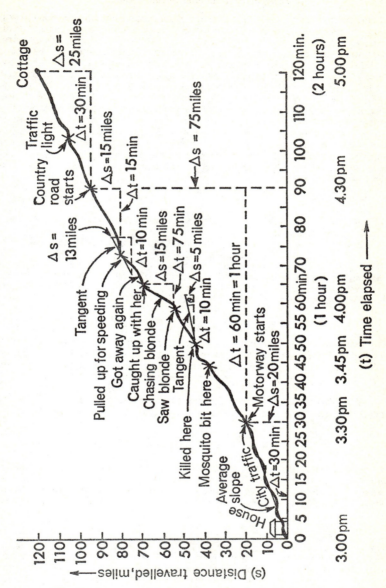

Fig. 13. Average and instantaneous speeds during car trip.

The average speed of 75 mile/h during the second lap on the motorway is not very informative, since many things happened to change the speed temporarily, as indicated on the graph. You will recall that a mosquito bit the driver at about 3.45 p.m., whereupon he rapidly accelerated and then slowed down, until he killed it at about 3.48 p.m. Suppose you wanted to know the instantaneous speed of the car at the moment the mosquito was killed. The problem could easily be solved if you could draw a line at the point that would have the same average rise or slope as the graph of the speed at that instant. This is accomplished by drawing a line tangent to the point of the graph where the mosquito was killed at 3.48 p.m. (You will recall from elementary geometry that only one line can be drawn tangent to a point on a curve.) To find the numerical value of the tangent line, we measure its slope during a convenient interval. Thus, we find from the graph that the tangent rises by 1 division or 5 miles (Δs) during a horizontal interval of 4 divisions of 10 min (Δt). Hence, expressed in mile/min,

$$\text{Slope of tangent line} = \frac{\Delta s}{\Delta t} = \frac{5 \text{ mile}}{10 \text{ min}} = 0.5 \text{ mile/min or 30 mile/h.}$$

(The actual slope is smaller, of course, since the vertical scale is only one-half of the horizontal scale.)

The value of the slope (30 mile/h) is also the desired instantaneous speed of the car at the moment the mosquito was killed.

You will recall from elementary trigonometry that the tangent of an acute angle in a right triangle is defined by the ratio of the opposite side to the side adjacent to the angle. Hence, the tangent of the angle (θ) formed by the tangent line with the horizontal in Fig. 13 is the ratio

$$\frac{\Delta s}{\Delta t} = \frac{1 \text{ graph square}}{4 \text{ graph squares}} = \tfrac{1}{4} \text{ or } 0.25$$

This corresponds to an angle, $\theta = 14°\ 2'$, or about 14 degrees. (The same vertical and horizontal scale must be used, of course, to define an angle.)

We now have a handy way to determine the speed at any instant by drawing a line tangent to the point on the graph for the desired instant. The slope of that tangent line is (numerically) the instantaneous speed at that moment. We can apply this method to other situations during the trip. At about 3.58 p.m. the driver accelerated (for reasons explained in Fig. 13) to the incredible speed of 120 mile/h.

$$\frac{\Delta s}{\Delta t} = \frac{15 \text{ mile}}{7.5 \text{ min}} = 2 \text{ mile/min or 120 mile/h}$$

A police car caught up with him at about 4.13 p.m. and gave him a deserved warning about speeding. What was his speed at the instant he was caught? We draw a line tangent to the point at 4.13 p.m. (73 min elapsed) and 80 miles out. The slope of this tangent line, in mile/min, is given by

$$\frac{\Delta s}{\Delta t} = \frac{13 \text{ mile}}{10 \text{ min}} = 1.3 \text{ mile/min or } 1.3 \times 60 = 78 \text{ mile/h}$$

Thus, the instantaneous speed of the car at the time he was caught, corresponding to the slope of the tangent, was 78 mile/h, which appears to have been about 18 mile/h over the legal speed limit. This assumption is confirmed by the subsequent leisurely ride from 4.15 (75 min) to 4.30 p.m. (90 min), during which time the car covered a distance of 15 miles (from 80 to 95 miles). Hence, his speed during this quarter hour was

$$\frac{\Delta s}{\Delta t} = \frac{15 \text{ mile}}{15 \text{ min}} = 1 \text{ mile/min, or 60 mile/h}$$

Meaning of 'Derivative'

You will recall that we undertook the discussion of that eventful journey in order to give concrete meaning to the term 'derivative' or 'instantaneous rate of change'. We had defined the instantaneous speed, $v = \dfrac{ds}{dt}$, where $\dfrac{ds}{dt}$ stood for the instantaneous rate of change or 'derivative' of distance (s) with respect to time (t). We now know how to obtain the instantaneous rate of change, or derivative, by graphical means. We simply draw a graph of the distance covered against the time elapsed, and whenever we want to know the instantaneous speed we draw a line tangent to the point in question. The slope of the tangent of that line will give us the instantaneous rate of change (derivative) of distance with respect to time and, hence the instantaneous speed. Moreover, this method holds true for any quantity (say, x) that varies as a function of time [written $f(t)$]; hence, we can define the derivative (instantaneous rate of change) of a variable, $x = f(t)$, at any point as the slope of the tangent of that function at that point. Mathematically, this can be written as follows:

If $x = f(t)$, then the instantaneous rate of change or derivative of x with respect to t $\dfrac{dx}{dt} = f'(t) = $ slope of *tangent* of $f(t)$ at any point. (The term $f'(t)$ is mathematical shorthand for 'derivative of a function of time'.) Though we have used the letter 't' to denote time, it could stand for any independent variable, i.e. any quantity not under our control. The method of drawing a tangent to a point on the graph of a function to obtain the derivative is useful primarily when the exact mathematical function is unknown. When the mathematical expression is known direct methods, known as differentiation, exist for evaluating the derivative at any time without the need for plotting the function or drawing tangents.

Since most physical quantities in nature vary non-uniformly with time, the instantaneous rates of change or derivatives of these quantities are of great importance. Usually the rates of change or derivatives are known and expressed in the form of differential equations.

What About Acceleration?

Whenever a driver wants to increase the speed of his car he steps on the 'accelerator'; acceleration, thus, is a change of velocity. If a car travels along at 30 mile/h and then increases its speed to 40 mile/h during 1 s its acceleration is 10 mile/h during that second. If it keeps increasing its speed by 10 mile/h every second thereafter it is said to have a uniform acceleration of 10 mile/h per second. Obviously, it can't keep accelerating at this uniform rate indefinitely, since within 7 s after it started to accelerate it will have reached a speed of 100 mile/h (from the initial speed of 30 mile/h). Thus, eventually, it will have to slow down. If it takes, say, another 3 s to slow down to a velocity of 80 mile/h and the car travels at this speed thereafter it will have taken a total of 10 s to effect a net change in velocity of 80−30 or 50 mile/h. We can say therefore, that the average acceleration (change in velocity) during that 10-s interval was 50 mile/h/10 s = 5 mile/h/s. This is true even though the acceleration during the first 7 s was 10 mile/h/s.

Hence, we can define average acceleration during any interval as the change in velocity divided by the change in time. Expressed as an equation, during any interval the

$$\text{Average acceleration} = \frac{\text{Change in velocity}}{\text{Change in time}} = \frac{\Delta v}{\Delta t}$$

where Δv and Δt stand for 'change in' velocity and time, respectively.

We are rarely interested, however, in 'average' acceleration, and the acceleration is practically never uniform. To find the value of a continually changing acceleration at any time we must shorten the time interval Δt more and more in order to obtain an average value for successively briefer periods. Since this is insufficient to determine the acceleration at a desired instant of time, we are finally driven to the definition of the

$$\text{Instantaneous acceleration} = \frac{\text{Infinitesimal change in velocity}}{\text{Infinitesimal change in time}}$$

Using the symbol a for acceleration and d for an 'infinitesimal change' (as before), we can translate this into the mathematical expression:

$$\text{Instantaneous acceleration, } a = \frac{dv}{dt}$$

which is read 'acceleration is the (instantaneous) rate of change, or the "derivative", of velocity with respect to time'. This can be shortened to 'acceleration is the time rate of change, or the derivative, of the velocity'.

We already have a definition of the instantaneous velocity, $v = \frac{ds}{dt}$, which we can substitute into the definition of acceleration. Doing this, we obtain acceleration,

$$a = \frac{dv}{dt} = \frac{d(ds/dt)}{dt}, \text{ which is written } \frac{d^2s}{dt^2}$$

In words, the acceleration is the 'derivative of the derivative' of the distance with respect to time. Instead of 'derivative of derivative' the term second derivative is used. In general, we can say, for a function $x = f(t)$ and its graph, the first derivative of x with respect to t,

$$f'(t) = \frac{dx}{dt}$$

and the second derivative of x with respect to t,

$$f''(t) = \frac{d^2x}{dt^2}$$

where the value of $f''(t)$ at any point (of the curve) is the slope of the tangent of $f'(t)$ at that point.

In practice, the value of the second derivative of a function $f(t)$ is found mathematically by differentiating the first derivative $f'(t)$; that is, by differentiating the function twice. Conversely, if the second derivative of a mathematical function is known the original function can be obtained by summing, or 'integrating', the function twice. Thus, the integral of the second derivative, $f''(t)$, is the first derivative, $f'(t)$, and the integral of the first derivative is the original function, $f(t)$. For example, the differential equation

$$a = \frac{d^2s}{dt^2} = \frac{dv}{dt}$$

(where the symbols are as before), can be rewritten: $a\,dt = dv$.

Integrating both sides of this equation (and using the symbol $\int$ for the integral),

$$\int a\,dt = \int dv$$

Performing the mathematical process of integration, this comes out $a\,t = v - v_0$, where v_0 is a constant of integration that represents the initial velocity at the beginning of the interval in question. Solving for the acceleration (a), we obtain

$$a = \frac{v - v_o}{t}$$

which is, of course, the same as the change in velocity divided by the time interval, t. Solving the expression above for the velocity:

$$v = v_o + at$$

which states that the final velocity, v, is the sum of the original velocity, v_o, and the product of the acceleration, a, times the interval, t.

Substituting $v = \frac{ds}{dt}$ in this expression, we can write

$$v = \frac{ds}{dt} = v_o + at$$

and multiplying through by dt:

$$ds = v_o\, dt + at\, dt$$

Again integrating both sides, as indicated by the symbol $\int$:

$$\int ds = \int v_o\, dt + \int at\, dt$$

When the indicated integrations are performed (which is very simple in this case) the expression for the distance covered

$$s = v_o t + \tfrac{1}{2}at^2$$

is obtained. This shows that the distance (s) covered by a body initially travelling at velocity v_o and then accelerating during a time interval t is the product of the initial velocity and the time, plus one-half of the product of the acceleration times the square of the time. If the body is initially at rest ($v_o = 0$) the distance covered during the accelerated motion is proportional to the square of the time interval, (i.e. $s = \tfrac{1}{2}at^2$).

The present excursion into the realm of mathematics, besides clarifying the relations between distance, velocity, and acceleration, should have given you some idea of what is involved in the concepts of 'rates of change' or 'derivative', and the formulation of a simple differential equation. Let us now apply these concepts to the solution of two analogous problems in the fields of mechanics and electricity.

A Problem in Mechanics

Consider the following simple mechanical problem. A weight or mass (m) is suspended from a spring of a certain stiffness, as is illustrated in Fig. 14.

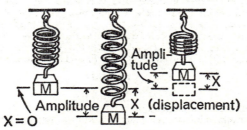

Fig. 14. A simple mechanical spring pendulum.

The weight is then pulled down with a certain applied force (F), quickly released, and is left free to oscillate thereafter. We would like to write the equation of motion of the weight, determine its frequency of oscillation, and, if possible, the displacement of the weight at any time (t).

Potential Energy. Let us analyse the forces involved in the motion of the spring pendulum. When the spring is extended by pulling the weight down, a certain amount of energy is expended, which is stored as potential energy in the tension of the spring. Disregarding friction for the moment, the potential energy stored in the spring must equal the energy originally expended in stretching (or compressing) the spring. Moreover, in most springs the force (F_1) that tends to restore the spring is proportional to the elongation or displacement, x (Fig. 14), of the spring from the normal (resting) position. Mathematically this restoring force (F_1) is expressed by

$$F_1 = -Kx$$

where K is a constant of proportionality known as the spring constant and the minus sign indicates that the displacement and the restoring force act in opposite directions.

Kinetic Energy. When the weight is let go, the potential energy stored in the spring is suddenly released and converted into energy of motion, called kinetic energy. During the motion the weight undergoes an acceleration, a, which depends on the inertia or mass, m, of the weight and upon the originally applied force, F. According to Sir Isaac Newton's famous second law of motion, formulated almost three centuries ago, the relationship between the force (F) acting on the mass (m) and the acceleration (a) attained by it is simply

$$a = \frac{F}{m} \text{ or } F = ma$$

that is, the acceleration is proportional to the force applied and inversely proportional to the mass. Since acceleration is the second time derivative of the distance, or the displacement x in this case, we can state Newton's law as follows:

$$F = ma = m\frac{d^2x}{dt^2}$$

where d^2x/dt^2 is the acceleration expressed as the second derivative of the displacement, x, with respect to time, t.

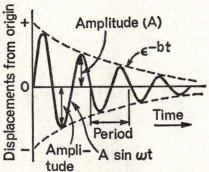

Fig. 15. Plot of equation in text, showing waveform of damped oscillation of a spring pendulum.

Force of Friction. After the spring has pulled the mass back to its original position ($x = 0$) the action doesn't stop there, because of the inertia or flywheel effect of the mass. The mass (weight) resists any sudden change in its motion; hence, it does not stop when the spring is slack again, but on the contrary continues to compress the spring until all the kinetic energy acquired during the motion of the weight is again stored as potential energy in the compressed spring. Now the spring releases its tension again, converting it into the energy of motion of the weight, and the process is repeated all over again. The action would continue indefinitely if it were not for the fact that some energy is lost during each oscillation because of the friction of the spring and support. Primarily due to the viscous friction of the spring, the maximum displacement (called amplitude) of the spring and weight is a little smaller during each oscillation, so that the oscillations finally die out. An oscillating motion, where the originally imparted energy is used up in the friction of the system, is known as a damped oscillation (see Fig. 15). If the weight were pulled during each downward swing with just sufficient energy to overcome the internal friction, an undamped oscillation would result (Fig. 16).

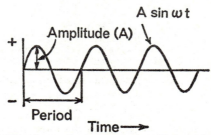

Fig. 16. Waveform of an undamped oscillation (sine wave).

It can be shown that the damping force (F_2) due to the viscous friction is proportional to the velocity (v) of motion of the spring. That is, the damping force due to friction is

$$F_2 = -\mu v$$

where μ is a constant of proportionality called the friction coefficient and the minus sign indicates that the damping force (F_2) opposes the motion. Again, remembering that velocity is the time rate of change, or time derivative, of the distance (displacement x), we can write damping force due to friction,

$$F_2 = -\mu v = -\mu \frac{dx}{dt}$$

Differential Equation of Motion. We have now expressed all the forces operating in the simple spring pendulum as a function of time. Since the applied force (F) unit equals the (algebraic) sum of the restoring and damping forces, we can equate $F = F_1 + F_2$, and, substituting the expressions derived above as a function of time, we obtain

$$F = m \frac{d^2x}{dt^2} = -Kx - \mu \frac{dx}{dt}$$

Finally, transposing the terms to the left side, we obtain the equation of motion:

$$m \frac{d^2x}{dt^2} + \mu \frac{dx}{dt} + Kx = 0$$

To the physicist, this latter expression describes the entire situation of the spring pendulum, including the forces acting and the motion resulting. It is known as an ordinary differential equation of the second order, because a second derivative is included, and it is easily solved for x by standard mathematical procedures. When this is done, the simplified expression for the displacement (x) as a function of time looks like this:

$$x = \varepsilon^{-bt} (A \sin \omega t)$$

Here the first term, ε^{-bt}, is known as an exponential function of time, since it involves the exponential ε ($\varepsilon = 2\cdot7183$, approximately). Because the exponential is raised to a negative power ($-bt$), the function is decreasing or decaying at a rate defined by the constant b.

The second term, $A \sin \omega t$, is the standard expression of a sine wave, which describes an undamped oscillation of constant amplitude, A (see Fig. 16). The symbol 'ω' stands for 2π times the frequency (f) of the oscillation; i.e. $\omega = 2\pi f$ ($\pi = 3\cdot14$, approximately). In the complete solution for x (not shown) the frequency of the oscillation is given approximately by the mathematical expression

$$\omega = 2\pi f = \sqrt{\frac{K}{m}} \text{ or } f = \frac{1}{2\pi}\sqrt{\frac{K}{m}}$$

In words, the frequency of oscillation is directly proportional to the square root of the ratio of the spring constant, K, to the mass, m.

When the exponential and sine-wave terms of the solution are put together, as shown above, the characteristic equation of a damped oscillation results, whose wave form is plotted in Fig. 15. Note that the amplitude (A) of the oscillations decays as shown by the dotted lines, which form the envelope of the sine wave. The equation of this envelope is the exponential function ε^{-bt}. The period of the wave is the time required to complete one complete oscillation or cycle (up and down and up again). The period is the reciprocal of the frequency ($1/f$). For comparison purposes the wave form of an undamped oscillation, the (sine) wave $A \sin \omega t$, is shown in Fig. 16. In such an oscillator energy has to be supplied continually to overcome the internal force of friction, as was previously mentioned.

A Problem in Electricity

Let us now consider a problem in electricity that is exactly analogous to the mechanical problem of the spring pendulum (see Fig. 17). A capacitor, C, that has been initially charged by a battery of voltage, E, is suddenly discharged (by placing switch S into 'discharge' position) through a coil of inductance L and resistance R which also includes the resistance of the connecting wires

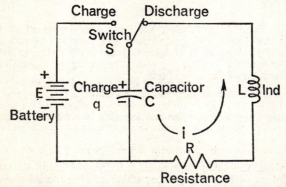

Fig. 17. An electrical oscillating circuit that is equivalent to the mechanical spring pendulum.

The schematic circuit diagram of the electrical set-up, shown in Fig. 17, is analogous to and could serve as an equivalent circuit for the mechanical pendulum shown in Fig. 14. If the resistance of the electrical circuit is small, electrical oscillations will take place that correspond to the damped mechanical oscillations of the spring pendulum. In the electrical case, we would like to know the current through the circuit and the charge on the capacitor at any time, t, after the switch is thrown to 'discharge', and we would also like to know the frequency of electrical oscillations.

Although we have boldly stated that the electrical circuit of Fig. 17 is analogous to the mechanical set-up of Fig. 14, this can be proved only if the form of the respective differential equations and their solutions are analogous. Before setting up these equations, let us explore some of the physical reasons that lead us to believe that the two situations are analogous. It is known from elementary electrical theory that the inductance of a coil resists any change in the current flowing through it. This is at least similar (and perhaps analogous) to the inertia or flywheel effect of a mass, that makes the latter resist any change in its motion. Similarly, the 'charging' of a capacitor appears to be physically analogous to the displacement or tensioning of a spring. Finally, the electrical resistance of the circuit seems to be analogous to the mechanical resistance, or friction, of the spring pendulum.

The Physical Analogy. When a capacitor is charged by an electromotive force (voltage), electrical energy is stored in the electric field between the plates of the capacitor. This is analogous to the storing of potential energy by the tensioning of a spring. If the capacitor is then suddenly discharged through an inductor (coil) the energy of the electric field is released by the motion of the charges (i.e. the current) through the coil, which build up a magnetic field about the coil. The energy in the electric field of the capacitor is, thus, temporarily converted to the energy of the magnetic field about the coil, corresponding to the conversion of potential energy into kinetic energy of motion in the pendulum. Furthermore, as in the mechanical set-up, the motion of the charges (current) does not stop when the capacitor is completely discharged, but, owing to the electrical inertia (inductance) of the coil, the current continues to flow until the capacitor is recharged in the opposite direction and the initial energy is again stored in the electric field. Evidently, then, the electrical inductance–capacitance circuit continues to oscillate by the alternate storage and release of energy in the fields of the capacitor and coil, just as the mechanical pendulum continues to oscillate by the alternate storage and release of mechanical energy in the spring and weight, respectively. The mechanical pendulum stops oscillating when all its energy is used up in internal friction; similarly, an inductive–capacitive circuit stops oscillating when all its electrical energy is dissipated in the resistance of the coil winding and conductors.

Finding the Voltage Drops. We have seen in the mechanical set-up that the applied force, F, must be balanced by all the counterforces of the pendulum, or equivalently, the sum of all forces must add up to zero. Similarly, in the electrical circuit the applied electromotive force (abbreviated e.m.f.) or voltage, E, must equal the voltage drops occurring throughout the circuit, or equivalently, the sum of all voltages (e.m.f.s and voltage drops) throughout the circuit must equal zero. We must therefore find the voltage drops across each of the components of the circuit and express them as a function of time, wherever possible.

Starting with the simplest, we know from elementary electricity that according to Ohm's law the voltage drop (V) across a resistance is simply the product of the resistance (R) and the current (i) through it. Expressing this in equation form, the resistive voltage drop

$$V_1 = iR$$

Evidently, this voltage drop depends only on R and i and not upon the *time* of current flow.

The voltage (called back e.m.f.) built up across a coil depends upon the rate of change of flux of the magnetic field about the coil. Since the magnetic flux is proportional to the current (i) flowing through the coil, it may be stated, equivalently, that the voltage across the coil is proportional to the time rate of change of the current. This proportionality is expressed mathematically as

$$V_2 = L \frac{di}{dt}$$

where L is the proportionality constant known as self-inductance.

Finally, the voltage drop across a capacitor follows from the definition of capacitance, which states that the ratio of the charge (q) on the capacitor to the voltage (V) across it is always the same (i.e. a constant). This constant, called capacitance, is given by

$$C = \frac{q}{V}$$

where C is expressed in farads, if q is in coulombs and V is the volts. By a simple transposition of this relation, we obtain for the voltage drop

$$V = \frac{q}{C}$$

that is, the voltage drop across a capacitor is the ratio of the charge to the capacitance. (The initial charge, Q, on the capacitor is, of course, the product of the applied voltage, E, and the capacitance, C, or $Q = CE$.) Although this relation is always valid, we would like to know the capacitor voltage as a function of time. This is easily ascertained by introducing the current. By definition, the current (i) is the time rate of change of (the flow of) charge (q); that is,

$$i = \frac{dq}{dt}$$

Conversely, an infinitesimal charge, dq, is the product of the current and an infinitesimal span of time, dt; that is,

$$dq = i\,dt \text{ (rewriting the above relation)}$$

To obtain the total charge q built up on the capacitor, we must sum up, or integrate, all the infinitesimal products $i\,dt$, so that

$$q = \int dq = \int i\,dt$$

Substituting for q in the relation for the voltage drop, we obtain

$$V = \frac{q}{C} = \frac{1}{C} \int i\,dt$$

Finally, summing up all the voltage drops in the circuit and setting them equal to the applied voltage (E), which is zero during discharge of the capacitor, we obtain

$$E = 0 = iR + L\frac{di}{dt} + \frac{1}{C}\int i\,dt$$

which is the fundamental equation of the discharge circuit shown in Fig. 17. Since this equation involves both integrals and differentials, we must 'differentiate' both sides with respect to time to convert it into an ordinary differential equation. Differentiating both sides and transposing terms, we obtain

$$L\frac{d^2i}{dt^2} + R\frac{di}{dt} + \frac{1}{C}i = 0$$

Charge Equation. This is the differential equation of the current (i) in the R–L–C circuit of Fig. 17 during discharge. As you can see, it is exactly analogous in form to the differential equation of the mechanical system we have explored earlier. By comparing the constant multipliers (coefficients) of the variable and its derivatives in both equations, we observe immediately that the inductance (L) is analogous to the mass (m), the electrical resistance (R) is analogous to the frictional resistance (μ), and the reciprocal of the capacitance ($1/C$) is analogous to the spring constant (K). However, the variables i and x in the two equations are not strictly analogous, since the current (i) is defined by the rate of change of the flow of charge (dq/dt), while we have previously stated that the spring displacement (x) corresponds directly to the charge (q) on a capacitor. Hence, to complete the analogy, let us find the differential equation for the charge, q, in the electrical circuit, using the previously developed relations

$$q = \int i\, dt; \quad \frac{dq}{dt} = i$$

and $\frac{d^2q}{dt^2} = \frac{di}{dt}$ (differentiating $i = \frac{dq}{dt}$)

Let us substitute these for the current in the fundamental equation

$$L\frac{di}{dt} + iR + \frac{1}{C}\int i\, dt = 0$$

Doing this, we obtain

$$L\frac{d^2q}{dt^2} + R\frac{dq}{dt} + \frac{1}{C}q = 0$$

Solution and Special Cases. This differential equation in terms of charge (q) is analogous in all respects to the differential equation of the displacement (x) in the mechanical system

$$m\frac{d^2x}{dt^2} + \mu\frac{dx}{dt} + Kx = 0$$

with the displacement (x) corresponding to the charge (q) and the corresponding coefficients being analogous, as previously outlined. The solutions of the differential equations for the current and charge are identical in form to the one previously shown for the mechanical system

$$i, q, \text{ or } x = \varepsilon^{-bt}(A \sin \omega t)$$

consisting of an exponentially decaying function (ε^{-bt}) and a sine wave of amplitude A and frequency $f = \omega/2\pi$. The frequency (f) of the electrical oscillations of the charge or current is given by the approximate expression

$$f = \frac{1}{2\pi\sqrt{LC}}$$

which holds when the resistance R is small compared to the quantity $2\sqrt{\frac{L}{C}}$. The wave form of the electrical oscillations is exactly as portrayed in Fig. 15 for the mechanical system.

$$\text{If } R = 2\sqrt{\frac{L}{C}}$$

only a single oscillation takes place, which decays exponentially, a situation known as critical damping. If the resistance R is greater than $2\sqrt{\frac{L}{C}}$ no electrical oscillations occur whatsoever, but the current or charge simply dies out exponentially when the capacitor is discharged. These same observations can be made, of course, in the analogous mechanical system, by substituting the corresponding quantities (m and K) in the frequency equation.

Table of Electrical and Mechanical Analogies

From the analogy of a spring pendulum to an R–L–C electrical oscillating system we can easily make up the following table of analogous electrical and mechanical quantities. This table could be extended to include many more analogous quantities from hydraulic, acoustic, magnetic, thermal, and other physical systems, but the few listed quantities illustrate how an analogue computer can solve analogous physical problems by solving just one basic equation.

Table of Analogous Mechanical and Electrical Quantities

MECHANICAL QUANTITY	SYMBOL	ELECTRICAL QUANTITY	SYMBOL
Applied force	F	Applied e.m.f. (volts)	E
Displacement	x	Charge (coulombs)	$q = \int i\, dt$
Velocity	$v = \dfrac{dx}{dt}$	Current (amperes)	$i = \dfrac{dq}{dt}$
Acceleration	$a = \dfrac{d^2x}{dt^2}$	Amperes/second	$\dfrac{di}{dt} = \dfrac{d^2q}{dt^2}$
Mass (Inertia)	m	Inductance (henries)	L
Spring constant	K	Reciprocal of capacitance (farads)	$1/C$
Frictional resistance	μ	Resistance (ohms)	R
Kinetic energy	$\tfrac{1}{2}mv^2$	Field energy	$\tfrac{1}{2}Li^2$
Potential energy	mx	Potential energy	Lq

REVIEW AND SUMMARY

General-purpose analogue computers provide solutions useful for many applications; fixed-purpose analogue computers serve as models for, or simulate, a specific problem.

Indirect analogue computers (usually general-purpose) solve the mathematical equations representing one or several physical systems; direct analogue computers (usually fixed-purpose) set up a direct analogy to the parameters involved in the problem (miniature dam or aeroplane).

General-purpose (indirect) or fixed-purpose (direct) analogue computers may be of the electrical (or electronic), mechanical, or fluid category, the electrical (electronic) type being most prevalent.

Speed (velocity) is the time rate of change, or derivative, of the distance ($v = ds/dt$).

The derivative (instantaneous rate of change) of a variable function at a particular point of its graph is given by the tangent or slope of the function at that point. [$f'(x)$ = tangent or slope of $f(x)$.]

Acceleration is the time rate of change, or derivative, of the velocity ($a = dv/dt$); it is also the second derivative of the distance ($a = d^2s/dt^2$).

In a body initially at rest the distance covered during accelerated motion is proportional to the square of the time interval ($s = \tfrac{1}{2}at^2$).

Spring Force. The restoring force of a stretched (or compressed) spring is proportional to the displacement ($F = -Kx$, where K is spring constant).

Introduction to the Analogue Computer

Inertia. Acceleration of body is proportional to applied force and inversely proportional to mass ($a = F/m$ or $F = ma$).

The damping force due to viscous friction is proportional to the velocity of motion ($F = \mu v$).

The oscillating action in a mechanical spring pendulum consists of the alternate storage of potential energy in the spring and its release in the form of kinetic energy (energy of motion). Because of the flywheel effect (inertia), the mass of the pendulum resists any change in its motion. The oscillation dies down because of friction. The differential equation of the displacement (x) in a mechanical oscillating system (spring pendulum) is given by:

$$m \frac{d^2x}{dt^2} + \mu \frac{dx}{dt} + Kx = 0$$

its solution is of the form $x = \varepsilon^{-bt} (A \sin \omega t)$. This equation describes a damped oscillation of frequency

$$f = \frac{1}{2\pi} \sqrt{\frac{K}{m}}$$

The oscillating action in an electrical R–L–C (resistance–capacitance–inductance) circuit consists of the alternate storage and release of energy in the electric field of the capacitor and magnetic field of the coil, respectively. The inductance of the coil resists any change in the flow of current. The oscillation dies down because of the resistance in the circuit. The fundamental equation of current flow in the R–L–C circuit during discharge is given by:

$$0 = iR + L \frac{di}{dt} + \frac{1}{C} \int i dt$$

The differential equation of the flow of charge in an R–L–C circuit (during discharge) is:

$$0 = L \frac{d^2q}{dt^2} + R \frac{dq}{dt} + \frac{q}{C}$$

its solution is of the form $q = \varepsilon^{-bt} (A \sin \omega t)$. This equation describes a damped oscillation of frequency,

$$f = \frac{1}{2\pi \sqrt{LC}}$$

A damped oscillation takes place if

$$R < 2 \sqrt{\frac{L}{C}}$$

critical damping occurs if

$$R = 2 \sqrt{\frac{L}{C}}$$

no oscillation occurs if

$$R > 2 \sqrt{\frac{L}{C}}$$

The following mechanical and electrical quantities are analogous: applied force (F) and applied e.m.f. (voltage E); displacement (x) and charge (q); velocity (v) and current (i); mass (m) and inductance (L); compliance (C_m) and the reciprocal of capacitance ($1/C$); mechanical friction (μ) and electrical resistance (R).

CHAPTER 3

BUILDING BLOCKS OF ANALOGUE COMPUTERS—I: MECHANICAL AND ELECTRICAL DEVICES

Present-day electronic analogue computers are highly sophisticated machines, which use few mechanically moving parts, and for that matter, few electrical devices of the ordinary variety. The silent efficiency and sophistication of these computers, however, rests upon a long evolution of relatively simple mechanical and electrical devices, some of which we have touched upon in our brief history of computers. We must now spend a little time with the underlying computational principles of these simpler devices, partially because they are still used in mechanical analogue computers and, also, because they provide an easily comprehensible introduction to the more 'elegant' electromechanical and electronic building blocks.

MECHANICAL COMPUTING DEVICES

Mechanical devices used for mathematical computations range all the way from simple pulley and cam arrangements to fairly complex gear assemblies and integrating devices. They can be made quite accurate, but they have in common the clumsiness of most mechanical devices and the lack of flexibility in changing over to different problems and operations (programming). Let us first look at some of the almost 'classic' devices available for addition and subtraction.

Addition and Subtraction

The simplest way of adding quantities is to represent them as lengths on a stick or ruler. By placing the lengths end to end, the addition of the individual lengths is automatically accomplished. The operation of the slide rule is based, of course, upon this principle, but the lengths added or subtracted represent logarithms, so that in actuality multiplication and division are accomplished. (To multiply two numbers their logs may be added; to divide, their logs may be subtracted.)

Pulley and Chain. The motion of two pulleys can be summed up by a third movable pulley, by means of a pulley-and-chain arrangement, such as that shown in Fig. 18. Three equal-sized pulleys are connected by a chain that is fixed at both ends. Pulley c responds to the motions of both pulleys a and b. When either pulley a or pulley b is moved by a certain amount pulley c is displaced by the same amount. When both pulleys a and b are moved pulley c is displaced by an amount equal to their *sum*; i.e. $c = a + b$. The arrangement can be extended to as many pulleys as are needed to add the desired number of quantities. It is not very practicable, however, to include pulleys and chains in an analogue computer.

Bar Linkages. Instead of pulleys and chains, bar linkages may be used to sum up linear displacements. A simple arrangement is shown in Fig. 19. Three rods that can slide freely within fixed guide blocks are linked at equal distances by a bar. The bar pivots around the centre rod c, and the motion of rods a and b is transmitted through the slots at the ends. By completing the similar triangles formed by the displacements of a and b you can easily determine that

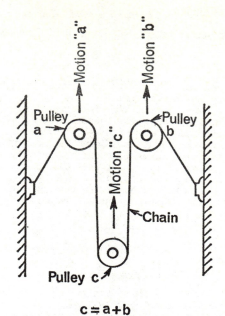

$$c = a + b$$

Fig. 18. Pulley-and-chain arrangement for adding two quantities.

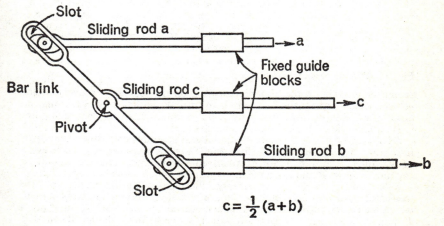

$$c = \frac{1}{2}(a + b)$$

Fig. 19. Bar linkages for adding two quantities.

rod c is always displaced by one-half the sum of displacements a and b; that is

$$c = \frac{a+b}{2}$$

As with the pulleys, this principle can be extended to include many more addends or change the ratio between rods.

The Differential Gear Assembly. We now turn to a more convenient and flexible means of adding or subtracting quantities—by shaft rotations. Since shafts always rotate in a circle, the additions or subtractions consist, of course, of angular displacements, rather than linear displacements, utilized in the previous devices. However, this is of no consequence, since angular displacements are easily converted into linear ones by multiplying the angle (θ) through which the shaft has rotated by its radius (r); i.e. displacement $s = r\theta$.

The first device of this category that we shall consider is the familiar (though little understood) differential gear assembly, illustrated in Fig. 20. In slightly

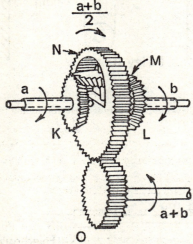

Fig. 20. Differential gear assembly for adding or subtracting two quantities (a and b).

modified form this is the 'differential' used on the rear axle of every motor vehicle, which permits the rear wheels to turn at different rates when driving around curves or getting stuck in mud. Its operation in an analogue computer is essentially the same as in a car, as we shall see.

As illustrated in Fig. 20, the differential gear assembly consists essentially of four gears, K, L, M, and N. The fifth gear, O, is not necessary to the operation of the differential, but performs a convenient auxiliary function, as will be explained. Gears K and L are the input gears and are coupled to input shafts a and b, respectively. (For convenience, the angular displacements are also measured by a and b.) These shafts are actually hollow collars, as shown, which are free to turn on the shaft of the central 'spider' gear, N. Input gears K and L both mesh with gear M, which is free to turn, though its shaft is axially mounted to the inside rim of 'spider' gear N. The latter gear (N) meshes with auxiliary gear O, which has half the number of teeth (one-half diameter) of gear N.

Assume for the moment that gear L is held fast and gear K is turned clockwise through some angle a. The side of gear M that meshes with gear K turns, of course, at the same rate as input gear K. The side of gear M that is in contact with gear L, however, is stationary, since L is held and cannot move. As a result, gear N in the centre is dragged along and moves in the same direction (clockwise) and through half the angle ($a/2$) by which gear K is turned. The reason for the factor $\frac{1}{2}$ is simple. With the left side of M turning at the same rate as gear K, while the right side (in contact with L) is stationary, the central shaft of M must turn at half the rate of K, and hence completes only half the rotation, $a/2$, in a given time. Since this shaft is coupled to spider gear N, the latter also completes only half the angle.

If gear K is held fast and input gear L is turned clockwise through some angle b the same action occurs, and gear N is dragged along through half the angle of rotation, $b/2$. Finally, if both gears K and L are turned through angles a and b, respectively, spider gear N will turn through an angle equal to half the sum of the two angles, or $\frac{a+b}{2}$. As we have mentioned, gear N meshes with output gear O, which has half the number of teeth or half the diameter of N. It will therefore turn at twice the rate and through twice the total angle of gear N, or through $\frac{2(a+b)}{2} = a+b$. Thus, the shaft rotation of output gear O represents the sum of the input rotations, a and b. Since a and b may have any values whatsoever, the differential gear assembly will add up angles without any mechanical limit.

You can easily visualize from Fig. 20 that turning one of the input shafts in the opposite (anticlockwise) direction will cause spider gear N to rotate through half the difference between the two shaft rotations; that is, it will subtract the two angles. For example, if gear K is turned clockwise through angle a and gear L is turned anticlockwise through angle $-b$, gear N will rotate through half their difference, $\frac{a-b}{2}$, and output gear O will turn through $a-b$. Thus, the differential gear arrangement can be used equally well for the subtraction of two angular displacements.

Multiplication and Division by a Constant

The principle of multiplication incorporated into the auxiliary (output) gear of the differential gear assembly is, of course, generally valid. If we mesh two gears, one of which has twice the pitch (number of teeth or diameter) of the other, the smaller gear will rotate at twice the rate of the larger, and hence, the angle of rotation will be multiplied by the factor 2. Consider, for example, the two gears shown in Fig. 21. The one mounted on shaft a has 72 teeth, while the other, mounted on shaft b, has only 36 teeth. If input shaft a is turned through one complete revolution (360°), the 72 teeth of the gear must mesh with 72 teeth on the other gear. Since only 36 teeth are provided, the gear and output shaft b must turn through two complete revolutions, or 720°. Thus, the angular displacement b (of the smaller gear) is twice that of a (the large gear). Conversely, if shaft b is turned through one full revolution, or 360°, shaft a will turn only through one-half revolution, or 180°, since only 36 teeth have been meshed. Thus, the arrangement works equally well for multiplication or division by the constant factor 2, depending upon which gear is used for the input.

The principle involved in the doubling mechanism discussed above can be extended to multiplication or division by any constant factor. Thus, by meshing two gears, one of which has three times the pitch, or number of teeth, of the other, you can multiply or divide by the factor 3, and so on. By coupling a number of gears in a gear-train or gear-box, any desired multiplication or division ratio can be attained.

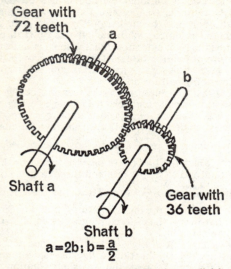

Fig. 21. Gear mechanism for multiplication or division by 2.

Generalized Multiplication and Division

If multiplication (or division) by a constant can be accomplished by fixed-ratio gears rotating at fixed speeds, multiplication by a continuously variable factor should be possible by means of *variable speed gears*. The disc-and-wheel device we shall look at next is one of a numerous category of variable-speed drives used in mechanical analogue computers. Essentially the same device (Fig. 22) is also used as an integrator.

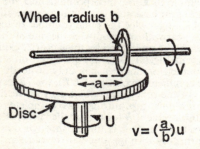

Fig. 22. Disc-and-wheel device for generalized multiplication.

A disc with an input shaft u drives a wheel with an output shaft v. The shaft of the wheel is parallel to a diameter of the disc and the wheel is free to turn on the shaft at any radius a. The rotation of the disc by shaft u also rotates the wheel, unless it is at the centre point.

Let us find the angular displacement (rotation) of the output shaft, v, for an angular displacement, u, of the input shaft. Turning the disc through an angle u rotates any point at a radius a from the centre of the disc by a distance a times u (au). A point on the circumference of the wheel at a radius a, therefore, turns through a distance au, when the disc is rotated through the angle u. The shaft of the wheel, however, rotates through an angle, v, equal to the distance travelled by a point on its circumference *divided* by the radius of the wheel, b. Since a point on the circumference turns through the distance au, the shaft turns through an angle

$$v = \frac{au}{b} = (a/b)u$$

In words, the input angle, u, is multiplied by the factor (a/b), where the radius a may vary anywhere from zero to many times b. Thus, the disc-and-wheel device permits the desired multiplication of any *multiplicand* (rotation u) by any multiplier (setting a/b) to produce the product [$v = (a/b)u$].

You can also see that *division* is achieved if a is smaller than b; that is, if the ratio a/b is smaller than 1. Moreover, in principle, the whole arrangement can be turned around by turning the shaft of the wheel through the angle v to produce an output rotation of the disk shaft, $u = (b/a)v$, thus reversing the multiplier ratio. In practice, this may not work too well, since it is difficult to obtain sufficient friction to rotate the massive disc by the relatively small wheel. The arrangement of Fig. 22 has the further drawback of not permitting continuous multiplication of previously obtained results by new multipliers. A change in the radius a will multiply further rotation of the disc, u, by the new multiplier (a/b), but will not affect a previously obtained result, since the device only multiplies increments in u. Hence, if a quantity must be multiplied by a new multiplier the disc must be reset to zero.

The disc-and-wheel device has been used in the past as a planetary transmission in motor cars to provide continuous speed control from full forward to full reverse. It is also useful as a mechanical integrating device, as we shall see presently.

Integration of Variable Functions

We have already mentioned that the disc-and-wheel device multiplies a quantity by the increments of a function. Without bothering to state the exact mathematical definition of integration at this time, let us just say that an integral represents the summation of the products of a variable quantity by the infinitesimal increments of another function. The disc-and-wheel device is, thus, well qualified to perform the operation of integration.

Fig. 23 illustrates the disc-and-wheel device of Fig. 22 with some minor modifications. We have mounted the disc with its input shaft (u) on a support, which can be moved horizontally at a controlled rate by means of a feed screw. This, of course, permits changing the radial distance, a, between the wheel and the centre of the disc. In addition, we have made the wheel quite small and given it a radius of unity in the system of measurements used. The multiplier by which u is multiplied, therefore, is $a \times 1$, or simply a. Thus, if the shaft of the disc is rotated by an infinitesimal angle, du, the output shaft of the wheel will turn through the infinitesimal angle $a \times du$, where a is, of course, the radius determined by the rotation of the feed screw. If a should change during the next infinitesimal increment du, the infinitesimal increment in v (i.e. dv) will be multiplied by the new value of a during this increment. If we continue to put in du's (by turning the disc) and vary the a's during each increment, we will get a series of dv's of varying size at the output. The sum of all these dv's is represented by the total rotation of the wheel, v. But this sum of all

dv's is exactly what is meant by the integral of a function. Thus, if $dv = a\, du$, the summation or integral of dv, which is written $\int dv = \int a\, du = v$, which is the desired function.

For example, suppose that the rotation of the feed screw through the radial (horizontal) distance a represents the velocity, v, of a car and that the rotation (u) of the disc represents increments in time, dt. The total rotation of the wheel will then represent the integral of the car's velocity with respect to time, which is the total distance, s, travelled by the car. In mathematical terms

$$\text{Distance, } s = \int v\, dt$$

which is obtained from the relation $v = ds/dt$, as we have seen earlier.

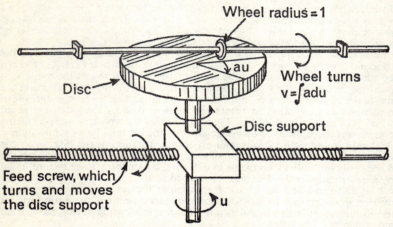

Fig. 23. Disc-and-wheel integrating mechanism.

If you now look back for a moment at the 'ball-and-disc' integrator, illustrated in Chapter 1 (Fig. 9) you will immediately note that this device operates exactly as the wheel-and-disc integrator, except for the substitution of a ball for the wheel. Since the latter device dates back to Lord Kelvin, living a century ago, you can see that nothing spectacular has happened in mechanical analogue devices.

Generation of Mathematical Functions

So far we have used purely mechanical devices to perform standard mathematical operations, such as addition, multiplication, and integration. Actual problems frequently require that some quantity be modified in accordance with a mathematical function, such as a sine curve ($y = A \sin x$), a logarithmic function ($y = b \log x$), an exponential function ($y = A\varepsilon^{-bx}$), and so on. Although these functions can be mechanized without too much trouble by the use of shaped cams and linkages, this is generally a fairly expensive procedure and hardly worthwhile, since there are many electrical devices that can do this economically and accurately. For example, the voltage available from the ordinary a.c. power line follows a sine wave exactly; that is, the instantaneous line voltage

$$E = E_{\max} \sin 2\pi ft$$

where $E_{\max}$ is the maximum or peak voltage and f is the frequency.

Mechanical function generation is used more frequently when the exact mathematical function is not known but when empirical data or curves are are available. For example, a curve may have been plotted, which shows the variation in the 'drag' of an aircraft wing as a function of the speed with which the wing is moving through the air (Fig. 24). If the drag behaviour of the wing

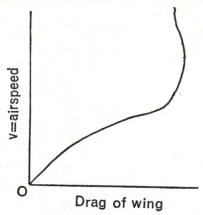

Fig. 24. Possible graph of wing drag *versus* air speed.

is to be simulated in an analogue computer the empirical drag curve must be 'mechanized'. A number of methods exist for doing this. We can simply paste or draw the curve on the body of a cylindrical drum, which can be rotated at a uniform speed by a hand crank or motor, simulating the speed of the aircraft. As the drum rotates, a mechanical pointer with some sort of sighting device can be made to glide along the curve, as is illustrated in Fig. 25. A mechanical adjustment is provided to keep the sighting device of the

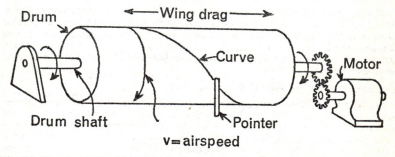

Fig. 25. Graph of Fig. 24 mechanized on a drum, with pointer following graph.

pointer centred on the curve at all times. As long as this is the case, the position of the pointer corresponds to the wing drag along the length (axis) of the drum and the speed of the aircraft around the body of the drum.

The function of the operator, who must keep the pointer centred on the graph, can easily be taken over by a photocell. If we paint the area below the graph black and keep the area above the curve white we can make the photocell respond to 'grey' to keep the pointer centred along the border line of the graph. We can also use the function generator in reverse to graph a function that the computer puts out. We only need to attach a pen to the pointer and wrap a sheet of paper around the drum. If the pointer is controlled by the dependent variable (y) as the drum is rotated in accordance with the independent variable (x), a graph of the function $y = f(x)$ will automatically be traced on the paper.

Cam and Follower. A more permament way of generating a function mathematically is to construct a cam whose shape represents the desired function. A two-dimensional model of a cam-type function generator is illustrated in Fig. 26. A cam is so shaped that its height (y) at any point along its horizontal

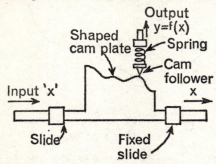

Fig. 26. Cam and follower for generating the function $y = f(x)$.

axis (x) represents the value of the function at that point. A cam follower is attached to slide along the edge of the cam. By moving the cam along the horizontal in accordance with the independent variable (x), the vertical motion of the cam follower will automatically represent the value (output) of the dependent variable (y); that is, $y = f(x)$.

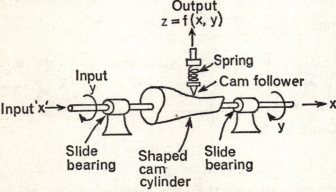

Fig. 27. Cam cylinder for generating a function $z = f(x, y)$.

The cam-and-follower principle can be extended to more complex functions. Suppose a dependent variable, z, is a function of two independent variables, x and y. Such a mathematical function is represented by a three-dimensional surface, rather than a plane. Fig. 27 illustrates a cylindrical cam that is shaped in accordance with some function, $z = f(x, y)$. The vertical output (z) from the cam follower is the desired function of the two inputs, x, along the horizontal axis of the cam, and y, around the body of the cylinder. The horizontal or x input is accomplished, of course, by sliding the cam cylinder horizontally in its bearing, while the y input is obtained by the rotation of the cam-cylinder shaft.

ELECTRICAL COMPUTING DEVICES—PASSIVE NETWORKS

Using voltages and currents in electrical 'networks', we shall now perform the same mathematical operations we have just carried out mechanically. A group of resistors connected together in some way is known as a passive network, since it can only dissipate the power applied to it and is incapable of generating any power. Later we shall connect amplifiers into these networks, thus making them 'active', to compensate for the losses incurred in the passive components. The main assumption in the following operations with resistance networks is that all resistors behave 'linearly', which means that the current through each resistor varies directly with the voltage across the resistor. If a graph of current flow versus applied voltage is drawn for such a resistor it turns out to be a straight line; hence, the term linear. Moreover, such linear behaviour is summed up mathematically by Ohm's law, which states that the current through a resistance (or resistor network) is directly proportional to the applied voltage (e.m.f.) and inversely proportional to the resistance. Mathematically, the current

$$I \text{ (amperes)} = \frac{E \text{ (volts)}}{R \text{ (ohms)}}$$

Most of our operations with electrical networks will be based on Ohm's law.

There are two additional principles related to linear behaviour and Ohm's law, which make possible arithmetical operations with electrical networks. One of these is called the principle of superposition, which can be summarized as follows: if several voltage sources are present in a linear electrical network, the currents and voltage drops due to each source may be computed separately as though all other voltage sources were absent (zero). The resultant currents and voltage drops may then be added separately to arrive at the overall effect. The implication of this principle is, of course, that the effects due to each voltage will superimpose as a simple sum and that there is no interaction between them. This is always true in a circuit made up of linear components.

The other principle consists of two laws, known as Kirchhoff's laws, which are an extension of Ohm's law. In brief, Kirchhoff's laws may be stated as follows:

1. The algebraic sum of the currents at a junction point is zero. Expressed mathematically:

$$\text{Sum } I = 0$$

In common-sense terms, this law simply states that as much current flows away from a point as flows towards it. (A plus sign is assigned to all currents flowing towards a junction; a minus sign to those flowing away from the junction.)

2. The sum of the e.m.f.s (battery or generator voltages) around any closed loop of a circuit equals the sum of the voltage drops across the resistances in that loop. Again, stated mathematically:

$$\text{Sum of e.m.f.s} = \text{Sum of } IR \text{ drops } (E = IR)$$

By assigning plus signs to e.m.f.s and minus signs to voltage (IR) drops, this may also be stated in algebraic form:

$$\text{Sum } E + \text{Sum } IR = 0$$

Multiplication and Division by a Constant

We shall start with electrical multiplication and division, since this happens to be electrically simpler than addition or subtraction. Consider the simple voltage divider illustrated in Fig. 28. An input voltage, x, is applied across a

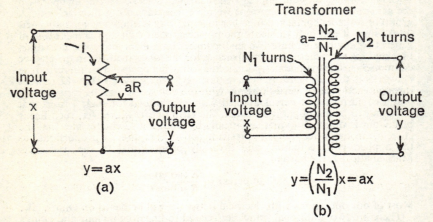

Fig. 28. Electrical multiplication or division by a constant: (a) voltage-divider potentiometer (coefficient potentiometer); (b) transformer.

resistor, R, which is tapped at some point. A certain fraction of the total resistance, aR, appears between the tap and the bottom end of the resistor. The output voltage, y, is taken between the tap and the bottom end. If the tap is fixed at the setting a this arrangement is known as a voltage divider; if the tap is adjustable (a slider on the resistance winding) it is a potentiometer. By Ohm's law, the current (i) flowing through the resistor (R) is:

$$\text{Current, } i = \frac{\text{Voltage}}{\text{Resistance}} = \frac{x}{R}$$

The voltage drop or output voltage (y) appearing between the tap and bottom end of the resistor is the product of the current and the tapped-off resistance (aR), or:

$$\text{Output voltage, } y = \left(\frac{x}{R}\right) aR = ax$$

In words, the output voltage equals the input voltage multiplied by the constant fraction (a) of the resistance tapped off. If the tap consists of an adjustable slider the fractional setting a can be varied between 0 and 1, but it can, of course, never be greater than 1, since at that point the output voltage equals the input voltage. A voltage divider therefore provides multiplication by a fraction, or equivalently, division by a constant coefficient. Accurately calibrated, adjustable voltage dividers for doing this job are known as coefficient potentiometers.

Transformer. An electrical transformer permits multiplication of an input quantity (voltage) by either a constant whole number or a fraction. Texts on elementary electricity explain that in an ideal transformer (that is, one without losses) the ratio of the voltage across the secondary winding to that across the primary equals the ratio of the turns in the secondary winding to the primary turns. Once a transformer is constructed, the ratio (a) of the secondary turns (N_2) to the primary turns (N_1) is fixed, of course. Hence, for any given transformer the mathematical relation holds:

$$\frac{\text{Output voltage }(y)}{\text{Input voltage }(x)} = \frac{N_2}{N_1} = a$$

This may be rewritten: $y = ax$, where a is the turns ratio. Since the number of primary and secondary turns can be anything we wish (within limits of good design), the turns ratio, a, can be any quantity from a small fraction to a whole number up to about 10. A transformer, therefore, is capable of multiplying or dividing by a constant.

Addition of Voltages by a Resistor Network

In an electrical analogue computer most physical or numerical quantities are represented by voltages. (Negative quantities or numbers can be represented by voltages of negative polarity.) Thus, to add several quantities of the same kind, the corresponding voltages must be computed, with proper regard to the sign. This can be done by applying the voltages across resistive networks.

For example, consider the summing network, shown in Fig. 29. Here three input quantities, voltages w, x, and y, are impressed upon a parallel resistive network consisting of resistors R_1, R_2, R_3, and R_4. The output voltage, z, is taken across load resistor, R_4. This network exemplifies the principle of superposition previously mentioned, and it can be worked out by using it. (That is, the effect of each voltage can be computed separately and then all effects can be added to obtain the output voltage.) But rather than using this somewhat strange-appearing technique, let us use Kirchhoff's first law to compute the output voltage (z). According to this law, the sum of all the currents flowing into the output junction, K, must equal zero. Hence, we must compute the current flowing through each of the resistors and then set their sum (flowing into K) equal to zero. To compute the currents we use Ohm's law; that is, the current equals the applied voltage divided by the resistance.

The voltage applied across resistor R_1 is the difference between the output voltage z and the input voltage w, i.e. $z-w$. Hence, the current through R_1 is

$$i_1 = \frac{z - w}{R_1}$$

Similarly, the current through R_2 is

$$i_2 = \frac{z - x}{R_2}$$

the current through R_3 is

$$i_3 = \frac{z - y}{R_3}$$

and the current through R_4 is

$$i_4 = \frac{z}{R_4}$$

(voltage z is across R_4). Applying Kirchhoff's first law, by setting the sum of the currents flowing into junction K equal to zero, we obtain

$$\frac{z-w}{R_1} + \frac{z-x}{R_2} + \frac{z-y}{R_3} + \frac{z}{R_4} = 0$$

By factoring out z from this expression and transposing we can compute the output voltage:

$$z = \left(\frac{w}{R_1} + \frac{x}{R_2} + \frac{y}{R_3}\right) \times \frac{1}{1/R_1 + 1/R_2 + 1/R_3 + 1/R_4}$$

This expression gives the output voltage as the sum of the input voltages multiplied by constant factors, but to see this more clearly let us multiply each of the factors in the parenthesis by the factor at the right. When this is done, we obtain

$$z = w\left(\frac{R_2 R_3 R_4}{R_2 R_3 R_4 + R_1 R_3 R_4 + R_1 R_2 R_3 + R_1 R_2 R_4}\right)$$
$$+ x\left(\frac{R_1 R_3 R_4}{R_2 R_3 R_4 + R_1 R_3 R_4 + R_1 R_2 R_3 + R_1 R_2 R_4}\right)$$
$$+ y\left(\frac{R_1 R_2 R_4}{R_2 R_3 R_4 + R_1 R_3 R_4 + R_1 R_2 R_3 + R_1 R_2 R_4}\right)$$

Replacing the constant factors in the parentheses above, using the constant coefficients k_1, k_2, and k_3, this expression may be written

$$z = k_1 w + k_2 x + k_3 y$$

The last expression, finally, tells us that the output voltage, z, is the sum of the input voltages w, x, and y, each being multiplied by a constant coefficient (k_1, k_2, k_3) that depends only upon the resistances of the network. The fact that each of the voltages to be added is multiplied by a constant is actually an advantage in practice, since most physical variables must be added in certain quantitative ratios, which can be expressed as constant coefficients.

If all the resistors in the network of Fig. 29 are made equal so that $R_1 = R_2 = R_3 = R_4 = R$ the solution for z is simplified to $z = \frac{1}{4}(w + x + y)$, as you can verify yourself; that is, the output voltage equals one-quarter of the sum of the input voltages. This result also points out the disadvantage of the passive resistive network, namely, that the output voltage can never be equal to the sum of the input voltages because of the considerable losses (called attenuation) occurring in the circuit. Moreover, when the resistances cannot be equal, the output voltage depends appreciably upon the load resistance (R_4) of the circuit, as is shown by the earlier mathematical expression. This 'loading effect' becomes progressively worse as more networks must be added for further calculations, and diminishes the attainable accuracy. In the next chapter we shall see how both the attenuation and the loading effect can be overcome by the addition of an active component—the operational amplifier.

Subtraction. A 'summing' network of the type shown in Fig. 29 can perform subtraction, rather than addition, simply by reversing the polarity of the quantities (voltages) to be subtracted. For example, if voltages x and y in Fig. 29 are to be subtracted from input voltage w equal but negative (—) d.c. voltages are inserted at the x and y inputs. Thus, for input voltages w, $-x$, and $-y$, the output voltage in an equal-resistor network will be

$$z = \tfrac{1}{4}(w - x - y)$$

This follows, of course, from the fact that Kirchhoff's first law states that the algebraic sum of the currents flowing into a junction is zero. A negative voltage sets up a negative current; that is, a current that is opposite in direction to the others.

Electronic Calculus with Resistance–Capacitance Networks. The mathematical relationships describing the charge and discharge of a capacitor through a resistance are almost naturally adapted to the calculus operations of integration and differentiation. We have already seen (in the last chapter)

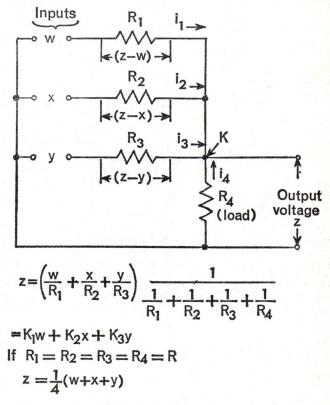

Fig. 29. Electrical addition of voltages by a resistor network.

that the voltage across a capacitor rises in direct proportion to the charge being supplied to it and that the total charge itself represents the summation or 'integral' of the flow of current over a certain time. It would appear, therefore, that we should be able to use the voltage developed across a capacitor during the charging period for the mathematical process of integration. This is indeed true.

Integration by Resistance–Capacitance Network. Fig. 30 (a) illustrates the circuit of a simple resistance–capacitance integrator. An input voltage, x, is applied across the R–C combination and the output voltage, y, is taken across capacitor C. Physically, what happens is simply this: the current (i) set up by the input voltage (x) charges the capacitor at a rate that depends on the values of the resistance and capacitance. The voltage (y) developed across the capacitor at any instant represents the total accumulated charge, which is the *integral* (summation) of the charging rate. By making the charging rate a function of the input voltage, the capacitor voltage will also represent the integral of the input voltage (x), which is the quantity to be computed.

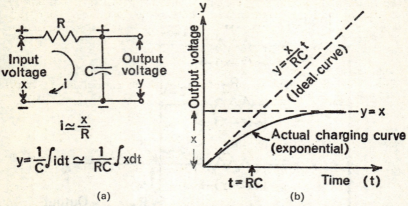

Fig. 30 (a) electrical integration by capacitor and resistor; (b) ideal and actual output of simple R-C integrator for constant voltage output.

Let us work out the approximate mathematical formula that demonstrates that the circuit of Fig. 30 actually integrates. When the input voltage, x, is first applied to the circuit there is initially no voltage across the capacitor, and the charging current, i, rushes in through resistor R. In the absence of a capacitor voltage this initial current is determined by the resistance in accordance with Ohm's law; that is, the charging current

$$i \simeq \frac{x}{R},$$

where the wavy equal sign ($\simeq$) stands for 'approximately equal to'. From the relations for the charge ($q = \int i dt$) and the capacitor voltage

$$y = \frac{q}{C}$$

developed in the previous chapter, we can express the voltage developed across the capacitor by

$$= \frac{1}{C} \int i \, dt$$

Finally, substituting the approximate expression above for the current, i, we obtain the approximate relation

$$y \simeq \frac{1}{RC} \int x \, dt$$

which states (in words) that the output voltage is approximately equal to the integral of the input voltage (with respect to time) divided by the product of the resistance and capacitance (RC). (The RC product is known as the time constant of the circuit, because it determines the time required for the capacitor to be charged or discharged.)

We can easily see why the expression obtained above is only approximate, and why the circuit of Fig. 30 (a) performs true integration but for a brief instant. As charge is built up on capacitor C, its voltage rises and starts to oppose the input voltage (x). This reduces the rate of flow of current through the resistor and the current is no longer given by the relation $i = x/R$ (Ohm's law). The capacitor voltage then builds up more slowly in exponential fashion, and it eventually reaches, but can never exceed, the input voltage (x). In contrast, true integration according to the formula developed above would result in a continually increasing output voltage. Assume, for example, that the input voltage has a constant value, x, such as provided by a battery or d.c. generator. With x being a constant we can remove it from the integral sign ($\int$) in the previous expression, and obtain

$$y \simeq \frac{x}{RC} \int dt = \frac{x}{RC} t$$

(since $\int dt = t$ by definition).

Thus, for true integration of a d.c. input voltage, the output voltage would increase linearly and without limit, as portrayed by the dotted straight line ($y = \frac{x}{RC} t$) in Fig. 30 (b). Actually, however, the output voltage increases exponentially to the value of the fixed input voltage ($y = x$), as is shown by the solid curve in Fig. 30 (b). Note that for a very brief interval, when the elapsed time is short compared to the time constant RC, the exponential charging curve (solid line) follows approximately the ideal (dotted) curve, and for this brief interval true integration is taking place. For this reason, the use of the simple R–C integrator is usually confined to the initial, linear portion of the charging curve. In practical analogue computers various circuits are employed for 'linearizing' the integrator curve. Generally an operational amplifier with 'feedback' is used for this purpose, as we shall see in the next chapter.

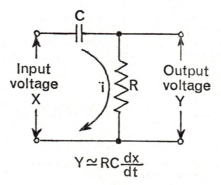

Fig. 31. Differentiation by capacitor and resistor.

Differentiation by R–C Network. Differentiation is the inverse of integration. A differentiating circuit, therefore, must produce an output voltage that is proportional to the rate of change of the input voltage. A simple R–C differentiating circuit is shown in Fig. 31; it differs from the integrator

(Fig. 30) only in that the output voltage (y) is taken across the resistor (R) rather than across the capacitor (C).

To see why the circuit differentiates let us develop the approximate relation for the voltage across the resistor (output voltage). By Ohm's law the voltage drop across the resistor at any time must equal the product of the current (i) and the resistance (R), or the output voltage $y = iR$. By definition (as developed in the last chapter), the charging current (i) is the time rate of change of the charge, or $i = \dfrac{dq}{dt}$. Substituting for i in the expression above, we obtain

$$y = R \frac{dq}{dt}$$

We can obtain the functional relation to the input voltage, x, by recalling that, by definition of capacitance, the charge on the capacitor,

$$q = Cx$$

The ratio of an infinitesimal change in charge (dq) to a correspondingly small change in time (dt) must, in accordance with the definition above, equal the capacitance times the ratio of an infinitesimal change in input voltage (dx) to a similar small change in time (dt), or expressed in equation form

$$\frac{dq}{dt} = C \frac{dx}{dt}$$

which is, of course, simply the derivative with respect to time of the relation $q = Cx$. Substituting for dq/dt in the previous equation, for the output voltage, we obtain, finally,

$$y = R \frac{dq}{dt} \simeq RC \frac{dx}{dt}$$

In words, the output voltage (y) is approximately equal to the product of the time constant (RC) and the rate of change of the input voltage with respect to time (dx/dt), which is the desired result.

We use the phrase and symbol ($\simeq$) for 'approximately equal to' since, for essentially the same reason as in the integrator, true differentiation does not take place in the circuit of Fig. 31. The voltage drop developed across the resistor opposes the flow of charge required by the formula, which causes the output voltage (y) to decay exponentially to zero (when the capacitor is charged), rather than to drop abruptly to zero, as would happen for true differentiation. You can easily see that for a constant (d.c.) input voltage (x) the time rate of change or 'derivative' of the input voltage, $\dfrac{dx}{dt}$ must equal zero. (The derivative of a constant is always zero.) Actually, for the reasons explained, the output voltage reaches zero only after a considerable time (equal to several times RC) has elapsed. As in the case of the integrator, the cure for this non-linear behaviour is an operational (feedback) amplifier.

REVIEW AND SUMMARY

Mechanical addition or subtraction can be accomplished by means of pulley-and-chain arrangements or bar linkages (Figs. 18 and 19). These operate by adding or subtracting linear displacements.

A differential gear assembly is essentially an arrangement of four operating gears and one auxiliary gear that can add or subtract the angular displacements (shaft rotations) of its two input shafts (Fig. 20).

Multiplication or division by a constant can be achieved by meshing two gears whose ratio of teeth or pitch (gear ratio) is the desired constant (Fig. 21).

Generalized multiplication and division by a continuously variable factor is accomplished by means of variable-speed drives, such as the disc-and-wheel device shown in Fig. 22. The output shaft rotation of a disc-and-wheel device is equal to the product of the input shaft rotation and the ratio of the disc radius to the wheel radius

$$(v = \frac{a}{b} u)$$

Division occurs if the disc radius (a) is smaller than the wheel radius (b).

The total output shaft rotation (v) of a disc-and-wheel (or ball-and-wheel) device represents the summation of the products of a variable quantity (disc radius a) by small (infinitesimal) increments of another quantity (input shaft rotation du); i.e. $v = \int a du$. The arrangement, therefore, is suitable for mechanical integration.

The generation of mathematical functions can be accomplished mechanically, either by placing the function (curve) on the drum of a curve follower or by causing a cam follower to 'ride' on a cam plate that is shaped in accordance with the desired function. By using a shaped cam cylinder, the relation between three variables can be represented mechanically (see Figs. 25, 26, and 27).

Passive electrical components (resistors, potentiometers, capacitors, inductors, and transformers) can be used for mathematical computations in accordance with the physical laws governing their behaviour. For arithmetic operations linear components (those that follow Ohm's law) are required.

Electrical multiplication or division by a constant is achieved by either a fixed or variable voltage divider (potentiometer) or a transformer whose turns ratio is the desired constant (see Fig. 28).

Electrical addition or subtraction of voltages, each multiplied by a constant coefficient, can be attained by means of resistor networks, whose behaviour is governed by Kirchhoff's laws (Fig. 29). Considerable losses (attenuation) and loading occur.

If an input voltage (x) is applied across a series resistance–capacitance combination, the output voltage (y), taken across the capacitor, varies approximately as the integral of the input voltage ($y \simeq \frac{1}{RC} \int x \, dt$). True integration takes place only for an interval that is short compared to the time constant (product $R\,C$), since the output voltage (charging curve) slopes off exponentially from the straight line representing the integral (Fig. 30) because of the back voltage developed across the capacitor.

If the output voltage (y) is taken across the resistance of a series R–C combination (Fig. 31) approximate differentiation of the input voltage (x) applied across the combination results

$$\left(y \simeq RC \frac{dx}{dt} \right)$$

True differentiation in an R–C differentiator takes place only for intervals that are long compared to the time constant (RC) of the circuit.

CHAPTER 4

BUILDING BLOCKS OF ANALOGUE COMPUTERS—II: OPERATIONAL AMPLIFIERS

We shall now add a sprinkling of electronic and electromechanical devices to the mechanical and electrical ingredients we have already studied, in order to broaden and perfect our knowledge of the analogue computing art. In the last chapter we saw that the fundamental operations of arithmetic and calculus are easily performed by simple electrical components and circuits containing only resistors and capacitors. However, each of these simple devices had basic flaws, which made their use rather limited and sometimes altogether impractical. The resistive summing networks had large losses, which made the output always less than the sum of the inputs, and moreover, any fair number of them 'loaded down' the circuits more than could be permitted. Our voltage dividers and potentiometers ('pots') multiplied and divided, but only by a constant; if you wanted to change the constant you either had to get a new set of resistors or change the setting of the coefficient 'pot' by hand. The $R-C$ integrators and differentiators also worked in some fashion, but true integration and differentiation was limiting to the 'linear' portion of the capacitor charging curve; that is, only for tiny output voltages and brief time intervals. The common flaw in all these devices was the inevitable loss of 'passive' electrical components, and in the case of the potentiometer multiplier, the inconvenience of changing settings by hand for multiplication by variable multipliers.

It appears that the cure for these defects would be the addition of 'active' components, such as electronic amplifiers, to overcome the losses of the passive devices, and of electric motors to drive the multiplier 'pots'. Since this is, indeed, correct and rather obvious, you might well wonder why these required additions were not made, say, in the early 1920s when amplifiers and electric motors were familiar items, rather than in the 1940s primarily as a result of Second World War developments. The answer is that any old amplifier or electric motor just will not do. The amplifiers of the 1920s lacked the all-important feedback circuits, which provide stability and low distortion. As the valves and other components got older, the amplification would drop off and the general performance would deteriorate badly. You can imagine what this would do to the accuracy of a computer, based upon constant, stable quantities. Similarly, while excellent, bulky electric motors were available many decades ago, the development of small, efficient fractional-horsepower motors is of more recent origin. Moreover, the use of an electric motor to change the setting of a multiplier pot is really not much better than doing the job by hand. You have to throw a switch to turn the motor on and then turn it off at the exact required setting of the potentiometer for the next multiplication; this would undoubtedly involve some adjusting and 'backing up'. You would not call this automatic operation. The concept of a self-correcting automatic control loop—the servomechanism—had to be developed first, before the motor could do the job automatically.

The next two chapters deal with some of these electronic and electromechanical refinements of electrical analogue devices, which gave rise to the

Operational Amplifiers

present-day automatic analogue computers. We shall not delve too deeply into the design and operation of feedback amplifiers and automatic control (servo-) mechanisms, since they enter the computer field only incidentally as building blocks (so-called 'black boxes') for performing mathematical operations. If you would like more information about the amplifiers and servomechanisms themselves you should refer to the many popular texts available.

FEEDBACK IN AMPLIFIERS

The operational amplifier used in analogue computers is a special type of high-gain d.c. amplifier utilizing large amounts of feedback. Let us assume for a moment that an amplifier is an electronic assembly that strengthens a voltage applied to its input without drawing appreciable current (or power) from the voltage source. A high-gain amplifier provides an output voltage that is many times (more than 100,000,000 in modern units) the value of the input voltage. A d.c. amplifier, finally, amplifies any signal of any frequency from d.c. on up to a certain maximum a.c. frequency, which depends upon the characteristics of the amplifier. Using these definitions and considering the amplifier as a 'black box' for the present, we must now turn to the feedback principle, which makes possible the use of high-gain d.c. amplifiers in analogue computers.

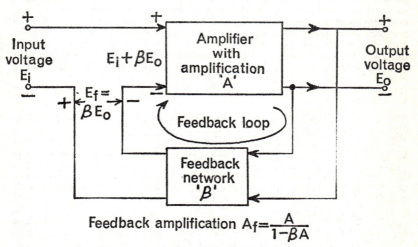

Fig. 32. Functional block diagram of feedback amplifier.

Basic Feedback Amplifier

Fig. 32 shows the schematic set-up of an amplifier incorporating a 'feedback loop'. In the absence of feedback, a voltage, E_i, applied to the input terminals is amplified A times by the amplifier, so that the output voltage

$$E_o = AE_i$$

This relation also defines the voltage amplification without feedback, which is known as the forward or 'open-loop' amplification $A = \dfrac{E_o}{E_i}$.

(It is sometimes, incorrectly, called 'the gain'.) Let us now feed a portion (β) of the output voltage back to the input of the amplifier by means of a feedback network inserted into the 'feedback loop'. In the simplest case, which is of interest here, the feedback network consists of two resistors which divide the output voltage of the amplifier in a certain ratio,

$$\beta = \frac{E_f}{E_o},$$ called the feedback factor,

where E_f is the voltage fed back to the input and β is the ratio or fraction of this feedback voltage to the total output voltage, E_o. If the feedback network contains reactive or active (amplifying) components instead of simple resistors, the feedback factor, β, is not a simple fraction, but will be a complex quantity with a phase angle; but the important feature is that β is always independent of any external load connected to the amplifier.

The feedback voltage, $E_f = \beta E_o$, adds to the externally applied input voltage (E_i), so that the total input voltage at the amplifier terminals consists of their sum, or the total input voltage (with feedback) = $E_i + \beta E_o$. This total input voltage multiplied by the voltage amplification, A, of the amplifier must, of course, equal the output voltage, E_o, or

$$E_o = (E_i + \beta E_o)A$$

Now, if we want to know the voltage amplification of any amplifier (whether or not it has feedback), we simply measure the output and input voltages and compute their ratio, since this is what is meant by amplification. Hence, to compute the voltage amplification in the presence of feedback, known as the 'closed-loop' amplification (symbol A_f), we must again form the ratio of the output to the input voltage, E_o/E_i. Solving the expression above for this ratio (E_o/E_i), we obtain for the closed-loop voltage amplification (with feedback),

$$A_f = \frac{E_o}{E_i} = \frac{A}{1 - \beta A}$$

where A is the open-loop amplification of the amplifier (without feedback) and $\beta = E_f/E_o$, as previously defined.

Positive and Negative Feedback. Our discussion so far has been based upon the assumption that the feedback factor, β, is a positive quantity, so that the denominator $(1 - \beta A)$ of the expression for A_f above is always less than 1. Consequently, with positive feedback ($\beta = +$), the voltage amplification in the presence of feedback, A_f, is always greater than the normal (open-loop) amplification without feedback. Since for positive feedback the voltage fed back directly reinforces the applied input voltage, it is also known as regenerative feedback. Unfortunately, positive or regenerative feedback increases the instability, noise, and distortion of the amplifier in the same proportion as it increases the voltage amplification, which is wholly undesirable in a computer-type amplifier. What we require, therefore, is negative feedback.

Elementary texts on electronics show that a single amplifier stage (valve or transistor) or an odd number of stages reverse the polarity (also called phase) of an output voltage with respect to that of the input voltage. This means that a positive (or positive-going) voltage applied to the input of an odd number of amplifier stages will automatically come out negative (or negative-going) at the output, and vice versa. Thus, by tapping off a fraction of the output voltage from an odd number of amplifier stages and feeding it back to the input, we automatically obtain negative feedback. The negative feedback voltage, $-\beta E_o$, subtracts rather than adds to the input voltage (E_i), so that the output voltage and amplification are decreased with respect to that available without feedback. Because of this action, negative feedback is also called degenerative feedback. To find the voltage amplification with negative (degenerative) feedback, we simply substitute a negative feedback factor, $-\beta$, in the previous formula and obtain the voltage amplification with negative feedback,

$$A_f = \frac{A}{1 + \beta A}$$

Stability. You can see from the formula above that the amplification with negative feedback is always considerably less than the amplification obtained without any feedback at all. For example, if the amplification without feedback (A) of an amplifier is 1000, and 10 per cent of the output ($\beta = 0.1$) is fed back to the input, the amplification with negative feedback

$$A_f = \frac{A}{1 + \beta A} = \frac{1000}{1 + (0.1 \times 1000)} = \frac{1000}{101} \simeq 10$$

That is, the negative feedback has decreased the amplification by a factor of 100 in this case. You may well wonder what the advantage of this procedure is. As a matter of fact, the advantages are manifold. It can be shown by simple physical and mathematical reasoning that in a negative feedback amplifier the distortion and noise are reduced, and the frequency response and stability are improved, in the same proportion as the amplification is reduced.

In a computer-type amplifier we are primarily interested in stability to assure the accuracy of computations. In particular, if the amplification (A) and the feedback factor (β) are large, so that their product (βA) is very large compared with 1, we can neglect '1' in the formula for voltage amplification with negative feedback, and write

$$\text{for } \beta A \gg 1, \ A_f = \frac{A}{1 + \beta A} \approx \frac{A}{\beta A} = \frac{1}{\beta}.$$

That is, the voltage amplification with feedback is simply the reciprocal of the feedback factor ($1/\beta$), and hence it is entirely independent of the open-loop amplification (A) of the amplifier. This is indeed a remarkable result. We already know that the feedback factor (β) depends only upon the ratio of two resistors across the output and is independent of any load connected to the amplifier. Now we learn in addition that, for high (open-loop) amplification and large feedback, the (closed-loop) voltage amplification is independent of the amplifier itself and is fixed by the feedback factor alone. You can easily see that this leads to a pronounced improvement in stability. Amplifiers are usually prone to a number of factors causing instability, such as voltage fluctuations, ageing of valves or transistors, differences due to replacements of components, and general deterioration of performance with age. Now we can see that none of these instability factors make any difference, as long as the amplification remains sufficiently high to make the βA product large compared with 1. (If the amplification drops too low, the approximate formula will, of course, be no longer valid.) Even for our example, with a relatively low open-loop amplification ($A = 1000$) and a feedback factor β of 0.1, the amplification with feedback is $1000/101 = 9.9$, or is almost equal to the reciprocal of the feedback factor, $1/\beta = 1/0.1 = 10$. In other words, the amplification is almost independent of the amplifier. In computer-type feedback (operational) amplifiers the open-loop amplification is more likely to be in the order of one hundred million.

Operational Feedback Amplifier

Let us now consider a special type of high-gain, negative-feedback amplifier, known as an operational amplifier because it permits a number of mathematical operations. Fig. 33 illustrates the block diagram and schematic representations of an operational amplifier. There are several significant differences between this and the standard voltage-feedback arrangements. Note that one (−) terminal of the input voltage source is common with the corresponding (−) amplifier output terminal. Furthermore, a feedback resistor (R_f) is bridged directly from the (+) input to the (+) output terminal of the amplifier, instead of the usual feedback arrangement consisting of a voltage divider across the amplifier output. As a result of this coupling between input

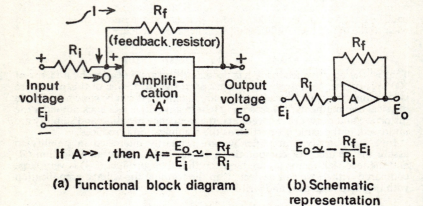

Fig. 33. The operational amplifier.

and output around the amplifier, an output signal will always appear even if the amplification drops to zero. (This is indicated by the path of the current, I, from input to output through R_i and R_f.) The amplifier proper consists of an odd number of stages, so that negative feedback is obtained. The amplification and the feedback are made very large for extreme stability. Because of the large degenerative (negative) feedback voltage, which blocks the input voltage (E_i), the actual input signal to the amplifier is quite small—almost zero. As a result, a virtual earth or short circuit exists across the amplifier input, although no current flows through this short. The current provided by the input voltage source, E_i, actually flows past the short through R_f.

Amplification. It is not too difficult to derive a formula for the closed-loop amplification (with feedback) of the operational amplifier in terms of the open-loop amplification (without feedback) and the two resistors, R_i and R_f. You can use either the principle of superposition or Kirchhoff's first law to work this out for yourself, as an exercise. The answer can be put into the following form:

Amplification with feedback,

$$A_f = \frac{E_o}{E_i} \simeq -\frac{R_f}{R_i}\left[\frac{1}{1 + \frac{1}{A}\left(1 + \frac{R_f}{R_i}\right)}\right]$$

where A is the open-loop amplification (i.e. without feedback).

In terms of the situation pictured above, where the open-loop amplification (A) and the feedback voltage are both very high, so that the actual input signal is practically zero and a virtual short circuit exists at the amplifier input, a simplified formula for the voltage amplification can easily be derived.

Since this situation prevails in all practical operational amplifiers, such a formula is more significant than the complete one given above. Thus, for the special case of high amplification and large feedback, the operation of the circuit is essentially independent of the amplifier and we may simply consider the current flow in the resistors R_i and R_f. By Ohm's law the current in R_i due to the input voltage E_i, is E_i/R_i. Similarly, the current in R_f, due to the output voltage E_o, is E_o/R_f. By Kirchhoff's first law, the sum of these two currents flowing into the input junction '0' [in Fig. 33 (a)] must equal zero. Hence,

$$\frac{E_i}{R_i} + \frac{E_o}{R_f} = 0$$

Solving for the output voltage, E_o, we obtain

$$E_o = -\frac{R_f}{R_i} E_i$$

which shows that the output voltage is simply the input voltage (reversed in polarity) multiplied by the ratio of the feedback resistance (R_f) to the input resistance (R_i). The voltage amplification with feedback (A_f) is, of course, the ratio of the output voltage to the input voltage, or

$$A_f = \frac{E_o}{E_i} \simeq -\frac{R_f}{R_i} \quad (A \gg 1)$$

where we use the 'approximately equal to' sign ($\simeq$) since the formula is valid only when the open-loop gain is very high compared to 1 ($A > 1$). (You can, of course, obtain the same result directly from the previous formula by letting A approach infinity.) This result shows that the feedback amplification depends only upon the ratio of the resistances and is independent of the open-loop amplification, provided the latter is very high compared to unity. By making the resistances equal (i.e. $R_f/R_i = 1$), you can see immediately that for this case the operational amplifier simply reverses the polarity (phase) of an input voltage, since then

$$E_o = -E_i \qquad (R_f = R_i)$$

Thus, for $R_f = R_i$, the operational amplifier performs the mathematical operation of sign changing (multiplication by -1). Moreover, sign changing will automatically be included in all the operations of the amplifier we shall study.

Basic Characteristics of Operational Amplifiers

The basic purpose of operational amplifiers is to overcome the losses (attenuation) and loading incurred when a number of passive computing networks are connected together. As each additional element is 'shunted' across a passive network, it further loads down the network, drags down the input impedance, and causes additional losses. The input impedance of a network represents the total opposition to current flow, and it must be high enough so as not to draw an appreciable current from the voltage source, which would change its value. At the same time the output impedance (or resistance) of a computing network must be quite low so that the network is not affected by a load resistance 'shunted' across the output. (The effect of a load upon a network is what is meant by 'loading'.)

The high-gain operational amplifier with large feedback admirably fulfils these requirements. Because of the feedback the output impedance is very low, and the closed-loop amplification is virtually independent of the load or the amplifier itself. At the same time the input impedance of the operational amplifier is very high (in the order of millions of ohms), so that the current drawn from an input voltage source is extremely low and its value is not affected. Moreover, the operational amplifier satisfies both important requirements without attenuating the signal, since its output depends only upon the arbitrary ratio of two resistors. In brief, the operational amplifier represents an ideal device for isolating a load from a computing network without introducing losses.

Practical Features. We shall not bother to show any operational amplifier circuits, since the complexity of their design may be confusing; for the purposes of this book we may simply think of them as 'black boxes' with certain necessary characteristics. The amplifier may consist of thermionic valves or may be transistorized. The forward amplification is very high, in the order of several hundred million (1 to 5×10^8), to attain the required stability. Practical amplifiers, though reversing the polarity of an input signal, do not introduce any appreciable phase shift over the entire operating frequency range. To reduce the drift usually present in high-gain d.c. amplifiers, the principle of chopper stabilization is generally utilized. In this arrangement the output of a conventional d.c. amplifier is combined with that of a 'modulated-carrier' type a.c. amplifier, which uses a synchronous vibrator (chopper) to convert the d.c. to an a.c. input, amplify, rectify, and filter it. The chopper-type amplifier is exceptionally free from drift, and combined with the excellent high-frequency response of the conventional d.c. amplifier, the combination is superior to either type alone.

COMPUTING WITH OPERATIONAL AMPLIFIERS

We now have a superb tool for performing a variety of mathematical operations—the operational amplifier. The output voltage, as we have seen, depends only upon the input voltage multiplied by the ratio of the feedback to the input resistances (or impedances). By deploying this ratio and the inputs in various ways, we can carry out almost any operation in algebra, and calculus, except efficient general multiplication and function generation. For these latter two functions we shall enlist the help of the servomechanism, to be explained in the next chapter.

1 MULTIPLICATION BY A CONSTANT (SCALE CHANGING)

The basic operational amplifier (Fig. 33) changes the sign (polarity) of the input voltage and multiplies it by a constant factor, as you will recall, since the output voltage of the amplifier is given by the approximate relation

$$E_o = -\frac{R_f}{R_i} E_i$$

By selecting the two resistors R_f and R_i in the required ratio, the input voltage may be multiplied by any desired constant coefficient, an operation that is also known as scale changing. The formula remains valid as long as the forward amplification is very high, and hence the accuracy of the scale change depends only upon the precision and physical stability of the two resistors. You can easily visualize that either resistor R_i or R_f may be made adjustable or continuously variable, in which case multiplication by a variable coefficient (variable scale change) is realized. If R_f is made variable the output voltage varies linearly with the resistance of R_f; if R_i is made variable the output voltage varies as the reciprocal of the resistance R_i. This is not the same, of course, as generalized multiplication by two variable functions, since the multiplier—though manually adjustable—remains fixed for each multiplication.

2 ADDITION OF VARIABLES—THE SUMMING AMPLIFIER

Fig. 34 shows two representations of a summing amplifier for the addition of three variable quantities (voltages E_1, E_2, and E_3). The principle illustrated can be extended, of course, to any number of voltages representing different mathematical quantities. You will realize immediately that the summing amplifier is simply a combination of the summing network of Fig. 29 and the operational amplifier of Fig. 33. In contrast to the simple summing network, however, the output voltage of a summing amplifier can be equal to the (algebraic) sum of the input voltages, except for a change in sign (polarity reversal). On the other hand, if the scale of any input voltage is to be adjusted by any desired constant multiplier before adding it to the other input quantities this operation, too, can easily be carried out with a summing amplifier.

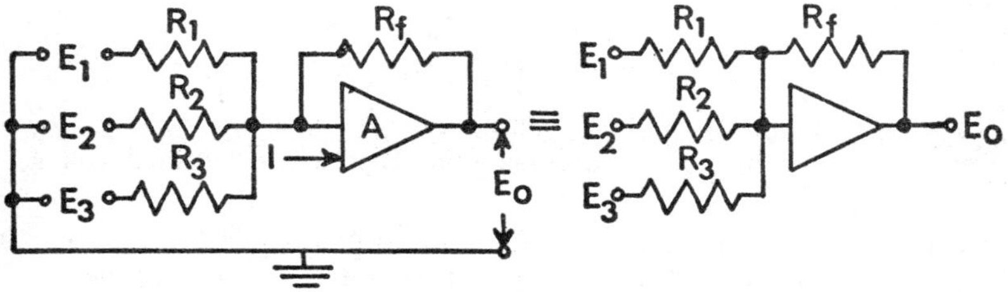

(a) Schematic diagram   (b) Equivalent symbol

$$E_O = -\left[\frac{R_f}{R_1}E_1 + \frac{R_f}{R_2}E_2 + \frac{R_f}{R_3}E_3\right]$$

Fig. 34. Addition with an operational amplifier (summing amplifier).

Basic Adder Equation. To obtain the basic equation for addition with a summing amplifier, we can use our previous approximate method of considering only the currents flowing in the resistors and ignoring the effect of the amplifier. You will recall that this procedure is valid as long as the forward amplification is high and a virtual short circuit exists across the amplifier input terminals. (These requirements are always fulfilled in any actual operational amplifier.) Thus, by Ohm's law, the currents in input resistors R_1, R_2, and R_3 are $\frac{E_1}{R_1}$, $\frac{E_2}{R_2}$, and $\frac{E_3}{R_3}$, respectively. Similarly, the current in the feedback resistor R_f due to the output voltage, E_o, is $\frac{E_o}{R_f}$. By Kirchhoff's first law, the sum of these currents (I), flowing into the common amplifier input junction, must equal zero. Setting the sum of the currents equal to zero, we obtain

$$\frac{E_1}{R_1} + \frac{E_2}{R_2} + \frac{E_1}{R_3} + \frac{E_o}{R_f} = 0$$

Solving this expression for the output voltage, E_o, gives the desired result:

$$E_o = -\left[\frac{R_f}{R_1}E_1 + \frac{R_f}{R_2}E_2 + \frac{R_f}{R_3}E_3\right]$$

In words, the output voltage of a summing amplifier is proportional to the sum of the input voltages (with the sign changed). Subtraction, rather than addition, may be performed by introducing input voltages that are negative (−) in polarity. The scale of each input voltage may be multiplied by a desired constant by selecting the appropriate value for the corresponding input resistor, as we shall see in some practical cases in a moment. If all resistors are made equal, pure addition (or subtraction) with a sign change is obtained. For, if $R_1 = R_2 = R_3 = R_f$, then

$$E_o = -(E_1 + E_2 + E_3)$$

In contrast, if all input voltages are to be multiplied by the same coefficient before addition, the input resistors are all made equal to a value R, and the ratio of the feedback resistor, R_f, to the value of R is chosen to give the desired constant coefficient (resulting in an equal scale change). The relation between input and output in this case is

$$E_o = -\frac{R_f}{R}(E_1 + E_2 + E_3)$$

Typical Practical Set-up. A typical problem from analogue computer practice is illustrated by Fig. 35. Six quantities, represented by input voltages

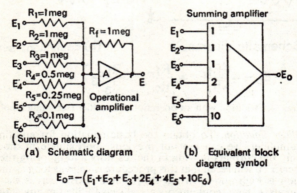

Fig. 35. Practical example of addition with summing amplifier and corresponding machine equation.

$E_1, E_2, \ldots E_6$, are to be summed by the amplifier. Voltages E_1, E_2, and E_3 are to be added directly, E_4 is to be multiplied by a factor of 2 before addition, E_5 by a factor of 4, and E_6 by 10. To obtain multiplication by 2, the input resistor for E_4 (i.e. R_4) must, of course, equal $\frac{1}{2}$ of the value of the feedback resistance R_f; for multiplication by 4, the corresponding input resistor must be $\frac{1}{4}$ of R_f; for multiplication by 10, the input resistor must be $\frac{1}{10}$ of R_f, and so on. Thus, a practical set of values might be as follows [see also Fig. 35 (a)]:

$R_f = 1$ megohm (1 million ohms)
$R_1 = R_2 = R_3 = 1$ megohm each
$R_4 = 0.5$ megohm
$R_5 = \frac{1}{4}$ or 0.25 megohm
$R = 0.1$ megohm or 100,000 ohms

Substituting these values in the general equation for the output voltage, we obtain the so-called 'machine equation' of the summing amplifier:

$$E_o = -(E_1 + E_2 + E_3 + 2E_4 + 4E_5 + 10E_6)$$

Fig. 35 (*b*) shows a commonly used block diagram symbol for such an arrangement, with the applicable constant multipliers (coefficients) indicated next to each input voltage terminal.

Parallel Inputs. Fig. 36 illustrates how the 'machine equation' of our previous example can be mechanized in another way, by means of parallel inputs. Inputs E_1, E_2, and E_3 are the same as before (see Fig. 35) and use 1-megohm resistors, each. Input voltage E_4, however, is applied to two different amplifier inputs through a 1-megohm resistor each. The quantity is, thus,

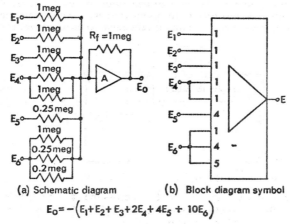

(a) Schematic diagram   (b) Block diagram symbol

$$E_O = -(E_1 + E_2 + E_3 + 2E_4 + 4E_5 + 10E_6)$$

Fig. 36. Example of Fig. 35 mechanized by means of parallel inputs.

added to itself, and $2E_4$ results with two parallel inputs. Input voltage E_5 is applied through a 0·25-megohm resistor, just as before. Input voltage E_6 is applied to three parallel amplifier inputs: one is through a 1-megohm resistor, resulting in E_6; the second is through a 0·25-megohm resistor, resulting in $4E_6$; the third is through a 0·2-megohm resistor, resulting in $5E_6$. Thus E_6 is added to itself in proportions of 1, 4, and 5, resulting in a total of $10E_6$, as shown. Fig. 36 (*b*) illustrates the block diagram symbol representing the parallel-input type of arrangement. Many other combinations are possible, of course, which may be arranged to suit the requirements.

3 THE INTEGRATION OF FUNCTIONS

We have already become acquainted (in the last chapter) with a slightly imperfect integrator, consisting of a simple resistance–capacitance network. You will recall that the passive *R–C* combination did integrate, in a fashion, for a brief instant, but as soon as a charge began to build up on the capacitor the rising 'counter-voltage' across the capacitor would oppose the charging current and make a shambles of our integrator. Instead of a linearly rising output voltage for a d.c.-voltage input, representing its integral, we obtained an exponential output curve that was by no means linear [see Fig. 30 (*b*)].

The Perfect Integrator. It would appear that the problem posed by the *R–C* integrator could be solved, if we could—somehow—overcome the effects of the capacitor countervoltage, that is, if we could always maintain a charging current that is proportional to the input voltage, regardless of the voltage built up across the capacitor. With a constant flow of charging current being maintained (for a constant input voltage), the voltage across the capacitor would rise in the linear fashion depicted in Fig. 30 (*b*), and we would have a perfect integrator. Ideally, then, we need some sort of active element, such as a current generator, which keeps pumping charging current into the capacitor against the rising countervoltage. Moreover, this current generator must be a perfect specimen, which does not waste any portion of the applied input voltage (i.e. has no losses) and generates a current that is exactly proportional to the input voltage.

Fig. 37 illustrates the scheme for such a perfect integrator. An input voltage, E_i, is applied to a perfect current generator, which produces an output current (I) that is proportional to the input voltage. (To eliminate the proportionality constant we have made the current equal to the input voltage; that is,

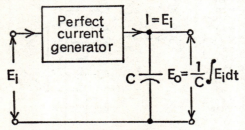

Fig. 37. The 'perfect' integrator.

$I = E_i$.) The output voltage, E_o, built up across the capacitor, is—as we saw earlier—the time integral of the charging current, or in mathematical form

$$E_o = \frac{1}{C}\int I\, dt = \frac{1}{C}\int E_i\, dt \text{ (since } I = E_i\text{)}$$

Thus, with this arrangement the output voltage is exactly proportional to the time integral of the input voltage. You may want to compare this ideal result of a theoretically perfect integrator with the performance of the various electronic integrators which follow.

The Buffer Amplifier. Since perfect current generators do not exist, we shall consider combinations of the *R–C* integrator with various amplifier arrangements. An amplifier with a high input resistance (consuming little of the applied power) and a low output resistance (unaffected by the load) is, of course, an approach towards a perfect current generator, since—if it is any good at all—its output will be approximately proportional to its input voltage. In fact, you will find that the insertion of any good amplifier between the output of an *R–C* integrator and the load vastly improves the integrator performance. Such a 'buffer amplifier', as it is called, isolates the integrator from the load and overcomes the counter e.m.f. of the capacitor to some extent, but its performance is still far from perfect. You will still need very large values of *R* and *C* (i.e. a large time constant *RC*) if the capacitor charging curve is to be relatively linear for any appreciable time. (You will recall that the charging curve of an *R–C* circuit is linear only during an interval that is short compared to the time constant, *RC*. True integration is obtained only during this interval. By increasing the time constant, the interval can be extended.) If *R* and *C* are made very large, however, the output voltage becomes very small, since it is inversely proportional to RC. Moreover, a large capacitance is usually accompanied by a measurable leakage resistance parallel with it, and which will 'load' the amplifier. Finally, grid current flowing through the leakage resistance may charge the capacitor and cause it to 'drift' from the correct voltage values.

The 'Bootstrap' Integrator. The addition of a feedback circuit to the buffer amplifier overcomes objections to it and makes it an almost perfect integrator. Fig. 38 illustrates this type of circuit, which is known as a 'bootstrap' integrator, since the capacitor voltage is lifted, so to speak, by its own bootstraps.

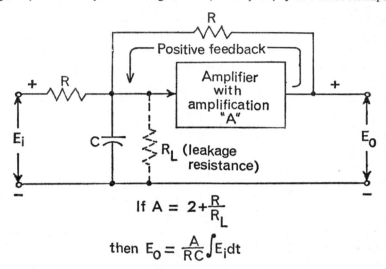

Fig. 38. Functional diagram of a 'bootstrap' integrator.

The bootstrap integrator depends for its operation on the application of positive (regenerative) feedback to overcome the effect of the capacitor countervoltage. To obtain positive feedback, the amplifier output voltage must have the same polarity as its input voltage, which means that an even number of stages must be used. The countervoltage built up on the capacitor (C) is applied to the input of a d.c. amplifier with a positive forward amplification A. The amplified output voltage is fed back through the resistor R to the integrating capacitor (C) and gives it just a sufficient voltage boost to permit it to charge in a linear manner, as required by a perfect integrator. If the amplification (A) and the feedback resistor (R) are correctly chosen the voltage across the capacitor, and hence also the output voltage E_o, are proportional to the time integral of the input voltage, E_i.

In the circuit of Fig. 38 it can be shown that for an amplification,

$$A = 2 + \frac{R}{R_L}$$

(R_L is the leakage resistance across capacitor C), perfect integration of the input voltage is obtained; that is, the output voltage,

$$E_o = \frac{A}{RC} \int E_i \, dt$$

If the capacitor leakage resistance, R_L, is very high compared to R, the voltage amplification, A, of the amplifier must equal approximately 2 for true integration.

A number of variations on the bootstrap principle are encountered in practice. For example, the positive feedback may be applied to the input voltage (E_i) to boost its value, rather than to the capacitor (C), with essentially the same effect. It can be shown that in the latter circuit the voltage amplification must be unity ($A = 1$) for correct integration.

The Operational Integrator. Let us now see how the operational amplifier can be used as a near-perfect integrator. The operational integrator, shown in Fig. 39, is almost universally used in analogue computers because it performs

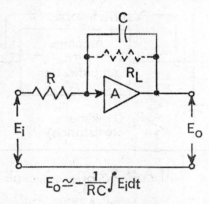

Fig. 39. The operational amplifier as integrator.

accurate integration with low losses and practicable values of the resistance and capacitance. Its operating principle is slightly different from the bootstrap integrator, as we shall see presently.

Note that in the operational integrator the capacitor of the R–C integrating network is not placed across the input of the amplifier, but is bridged between the amplifier input and output, and thus forms part of the feedback loop. The capacitor leakage resistance is also part of the feedback path. The operational amplifier inverts, of course, the polarity of an input signal, so that negative feedback results. An input voltage, E_i, will attempt to charge the capacitor plate, connected to the amplifier input, through resistor R. Simultaneously, however, the output voltage (E_o) of the amplifier tends to charge the other capacitor plate in a direction that neutralizes the charge on the input side. As a result, the capacitor does not charge as rapidly as it normally would, and the countervoltage (opposing the input voltage) is kept quite small. The net effect of this approximate action is that the amplifier feedback voltage just neutralizes the capacitor countervoltage at the input, while the output voltage (E_o) increases proportionately with the negative time integral of the input voltage (E_i). When the voltage amplification is very high ($A \gg 1$), this approximate relation may be stated mathematically as

$$E_o \simeq -\frac{1}{RC} \int E_i \, dt$$

4 DIFFERENTIATION BY AN OPERATIONAL AMPLIFIER

Although, for practical reasons, differentiation is generally avoided in analogue computers, let us consider for the sake of completeness the use of an operational amplifier as a differentiator. Fig. 40 shows the connexion of an R–C differentiating network to an operational amplifier. Note that the resistor (R) and capacitor (C) have simply been interchanged with respect to the integrator of Fig. 39. In general, the operation of this circuit corresponds to that of the operational-amplifier integrator, with the countervoltage just being neutralized by the negative feedback of the amplifier. Again, if the

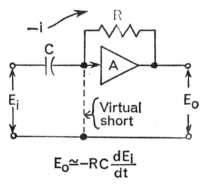

Fig. 40. The operational amplifier as differentiator.

amplification (A) is very high, so that a virtual short circuit exists across the amplifier input, the effect of the amplifier itself may be neglected. The input voltage (E_i) may then be considered to be across the capacitor, the output voltage (E_o) across the resistor, and a current (i) may be assumed to flow through C and R. (The virtual short does not consume any current.) From the definition of capacitance, the charge on the capacitor is

$$q = CE_i$$

The charging current, i, is the time rate of change of charge, or

$$i = \frac{dq}{dt} = C\frac{dE_i}{dt} \quad \text{(by substitution)}$$

Finally, the output voltage, E_o, must by Ohm's law equal the product of the current (i) and the resistance (R), so that

$$E_o = -iR = -RC\frac{dE_i}{dt} \quad \text{(by substitution)}$$

(The minus sign appears because of the amplifier's polarity reversal.) Thus, we see that the output voltage of the operational differentiator is approximately equal to the negative time derivative of the input voltage multiplied by the time constant (RC) of the differentiator.

5 COMBINED OPERATIONS

The operational-amplifier table (Fig. 41) reviews the common mathematical functions carried out with operational amplifiers in conjunction with simple input networks, and also illustrates some combined operations that can easily be undertaken. We have already considered the summing amplifier, which permits combined addition and multiplication by a constant coefficient (scale changing), as well as the operational integrator and differentiator. Because of the amplifier's polarity reversal, all these operations are accompanied by a change in sign, equivalent to multiplication by -1.

Summing Integrator. Let us now look at a few of the many operations possible by means of a combination of simple elements. The summing integrator shown in Fig. 41 is a logical combination of a summing amplifier with an

Operational amplifier table		
Schematic diagram	Description	Operating equation
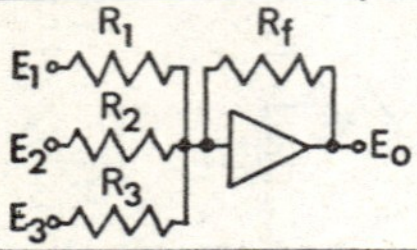 R_1, R_f inputs E_1, E_2, E_3 → E_o	Summing amplifier	$E_o = -\left(\dfrac{R_f}{R_1}E_1 + \dfrac{R_f}{R_2}E_2 + \dfrac{R_f}{R_3}E_3\right)$
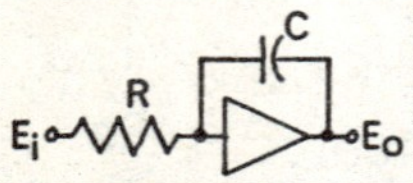 R, C feedback, E_i → E_o	Integrator	$E_o = -\dfrac{1}{RC}\int E_i\, dt$
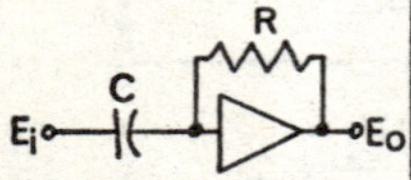 C, R feedback, E_i → E_o	Differentiator	$E_o = -RC\dfrac{dE_i}{dt}$
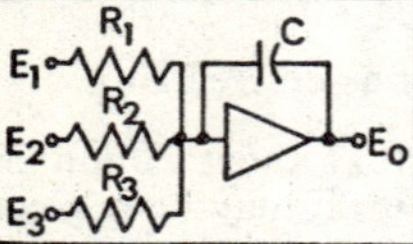 R_1, R_2, R_3 inputs, C feedback, E_1, E_2, E_3 → E_o	Summing integrator	$E_o = -\left(\dfrac{1}{R_1C}\int E_1\, dt + \dfrac{1}{R_2C}\int E_2\, dt + \dfrac{1}{R_3C}\int E_3\, dt\right)$
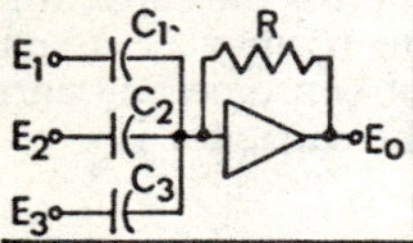 C_1, C_2, C_3 inputs, R feedback, E_1, E_2, E_3 → E_o	Summing differentiator	$E_o = -\left(RC_1\dfrac{dE_1}{dt} + RC_2\dfrac{dE_2}{dt} + RC_3\dfrac{dE_3}{dt}\right)$
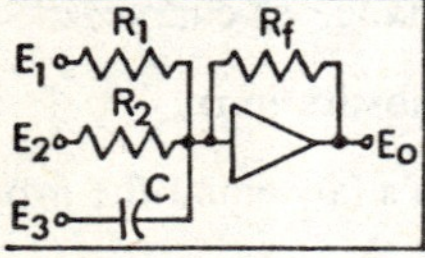 R_1, R_2, C inputs, R_f feedback, E_1, E_2, E_3 → E_o	Combination Summation of algebraic and differentiated quantities	$E_o = -\left(\dfrac{R_f}{R_1}E_1 + \dfrac{R_f}{R_2}E_2 + R_fC\dfrac{dE_3}{dt}\right)$

Fig. 41. The operational-amplifier table.

operational integrator. In the illustration three input voltages, E_1, E_2, E_3, are applied to the input of the amplifier through resistors, R_1, R_2, and R_3, respectively, which form a simple summing network. The combined (summed) input is integrated by capacitor C in conjunction with the amplifier, as described earlier. We can, therefore, write without further ado the equation of the amplifier output voltage

$$E_o = -\left(\frac{1}{R_1C}\int E_1\, dt + \frac{1}{R_2C}\int E_2\, dt + \frac{1}{R_3C}\int E_3\, dt\right)$$

which is seen to be the addition of the three integrated input quantities divided by the respective time constants (RC) of the inputs.

Summing Differentiator. A summing differentiator (Fig. 41) is obtained by applying each of several inputs through a capacitor to the operational amplifier. For the illustrated case of three input voltages, E_1, E_2, E_3, the output voltage (E_o) is simply the *sum of the three differentiated quantities*, each multiplied by the respective input time constant (RC). In equation form,

$$E_o = -\left(RC_1 \frac{dE_1}{dt} + RC_2 \frac{dE_2}{dt} + RC_3 \frac{dE_3}{dt}\right)$$

Finally, the operational-amplifier table (Fig. 41) gives an illustration of a combined operation, where two of the input voltages (E_1 and E_2) are to be summed, while the third input voltage (E_3) is to be differentiated and added to the result. Even this type of operation is possible with a single operational amplifier and the appropriate input network. Input voltages E_1 and E_2, which are to be added only, are applied through resistors R_1 and R_2, respectively. The third input, E_3, which is to be added and differentiated, is applied through input capacitor C. As in all other operations, the scale of the input voltages can be adjusted by choosing a suitable ratio for the input resistors and the feedback resistor, R_f. For the case illustrated, the output voltage then represents the addition of the two scale-adjusted input voltages E_1 and E_2, plus the differentiated third input voltage, E_3, multiplied by its time constant (R_fC). We may therefore write the equation of the output voltage as follows:

$$E_o = -\left(\frac{R_f}{R_1} E_1 + \frac{R_f}{R_2} E_2 + R_fC \frac{dE_3}{dt}\right)$$

As we shall see in Chapter 5, much more complicated mathematical operations may be carried out by inserting complex impedances (combinations of resistors and capacitors) into the feedback loop of the operational amplifier.

REVIEW AND SUMMARY

An operational amplifier is a high-gain d.c. amplifier using large amounts of feedback.

The forward or open-loop voltage amplification (without feedback), A, is the ratio of the amplifier output voltage (E_o) to the input voltage (E_i).

The closed-loop amplification (A_f) is the ratio of the amplifier output voltage to the input voltage in the presence of feedback. It is given by the formula

$$A_f = \frac{A}{1 - \beta A}$$

where the feedback factor $\beta = \frac{E_f}{E_o}$, and E_f is the voltage fed back to the input (see Fig. 32). The factor β may be positive or negative.

If the feedback factor is positive ($\beta = +$), positive feedback takes place. Positive or regenerative feedback reinforces the input voltage and increases the voltage amplification, but it also increases noise and distortion, and leads to instability or oscillations.

If the feedback factor is negative ($\beta = -$), as is the case for an odd number of amplifier stages, negative, degenerative, or inverse feedback takes place. The closed-loop voltage amplification (A_f) is then reduced with respect to the open-loop amplification (A) in accordance with the relation

$$A_f = \frac{A}{1 + \beta A}$$

If βA is very large compared to 1 the amplification is given approximately by $1/\beta$; that is, it is independent of the open-loop amplification, as well as of the load. This results in a pronounced increase in stability. Negative feedback also reduces distortion and noise, in the same proportion as it reduces amplification, and it improves the amplifier frequency response.

In an operational amplifier (Fig. 33) with very high forward amplification and a large amount of negative feedback, the actual input voltage is almost zero, so that a virtual short circuit is said to exist across the input. In such an amplifier, with an input resistor R_i bridged by a feedback resistor R_f, the closed-loop amplification A_f is given by the approximate relation

$$A_f \simeq - \frac{R_f}{R_i}$$

and the output voltage is

$$E_o \simeq \frac{R_f}{R_i} E_i$$

The feedback amplification, thus, depends only on the ratio of the input and feedback resistances; the output voltage is the product of this ratio and the input voltage, reversed in polarity.

If the feedback resistance equals the input resistance ($R_f = R_i$), the amplifier operates as a sign changer; that is, $E_o = -E_i$.

Operational amplifiers are used in analogue computers to isolate a load from a computing network without introducing losses; the amplifiers must have minimum phase shift and no drift.

Scale Change. The basic operational amplifier (Fig. 33) multiplies the input voltage by a constant coefficient, so that

$$E_o = - \frac{R_f}{R_i} E_i$$

Thus a change of scale and a change in sign are produced.

Addition. By adding a resistive summing network to an operational amplifier, a summing amplifier (Fig. 34) is obtained. If E_1, E_2, and E_3 are the input voltages, and R_1, R_2, and R_3 the respective input resistors, the output voltage, is

$$E_o = - \left(\frac{R_f}{R_1} E_1 + \frac{R_f}{R_2} E_2 + \frac{R_f}{R_3} E_3 \right)$$

Simultaneous scale change of the input voltages is accomplished by adjusting the input resistors and by paralleling the inputs.

The 'perfect' integrator consists of a current generator that pumps a charging current into a capacitor exactly proportional to the input voltage, regardless of the back e.m.f. built up on the capacitor

$$(E_o = \frac{1}{C} \int E_i \, dt)$$

The 'bootstrap' integrator (Fig. 38) uses positive feedback to overcome the effect of the capacitor countervoltage. It provides a sufficient voltage boost to allow the capacitor to charge linearly

$$(E_o = \frac{A}{RC} \int E_i \, dt)$$

In the operational integrator (Fig. 39) the integrating capacitor is bridged between the amplifier input and output and forms part of the negative feedback loop. The feedback voltage just neutralizes the capacitor countervoltage, so that the output voltage increases proportionately with the negative time integral of the input voltage

$$(E_o \simeq - \frac{1}{RC} \int E_i \, dt)$$

Operational Amplifiers

In the operational differentiator (Fig. 40) the positions of the capacitor and resistor are interchanged with respect to the integrator. The output voltage of an operational differentiator equals (approximately) the negative time derivative of the input voltage multiplied by the time constant (RC); that is

$$E_o \simeq - RC \frac{dE_i}{dt}$$

A summing integrator (Fig. 41) integrates the sum of a number of input voltages; it is a combination of a summing amplifier and integrator.

A summing differentiator (Fig. 41) computes the sum of a number of differentiated input voltages; it is a combination of differentiator and summing amplifier.

By combining the input network with appropriate feedback components, combined mathematical operations (addition, integration, etc.) may be carried out (see operational-amplifier table, Fig. 41).

CHAPTER 5

BUILDING BLOCKS OF ANALOGUE COMPUTERS—III: SERVOMECHANISMS AND FUNCTION GENERATORS

In our discussion of the operational amplifier and its uses, we have omitted two important mathematical operations for which the operational amplifier is not well suited. These are generalized multiplication of two variables and the generation of mathematical functions. In analogue computers these operations are usually performed by means of servomechanisms. A servomechanism is essentially an extension of the feedback principle to the field of automatic control. We have seen how a portion of the output of a feedback amplifier is fed back to its input, thus making the input dependent upon the output. We shall now apply this same idea to obtain automatic control of processes or mechanisms. By correcting the input of an operation in accordance with the effectiveness of the result (output), we can obtain very precise control over the operation.

Open- and Closed-loop Control. We have talked about 'open-loop' and 'closed-loop' amplification in connexion with feedback amplifiers. We now apply this concept more generally to control systems. In an open-loop or manual control system, operation is independent of the result (output) obtained. Any 'straightforward' control operation (one that does not employ feedback) is of the open-loop type. Examples are legion. Switching a motor on and off to drive some load is an open-loop operation. Turning on a heater or toasting a piece of bread are examples of open-loop control operations. Of course, you can 'close the loop' by switching off the heater when the desired temperature is reached, and by frequently checking the readiness of the toast. By making the control operation dependent on the end result you have—manually—closed the control loop. This is still manual control, however, and doesn't interest us.

When a control operation is a function of the result, or output, you have a closed-loop (automatic) control system. To obtain closed-loop operation you need some sort of sensing device that determines the effectiveness of the result (output) of the operation and feeds back the appropriate corrective information to the input of the system. Feedback is, thus, always present in a closed-loop control system. You can convert an open-loop into a closed-loop control by adding such a sensing device. Thus, by adding a thermostat to the heater which will turn it on and off in accordance with the desired room temperature, you have converted it into a closed-loop control system. Similarly, by installing in the toaster a device that can sense the temperature or colour of the piece of toast and control the operation in accordance with this output information, you have converted to closed-loop or automatic toasting.

CLOSED-LOOP (SERVO) CONTROL SYSTEM

A servomechanism is a form of closed-loop control system. The purpose of the servomechanism is to position some load, coupled to the output of the mechanism, in accordance with the position selected at the input. Moreover, the remotely located load must follow quickly and precisely any changes of position placed at the input of the system. The essential elements of such a servo control system are illustrated in Fig. 42.

The functions of the elements illustrated in Fig. 42 are as follows:

1. The input, which may be the position of a shaft or a control signal, constitutes a command to the system to carry out a desired operation.

2. The controller comprises the necessary mechanical and electrical driving means (an amplifier and a motor) for positioning the load in accordance with the input command.

3. The output of the system, or load, is the device which is to be positioned in correspondence with the input position (command).

4. The comparer, which may be some sort of sensing element, compares the position of the output with that of the input and develops an error signal that is proportional to the discrepancy, or error, between the input and output positions.

You can see that a feedback loop comprises an essential part of a closed-loop or servo control system. Information concerning the position of the load is fed back from the output of the system to the comparer (error detector), which then sends an 'error' signal to the controller that is proportional to the difference (error) between the input and output positions. This error signal is amplified to a level sufficient to actuate the motor by the servo amplifier in the controller. The motor then drives the load in such a direction as to decrease the error between the desired (input) and actual (output) load positions. When the load has been positioned exactly in accordance with the input position (command), the error signal drops to zero and the servo system stops. Thus, a servo system may be said to 'null out' (zero out) the error signal.

A Gun-turret Positioning Servomechanism

Servomechanisms were developed during the Second World War specifically for military applications in connexion with the positioning of radar aerials, anti-aircraft and anti-submarine fire control, steering of ships, etc. As an example, let us look at a typical gun-training servomechanism. You should be able to detect in the block diagram of this application (Fig. 43) all the basic elements of a closed-loop servo system shown in Fig. 42.

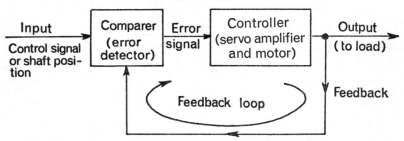

Fig. 42. The elements of a closed-loop servo control system.

A remote gun turret is to be trained on its target in accordance with a sighting mechanism and hand crank (input) operated by an artillery observer. The turret (output) is driven by a servo motor through a gearbox. The position of the turret at any time is indicated by means of a gear on the output shaft, which drives a synchro transmitter. The synchro transmitter is an electromechanical device similar to an a.c. generator, which generates a voltage that is proportional to the angular position of its input shaft. This voltage, which represents the gun-turret position, is the feedback signal that is sent back to the 'error' detector at the input. You may think of the synchro transmitter as a sort of electrical flexible shaft which can transmit positional information to a remote location. It is one of a whole family of synchro follow-up devices.

The error detector in this case is a control transformer, another type of synchro device. It is a transformer with a rotatable winding (rotor), which is coupled to the input shaft. The feedback signal is applied to the other, fixed, winding (stator) of the transformer. The output voltage from the control transformer is a function of the position of the movable (rotor) winding and the voltage applied to the fixed (stator) winding. Since the position of the rotor winding is determined by the angle of the input shaft (crank) and the

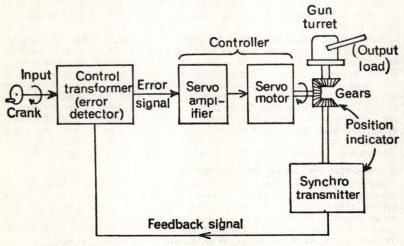

Fig. 43. Block diagram of a gun-turret positioning servomechanism.

stator voltage represents the gun-turret position, you can see that the control transformer output represents the difference, or error, between the crank (input) position and the gun-turret (output) position. By comparing the input with the output and generating an error signal in accordance with the difference between them, the control transformer thus fulfils the function of a comparer or error detector, described earlier. The error signal from the control transformer is boosted by a servo amplifier to the magnitude required for driving the servo motor. (The servo amplifier and motor together constitute the controller.) The motor corrects the position of the gun-turret so that it will follow the motion of the hand crank and decrease the error. When the error signal has been 'nulled out', the motor stops with the gun turret positioned in accordance with the angle selected by the hand crank.

A Computer Positioning Servo

Since we are interested in analogue computers rather than gun-turrets, we must now apply the servo concepts we have learned to the solution of mathematical problems. You will recall from the last chapter that we were able to multiply some quantity (voltage x) by a constant (a) by applying it to the input of a variable voltage divider (potentiometer) and adjusting its setting so that the fraction dx of the input voltage would be tapped off by the slider and made available as output [see Fig. 28 (a)]. Although we can vary the setting of the potentiometer and thereby change the value of the constant a, for any par-

Servomechanisms and Function Generators

ticular setting it will always remain a constant (less than 1). Now imagine that we can, in some way, vary the potentiometer setting rapidly and continuously, to represent the value of some other function (y). With the potentiometer setting being a function of y, the fraction of the input voltage tapped off by the slider is, of course, also a function of y, so that the output voltage equals xy. Thus, by making the potentiometer setting continuously variable in accordance with some quantity, we can multiply the input voltage (x) by that quantity (y), rather than by a constant (a). This, then, is what is required for generalized multiplication of two variables.

It would obviously not be practicable to vary the potentiometer setting manually in accordance with the variable to be multiplied, since no human operator could do this with sufficient accuracy and speed. We need, therefore, a positioning mechanism which will automatically adjust the setting of a potentiometer in accordance with a variable quantity or voltage. The computer positioning servomechanism, illustrated in Fig. 44, is designed to do this job.

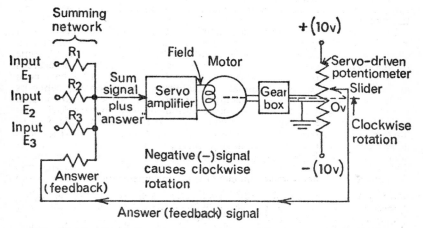

Fig. 44. Functional diagram of a computer positioning servomechanism.

Before considering the problem of the multiplication of two variables by means of a servomultiplier, let us see first how a servomechanism can position the movable arm (also called a slider, wiper, or brush) of a potentiometer in accordance with the value of a variable. In Fig. 44 we have shown three input voltages, E_1, E_2, and E_3, being applied to the servomechanism through respective resistors R_1, R_2, and R_3. This portrays a realistic situation, since usually a number of physical quantities (forces, speeds, etc.) must first be summed up before they can be multiplied by another quantity with which they are in a functional relation. Thus, resistors R_1, R_2, and R_3 represent an ordinary summing network, of the type we have met before, for adding the three input voltages E_1, E_2, and E_3. Their sum represents the actual input to the servo. The sum signal is applied to a servo amplifier, which strengthens it to a relatively high voltage of the same polarity as the input and with sufficient power to actuate the motor. The greater the input signal (or sum of the input voltages), the greater the output from the amplifier applied to the field winding of the motor, and hence the greater the speed with which the motor runs. The shaft of the servo motor is coupled through reducing gears (gearbox) to the movable slider of a potentiometer, usually wound on a flat card (and hence just called card). An input sum signal of one polarity will drive the motor and potentiometer slider in one direction, while a signal of the opposite polarity will drive the motor and slider in the opposite direction.

While we now have a means for automatically changing the setting of a potentiometer, we have not yet made sure that the amount of shaft and slider rotation is exactly proportional to the input quantities. Unless this is done, the whole arrangement is of doubtful value. In other words, we must feed back information to the input as to just how far the output shaft has turned and adjust the input to stop the servo motor when the correct amount of rotation has been attained. We will then have the kind of self-correcting, closed-loop servo system illustrated in Figs. 42 and 43. The special type of feedback used in a computer servo is known as an 'answer' signal, and it is obtained from the potentiometer itself. One potentiometer, the 'answer card', on the servo motor output shaft is devoted to the sole purpose of positioning the servo correctly; other cards are then attached to the shaft to perform multiplications and various other required mathematical operations. Let us see how this works out.

The Answer. To take a practical example, let us assume that a negative (−) sum signal (in Fig. 44) will drive the motor and potentiometer slider in a clockwise direction, while a positive signal will drive it anticlockwise. Assume further that the sum of the input voltages is negative and never exceeds −10 V. As shown in the diagram, voltages of +10 and −10 V have been applied to the fixed input terminals of the 'answer' potentiometer, while its centre tap has been earthed, placing it at zero volts (0 V). With the sum signal being negative, the slider of the potentiometer will move clockwise and pick off a positive voltage in the region above the centre tap. This positive 'answer' signal is fed back to a resistor in the input summing network and is added to the negative sum signal. As a result, the combined sum + answer signal applied to the servo amplifier is somewhat less negative, and the motor and slider run more slowly in a clockwise direction. You can see what will happen. As the potentiometer slider of the potentiometer turns clockwise and picks off an increasingly more positive answer voltage, it will cancel out more and more of the negative input (sum) signal. When the positive answer signal and the negative sum signal are exactly equal the combined sum + answer voltage input to the servo will be zero, and the motor stops. The potentiometer slider is then positioned at a point where the voltage picked off is equal in magnitude, but opposite in polarity, to the sum of the input voltages. Here, then, is another illustration of a servo 'nulling itself out' to such a position that its input sum equals zero volts.

As a concrete illustration assume that (in Fig. 44) $E_1 = +1$ V, $E_2 = +4$ V, and $E_3 = -13$ V. Ignoring, for simplicity, the constant coefficients imposed by the summing network, the total sum signal, therefore, will be $1 + 4 - 13 = -8$ V. This negative sum signal is applied to the servo amplifier and causes the output shaft and potentiometer slider to turn clockwise into the positive region of the answer pot. The positive answer signal is fed back to the input summing network and cancels out a portion of the negative input. When the potentiometer slider has been positioned to pick off an answer of $+8$ V the entire negative sum signal is cancelled out (i.e. the input sum equals zero) and the servo stops at that point.

Complications. To simplify the discussion we have taken a number of things for granted, which we must touch upon briefly now. For instance, we have assumed that the entire system—input signals, servo amplifier, motor, potentiometer, etc.—operates on direct current. Only by assuming that a d.c.

Servomechanisms and Function Generators

voltage of a certain polarity, amplified by a d.c. servo amplifier, would rotate a d.c. motor in a certain direction, while a signal of opposite polarity would cause the motor to run in the opposite direction, could we explain the self-correcting action of the servo. As a matter of fact, efficient direct-current servo amplifiers and motors are difficult to make, and alternating-current (a.c.) components are almost always used. This being the case, we have the choice of either converting the entire system, including the input signals, to a.c., or of using, somehow, d.c. input signals to operate a.c. servo amplifiers and motors. (The resistive summing networks, fortunately, give no trouble, since they obey Ohm's law for either a.c. or d.c.) Whichever method we use, we will have to solve the problem of dealing with voltages that have no fixed direction, but continually change their polarity. Let us take the all-a.c. system first.

An A.C. Positioning Servo

Phase. The distinguishing characteristic of d.c. voltages is polarity, while that of a.c. voltages is phase. In Fig. 45 (*a*) we have drawn three sine waves representing a.c. voltages of the type normally produced by commercial alternating-current generators. For the usual 50-cycle house current, each sine wave completes a cycle in $\frac{1}{50}$ second and then repeats itself. Such a complete cycle of the sine wave is said to consist of 360°, since the sine function runs through the complete range of its values every 360° (one rotation). We have, therefore, also marked the 'time axis' in degrees, 360 to each cycle.

Note that the three sine waves in Fig. 45 (*a*) are exactly alike, except that each starts at a different time. Wave 1 is zero volts in magnitude at zero time and zero degrees. It reaches its maximum positive amplitude at 90°, drops back to zero (volts) at 180°, reaches its maximum negative amplitude at 270°, and again drops to zero at 360°, or after $\frac{1}{50}$ second. Wave 2 has already reached its positive maximum at 0° and zero time (i.e. the time we started to look at it). It then drops back to zero at 90°, reaches its negative peak at 180°, drops to zero again at 270°, and again rises to a positive maximum at 360° or the end of $\frac{1}{50}$ second. It appears that wave 2 is always ahead of wave 1 by one-quarter of a cycle, or 90°. In the vocabulary of electricity this fact is expressed by the statement that 'wave 2 leads wave 1 by 90° in phase'.

Now look at sine wave 3. At the start of our observation (zero time and 0°) it is just at its negative peak and it does not reach zero (volts) until 90° or $\frac{1}{4}$ cycle have elapsed. It then rises to its positive peak (at 180°), drops back to zero (at 270°), and reaches its negative maximum again at 360°, after $\frac{1}{50}$ second. This wave, apparently, is always behind wave 1 by $\frac{1}{4}$ cycle or 90°, and hence it is said to lag wave 1 by $\frac{1}{4}$ cycle or 90° in phase.

Finally, consider the two sine waves shown in Fig. 45 (*b*). Wave 1 is exactly the same as that shown in Fig. 45 (*a*). It starts from zero at 0° and completes its cycle again at zero (volts) after 360° or $\frac{1}{50}$ second. Wave 2 also starts at zero (volts), but reaches its negative maximum when wave 1 rises to its positive maximum, both reaching zero again after 180°. Their roles are then reversed: wave 2 climbs to its positive maximum, while wave 1 reaches its negative peak, and at 360° both reach zero again. You cannot really tell which wave is ahead, but all you can say is that the two waves are displaced, or out of phase with each other, by one-half cycle or 180°.

A.C. Voltage Addition and Subtraction

Note also in Fig. 45 (*b*) that the two 180° out-of-phase voltages are always equal in amplitude (height) but opposite in polarity to each other. When wave 1 is at its positive maximum wave 2 is at its negative maximum, and vice versa. This is not only true at the maximum amplitudes but the waves are also equal

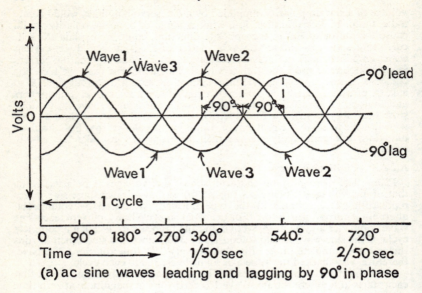

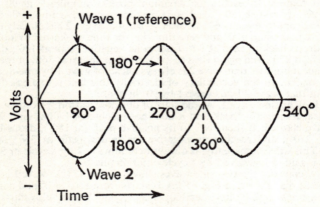

Fig. 45. (a) Sine waves leading and lagging by 90° in phase. (b) Two sine waves 180° out of phase with each other.

and opposite (in sign) to each other at every point along the time axis. You can guess what will happen if you apply two such 180° out-of-phase voltages to a resistor or a summing network. They will cancel each other out completely, and it will be impossible to detect a trace of voltage. Of course, if one voltage is larger in amplitude than the other they will not cancel out completely, but the net voltage will be the difference between the two voltages, and it will have the same phase as the larger voltage. It is obvious now that we can subtract one a.c. voltage from another simply by applying it with opposite phase to a resistor-summing network or summing amplifier. In contrast, if two a.c. voltages are applied which are in phase with each other (i.e. have the same phase at every point along the time axis) they will add up in the same way as d.c. voltages.

Phase Convention. We now have the perfect tool for the addition and subtraction of a.c. voltages, namely, their phase. In-phase voltages add, opposite-phase voltages subtract. To know throughout a large computer which voltages are in phase and which are opposite in (180° out of) phase, we must have some standard or reference voltage with which to compare the phase of a voltage. The reference voltage can be anything, such as the a.c. line voltage, wave 1 in Fig. 45 (*b*), etc. Let us now agree arbitrarily, for the sake of brevity, to label voltages in phase with the reference voltage as positive in phase (+), and all voltages opposite to, or 180° out of phrase with the reference as negative in phase (−). Thus, the designation '+ 100 V a.c.' is not a contradiction in terms, but means an a.c. voltage of 100 V magnitude that is in phase with a reference voltage, while '−25 V a.c.' is an a.c. voltage of 25 V magnitude that is opposite in phase to a reference voltage.

Two-phase A.C. Motor. We can now add and subtract a.c. voltages with ease. Furthermore, we have no trouble applying a plus or minus phase sum signal to the input of an a.c. servo amplifier. If the amplifier is a good one it will strengthen the magnitude of the signal to the required level, but it will not change the phase of the signal with respect to that of the reference voltage; that is, the amplifier must have absolutely no phase shift. The output from the servo amplifier is applied to a motor, as before, but unlike the d.c. motor, the required motor must respond to the phase rather than to the polarity of the input voltage; in other words, the motor must turn in one direction when a signal of positive phase is applied, and in the other direction when a negative phase signal is applied. The two-phase a.c. induction motor is made to order for this task. Although we cannot go into the details here, let us briefly summarize the operating principle of the two-phase a.c. motor.

Fig. 46 illustrates in schematic form the operating parts of the two-phase a.c. motor. Two field windings are arranged at right angles to each other, known respectively as the control field and the fixed field. (The term 'fixed' refers to the excitation voltage; both windings are mechanically stationary.) A metallic rotor, which has no winding at all, is placed in the magnetic field of the two windings. If you 'excite' either of the two windings with an a.c. voltage nothing happens. However, if you apply two a.c. voltages of the same frequency, but 90° out of phase with each other, to the control and fixed field windings, respectively, the rotor will start to turn. The diagram (Fig. 46) explains why. Here the fixed field voltage is shown to be 90° ahead in phase of the control field voltage (but has the same frequency). At zero time (0°), therefore, the control field voltage is zero (volts), while the fixed field voltage, is at its positive maximum. This can be summarized by drawing an arrow (called a vector) to the right in the direction of the fixed-field voltage and having the same length as the amplitude of the wave at zero time. Thus, the arrow shown within the rotor (in Fig. 46) and labelled 0° represents both the voltage and field in direction and magnitude at that instant. A moment later, at 90°, the fixed field is zero, while the control field voltage reaches its maximum positive value. This is shown by the 90° vector, drawn straight up in the direction of the control field voltage and having the same magnitude. One ¼ cycle later, at 180°, the control field is zero, while the fixed field is at its negative maximum, as shown by the 180° vector, drawn to the left. At 270°, or

¾ cycle, the fixed field is zero and the control field is at its negative maximum, as indicated by the downward 270° vector. Finally, at 360°, after one cycle has been completed, the original situation is repeated, with the fixed field at its positive maximum and the control field at zero.

Although we have shown it only at four instants of time, you can easily see that the vector (arrow) representing the net voltage and field direction rotates anticlockwise in a complete circle; that is, the two 90° out-of-phase voltages generate an anticlockwise rotating magnetic field. If you reverse the connexions of either field you reverse the phase of the corresponding excitation

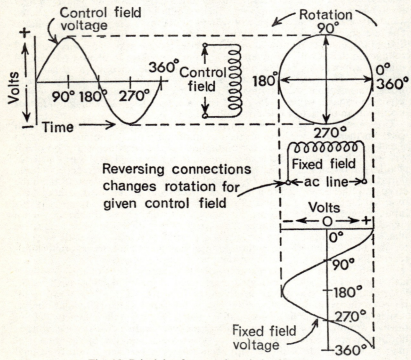

Fig. 46. Principle of a two-phase induction motor.

voltage, causing the field to rotate in the opposite, or clockwise, direction. You can easily convince yourself of this by tracing out the circle of vectors with one of the field-voltage waveforms turned upside down (i.e. reversed in phase). Whichever way you connect them, as long as the two field voltages are 90° out of phase with each other, the resulting magnetic field will rotate either clockwise or anticlockwise. As the field rotates, it induces a counter-field in the rotor and 'drags' it along. Thus, the rotor and shaft of the motor turn in the same direction as the magnetic field.

We now have the means for turning the motor in accordance with the servo amplifier output. By applying the amplifier output voltage to the control field winding of the motor, the rotor shaft will turn in one or the opposite direction, depending upon the (+ or −) phase of the voltage. Furthermore, the motor will rotate at a speed that is proportional to the magnitude of the amplifier output voltage, and hence to that of the servo input signal. Finally, you can see that with this arrangement the a.c. line voltage cannot be used directly as reference voltage. For, assuming that the control field voltage (in Fig. 46) is in phase with the reference voltage (i.e. +), the a.c. line voltage applied to the fixed field must be 90° out of phase with the control field voltage, and hence with the reference voltage. Thus, the a.c. line voltage feeding the fixed field cannot be used as reference voltage. The same reasoning applies if the control field voltage is negative in phase. However, this difficulty is easily removed by a phase shifter, which shifts either the fixed field or the reference voltage by 90° with respect to the a.c. line voltage.

The Complete A.C. Positioning Servo. We finally have the essential components of an a.c. positioning servo for analogue computers, as is illustrated in the schematic diagram below (see Fig. 47). By comparing this diagram with

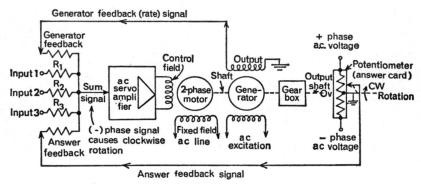

Fig. 47. Schematic diagram of an a.c. computer positioning servo.

the d.c. servo shown in Fig. 44, you will note that they are almost the same. Again three inputs are summed up, but this time with respect to phase, and the plus or minus phase sum signal is applied to an (a.c.) servo amplifier. The output voltage of the amplifier, strengthened in magnitude but unchanged in phase, is applied to the control field of a two-phase motor, whose fixed field is fed from the a.c. line. The output shaft of the motor drives a generator, whose function we shall explain in a moment, and the slider of a potentiometer card through reducing gears. In the system shown, a negative (−) phase signal causes clockwise rotation of the motor shaft and potentiometer and, hence, the slider picks off a plus (+) phase a.c. signal for a minus (−) phase servo input (sum) signal. The plus phase signal from the potentiometer is fed back as the answer to the input summing network and cancels out a portion of the minus phase input sum signal. The servo stops when the potentiometer slider has picked off a positive (+) answer voltage exactly equal to the negative (−) input sum voltage. If you replace 'phase' by 'polarity' you have essentially the same description as for the d.c. positioning servo.

A Modified D.C. Servo. We mentioned earlier that we have a choice of using either an all-a.c. servo system or d.c. input signals to operate a.c. servo amplifiers and motors. If the remainder of the analogue computer is operated by direct current we might prefer the latter alternative. This is none too easy in practice. We cannot go into the details in this book, but the method consists essentially of superimposing the d.c. input signal upon the a.c. system by a process of 'modulation' similar to radio broadcasting. The d.c. sum signal (including a d.c. answer feedback voltage) is 'chopped up' to a.c. by a synchronous vibrator (chopper), and the resulting signal modulates another voltage obtained from the a.c. line. The combined signal is applied to an a.c. servo, whose output is provided with a d.c. answer potentiometer for feeding back a d.c. answer signal to the input.

Another Complication: Hunting. What about the generator shown coupled to the two-phase motor in Fig. 47? To understand its purpose we must discuss a defect that is common to all servos, not just the a.c. variety. This defect is called 'hunting'. It means, quite literally, that an uncorrected servo will not go straight to its final resting position, but will 'hunt' for it, back and forth, sometimes breaking out into uncontrollable oscillations around its ultimate stopping position. It is not difficult to see why this happens. Suppose a fairly large negative (phase) input voltage (say −8 V) is applied to the servo shown in Fig. 47 (or to that in Fig. 44). With a negative input, the servo will start to run in a clockwise direction, and since the signal is large, the motor will attain a high speed, perhaps 3000 rev/min or more. As the slider of the answer card (potentiometer) approaches the +8 V point, where it will balance out the input signal, the motor is still running very fast. When the input signal actually reaches zero and there is no driving force, the motor still coasts along on its built-up momentum, and the slider overshoots the +8 V point on the card. When this happens, the input signal to the servo amplifier becomes positive, and the servo will start to run in the reverse (anticlockwise) direction. Again the motor may reach a sufficiently high speed to cause the slider to overshoot its goal and move past the +8 V point. It will then, of course, pick off less than +8 V, and the net input signal becomes negative once more. The servo again reverses its direction, and so on. Under some conditions this hunting may continue for ever, and the oscillations may even increase.

The Cure for Hunting: An Oppositely Phased Generator Signal. The cure for hunting is to slow the motor down as it reaches the 'no signal' (zero input voltage) point so that it comes to a smooth stop. Various mechanical means have been used in the past, but the best way is to feed back to the input of the servo an oppositely phased signal that is proportional to the speed of the motor. Such a signal is easily generated by an induction or 'tachometer' generator. With a fixed a.c. excitation (field) voltage, the output of such a generator is exactly proportional to the speed (rate) of its shaft rotation. (A tachometer indicates shaft speed.) Hence, we simply couple a tachometer generator to the motor shaft and feed back the generator output voltage to a resistor in the input summing network, so that it will oppose the phase of the servo input signal.

Although this procedure may sound silly, since it partially prevents the servo from doing what it is supposed to do, it really works, preventing the servo only from doing too much. The feedback or 'rate' signal from the generator tends to run the servo in a direction opposite to that of the input signal. Since the rate signal is proportional to the motor speed, it will cause the motor to slow down as it approaches the zero signal point at high speed. As the motor slows down, the rate signal decreases, of course, and the braking effect is diminished. As the motor finally reaches the zero signal point (+ 8 V), it will be moving very slowly and may simply coast to a stop at this point. The rate signal from the generator then also drops to zero and there is no remaining 'counter' signal tending to force the motor to run in the opposite direction. In other words, the brake operates only while the motor is running. If the slider of the answer card should slightly overshoot its goal and the motor starts to run in the opposite direction the generator 'brake' will go on again, supplying once more a signal of a phase opposite to that of the input signal. Any prolonged hunting is thereby prevented.

We shall presently see how this ingenious generator rate signal can be used in yet another way: for the integration of mathematical functions.

Integrating (Velocity) Servos

The type of servomechanism we have dealt with to this point is known as a positioning servo, since its output shaft and answer card (potentiometer) always assume a position proportionate to the servo input. Previously, however, we mentioned that the two-phase servo motor runs at a speed proportional to its control-field excitation, and hence proportional to the servo input signal. We have also seen that a tachometer (induction) generator produces a feedback signal that is always exactly proportional to the rate (velocity) of servo rotation. These are the two characteristics required to make the rate of servo output shaft rotation exactly proportional to the input signal, or equivalently, make the total output rotation proportional to the time integral of the input quantity. All we need do to obtain an integrating (velocity) servo, therefore, is to leave off the answer card that stops the servo in a certain position. The servo will then speed up to a velocity that is proportional to the input signal, and the total amount of shaft rotation will automatically represent the integral of the input signal.

What is the self-correcting mechanism (error detector), in this case, that will assure that the servo runs exactly at the required rate? The generator feedback (rate) signal, of course. The rate feedback becomes the equivalent of the 'answer' feedback signal in the positioning servo. As the servo speeds up under the influence of the input 'excitation', the tachometer generates a feedback signal proportional to the rate of rotation and with a phase opposite to that of the input signal. This feedback signal cancels a part of the input signal. As the motor continues to speed up, the feedback voltage cancels out more and more of the servo input voltage, and when the two balance each other exactly (i.e. are equal and opposite), the servo stops speeding up. It then continues to run at a constant rate, exactly proportional to the input signal. If the input signal should increase, the servo will speed up again until the generator feedback voltage once more matches the (increased) value of the input voltage.

You must not assume, however, that this self-balancing process is as exact in a continuously running (integrating) servo as in the positioning servo. If the input and generator feedback signals cancelled each other out exactly there would be no signal left over at the input to drive the motor and generator, that is, to overcome the friction of the bearings, gears, brushes, and whatever output 'load' may be attached to the servo shaft. (Although there is no answer card in an integrating servo, other continuously rotating potentiometers, resolvers, function generators, etc., are frequently attached, as we shall see in the next section.) Thus, there must always be a sufficiently large 'error' signal left over at the input to overcome the friction of the motor and output load. An integrating servo, therefore, will speed up until it drives the generator sufficiently fast to provide a feedback signal that will neutralize all of the input signal, except that amount required to overcome the friction of the motor, generator, and load.

An Example. Suppose you wanted a servo to indicate the ground speed of an aeroplane; that is, the rate at which it flew with respect to the ground underneath. Contrary to what you've probably just been thinking, a positioning servo could do this job very well. All you need do is obtain a voltage proportional to the ground speed of the aeroplane and apply this voltage to the input of a positioning servo. The output shaft and answer 'pot' of the servo will then turn to a position proportional to its input, i.e. to the ground speed. You could calibrate the shaft over the required range of speed, say, from 0 to 600 mile/h. If the plane flew at 300 mile/h the servo shaft would turn one-half of its full rotation; for 500 mile/h, the shaft would turn $\frac{5}{6}$ of its full rotation, and so on.

Suppose that you wanted the servo to indicate the total distance covered by the aeroplane in a certain time. Well, if you were absolutely sure that the plane was flying at all times at its advertised rate of perhaps, 575 mile/h you need only multiply this velocity by the total elapsed time to obtain the distance, since the distance (s) covered is the product of velocity (v) and time (t) ($s = vt$). In this case you could still use your speed-indicating positioning servo with a suitable multiplier. More likely than not, however, the rate at which the aeroplane is flying varies continually with time and, thus, the total distance covered is the time integral of the varying velocity, or as we have learned to write:

$$s = \int v\, dt$$

To show the total distance covered in this case, therefore, you will require an integrating servo, to which you apply the ground speed as input signal. The servo then automatically runs at a rate proportional to the plane's ground speed, and this rate is multiplied at each instant by an infinitesimal change in

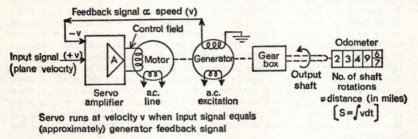

Fig. 48. Example of integrating servo for indicating total distance covered as function of ground speed and time.

time (dt), represented by the amount of shaft rotation during this interval. The total number of shaft rotations completed by the servo during the total time interval then represents the required integral; i.e. the total distance covered. You could easily attach a counter (odometer) to the servo output shaft, which would count the total number of shaft revolutions directly in miles covered (see Fig. 48).

Servomultipliers and Servodividers

We now understand positioning servos and integrating (velocity) servos; with these tools we can do quite a few things, such as multiply and divide, integrate, generate mathematical functions, and so on. Multiplication is accomplished by attaching potentiometer 'cards' to the output shaft of a positioning servo. You can add on as many cards as you wish—as long as you do not overload the servo, of course—and the voltage applied to each card will be multiplied by the servo input. A typical potentiometer card looks something like the illustration in Fig. 49 (*a*). The card itself is a rectangular insulating form, made of plastic, which is bent into a circle and held in place by an insulated strap and turnbuckle. Resistance wire is wound very evenly upon this card, so that the resistance tapped off by the slider (also known as wiper or brush) is exactly proportional to the amount of rotation. A voltage is applied to the free ends of the resistance winding, and as the servo shaft rotates the pot wiper, a proportionate fraction of this voltage is picked off by the wiper. The winding may have a number of additional fixed taps, such as an earthed centre tap (zero voltage) and other fixed voltages. The equivalent schematic symbol of a potentiometer card is shown in Fig. 49 (*b*).

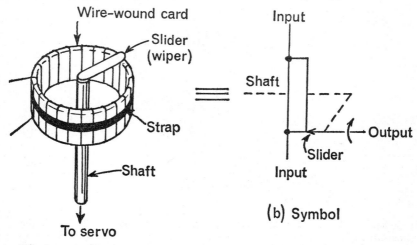

Fig. 49. (*a*) Potentiometer card and (*b*) equivalent symbol.

Multiplication of Two Variables. Fig. 50 illustrates two commonly used types of servomultipliers. Since we are not interested now in the servomechanism but only in the inputs and outputs, we have shown the servo in 'abbreviated' form by the amplifier symbol with a variable input voltage x and a (dotted) output shaft. Attached to this shaft are two potentiometer cards, whose input voltages are to be multiplied by the servo function (x). The first card has a positive input voltage, w, applied to one (top) end, while the other (bottom) end of the card is earthed, or at zero volts. (In electronic assemblies all voltages are measured with respect to the common chassis, which is usually earthed and thus represents zero voltage.) As the servo rotates, the voltage, w, is multiplied by the setting, a, of the potentiometer wiper, as we have seen in the preceding chapter [see Fig. 28 (*a*)]. Since this setting, however, is a direct function of the input variable, x, of the servo, the output voltage (E_o) picked off by the wiper is

$$E_o = aw = xw \quad (0 \leq x \leq 1)$$

That is, the output voltage is the product of the servo and potentiometer input voltages. However, since the minimum voltage picked off by the pot wiper is zero (0 V), and the maximum voltage is w, the value of the multiplier x can vary only between 0 and 1; i.e. x can have only positive values.

The second card (in Fig. 50) is also driven by the servo in accordance with its input function, x. This card, however, has its centre tap earthed (0 V) and has input voltages of $+y$ and $-y$ applied to its two free ends. Again the output voltage picked off by the wiper (E_o) is the product of the servo input (x)

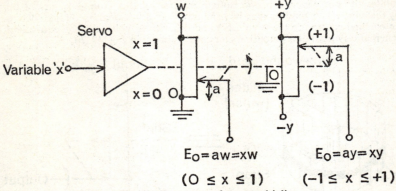

Fig. 50. Two types of servomultipliers.

and the card input ($\pm y$), but in this arrangement the wiper voltage can vary between $-y$, 0, and $+y$ V. Thus, the multiplier function, x, may have both positive and negative values between -1 and $+1$. This may be summarized mathematically by:

$$E_o = ay = xy \quad (-1 \leq x \leq +1)$$

A word about the mechanization of positive and negative multipliers in the diagrams: if the servo is a d.c. type the plus ($+$) and minus ($-$) signs next to the quantities (voltages) to be multiplied refer to the actual polarities of the voltages; if the servo is of the a.c. variety the plus ($+$) and minus ($-$) signs refer to the phase of the voltages with respect to a reference voltage, as we have explained earlier. In either case, multiplication by quantities with the indicated sign is accomplished.

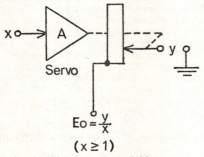

Fig. 51. A servodivider.

The Servodivider. Division is frequently only a special case of multiplication. You will recall from elementary arithmetic that dividing A by B ($A \div B = A/B$) is the same thing as multiplying A by the fraction $1/B$; that is $A/B = A \times 1/B$. Thus, since the servomultipliers of Fig. 50 actually multiply by a fraction of the servo input (x), they may be considered dividers rather than multipliers. For example, if the input function (in Fig. 50) x has the value $\frac{1}{2}$, corresponding to half of the shaft rotation, the output from the first card is $\frac{1}{2}w$ or $w/2$, while that of the second card is $\frac{1}{2}y$ or $y/2$; that is, w and y may be considered to have been divided by a factor of 2.

Division by a divisor greater than 1 can be handled more directly, as is illustrated by the servodivider (Fig. 51) on the opposite page. Here a servo with an input voltage x drives a potentiometer card that has no voltages applied to its fixed ends. A voltage y is applied to the wiper of the pot and the output voltage, E_o, is taken from the bottom end. As you can see, when the wiper is at this bottom end the output voltage $E_o = y$, in which case division by 1 has taken place. As the wiper rotates towards the top end of the card the voltage, y, is divided by the increasing card resistance, which is a function of the servo input, x. Hence, we can write for the value of the output voltage

$$E_o = \frac{y}{x} \quad (x \geq 1)$$

where x must be at least equal to 1. The maximum value of x depends upon a number of factors, but primarily on the resistance of the card for full rotation (top end).

SERVO FUNCTION GENERATORS

What if the card of a servomultiplier was not evenly wound or was of an irregular shape, so that the resistance picked off by the wiper would not increase linearly with the amount of rotation? The input voltage to the card

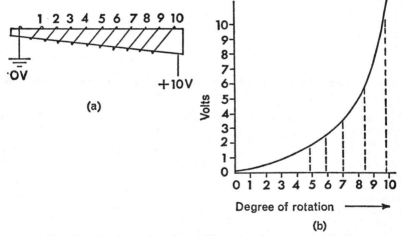

Fig. 52. (*a*) A tapered card, and (*b*) the function generated by it.

would still be multiplied by the servo, but with the resistance not directly proportional to the servo position, the output from the pot would be some function of the servo input rather than the simple product of the servo and card inputs. Thus, if x represents the input to the servo, and y the input to the card, you could write for the output voltage (E_o) with a non-linear pot

$$E_o = f(x)\, y$$

that is, the output is the product of *y* and a function of *x*. This would be highly undesirable for the ordinary servomultiplier, but if you wanted to generate some special type of mathematical function it would be just the thing to do.

As an example, consider the card shown in Fig. 52 (*a*). Here we have 'opened up' a round potentiometer card, so that its shape becomes apparent. Note that the card becomes increasingly thicker towards the right (+ 10 V) end, instead of being of uniform width, as the linear card shown in Fig. 49 (*a*). Each turn of resistance wire (ten numbered turns are shown), therefore, has an increasingly greater length and, hence, the resistance increases non-linearly with the number of turns. The voltage picked off by the wiper is

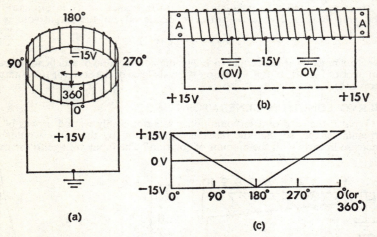

Fig. 53. Discontinuous function card. (*a*) Actual card, (*b*) 'opened' card, and (*c*) function generated.

therefore also a non-linear function of the amount of rotation. You can easily demonstrate this by drawing a simple graph of the voltage picked off by the wiper (from 0 to 10 V in this case) against the number of turns (1–10), or equivalently, the degree of shaft rotation. As you can see, this graph is anything but linear. Instead of tapering the width of the card, we could have obtained similar results with a card of uniform width by either changing the spacing of the turns in a non-uniform manner or changing the wire size or composition. By using these methods, the resistance of the card (or wiper voltage) can be made almost any desired function of the number of turns (or degrees of shaft rotation). Square and square-root cards are commonly used.

If the function to be generated is discontinuous, which means that it does not follow a smooth curve, but has a sharp break or sudden changes in value, this can be handled by placing voltages of the required value at fixed taps on the card. For example, consider the function card illustrated in Fig. 53. The actual card is shown in (*a*) of the figure, while the flattened-out (cut-open) version is presented in (*b*). The triangular function generated by this card is shown in (*c*). Remember, since the pot wiper can turn round in a complete circle, the two ends of the flattened card (0° and 360°) are really the same point, as shown by the mounting points 'A'.

Note in Fig. 53 that fixed taps are placed on the winding every 90° or quarter-revolution. Voltages of +15 V and −15 V are applied to the taps at 0° (or 360°) and at 180°, respectively. The 90° and 270° taps are connected to 'ground', thus placing them at zero volts potential. The resistance winding itself is 'linear', so that the wiper picks off a voltage proportional to the amount of rotation. Starting at 0°, the wiper picks off +15 V applied to the tap and the voltage then decreases smoothly, passing through zero volts at the 90° (grounded) tap and reaching −15 V at the 180° tap. For the second half of a complete rotation, from 180° to 360° (0°), the wiper voltage increases smoothly from −15 V to +15 V, passing through zero at the 270° point [see Fig. 53 (c)]. Thus, in this example a linear pot with fixed voltage taps has been used to advantage to generate a discontinuous function.

Sine-wave Function Generator. Sine and cosine functions enter into mathematical computations very frequently, and we must have devices to generate

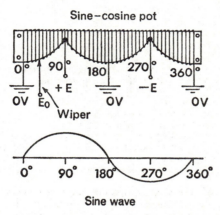

Fig. 54. Sine-cosine potentiometer and function generated.

them whenever needed. If you look back at Fig. 45 (*a*) you will see the appearance of a sine wave portrayed by wave 1. A cosine wave looks exactly the same as a sine wave, except that it leads the latter by 90° in phase; wave 2 in Fig. 45 (*a*) thus illustrates a cosine wave.

There are a number of ways of generating a sine or cosine wave by means of function potentiometers, but one of the simplest is illustrated by the specially shaped card in Fig. 54. Note that the card is tapped every 90° or quarter-revolution. Corresponding to the zero-amplitude points of a sine wave, the function card is earthed (zero volts) at 0°, 180°, and 360° (which is the same point as 0°). Since a sine wave reaches its maximum positive and negative amplitudes at 90° and 270°, respectively, positive and negative voltages ($\pm E$) of the desired magnitude are placed, respectively, on the taps at 90° and 270° rotation. Thus, the function is assured to pass through these extreme points at the correct degree of rotation. Between these values the card is tapered to make the resistance winding follow a sine wave. Note also in Fig. 54 that a cosine function is generated if the wiper is started at the 90° point for 0° rotation. By placing two wipers on the pot, separated by 90° or ¼ revolution, sine and cosine waves can be generated simultaneously.

Resolvers. There is another type of device, known as a resolver, which can generate simultaneous sine and cosine functions more elegantly. Like the control transformer, a resolver is essentially a rotary transformer with two fixed stator windings and one or two rotor windings, located in the magnetic field of the stators. The stator windings and the rotor windings (if there are more than one) are arranged at right angles to each other so that their voltages and currents are always 90° out of phase with each other. This is shown by the crossed rotor symbol in Fig. 55. Since the resolver is one of the family of synchro devices, of which we have met the synchro transmitter and control transformer, it is often referred to simply as 'synchro'. Conventional synchro transmitters and receivers, however, have three stator windings, spaced 120° apart, and only one rotor winding.

By virtue of its crossed stator windings, a resolver can 'resolve' an a.c. voltage applied to the rotor into stator output voltages that are sine and

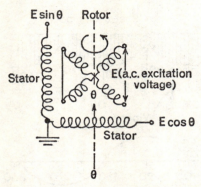

Fig. 55. Resolver giving outputs proportional to sine and cosine of input voltage.

cosine functions, respectively, of the angular position (θ) of the rotor. In other words, if you 'excite' the rotor winding with an a.c. voltage, E, the stator output voltages will be $E \sin \theta$ and $E \cos \theta$, respectively. Conversely, if you apply two 90°-out-of-phase voltages, $E \sin \theta$ and $E \cos \theta$, to the stator windings of a resolver, the voltage induced in the rotor winding will be of constant amplitude E and will have a phase angle that depends only on the angular position, θ, of the rotor.

A word of caution here. You cannot connect a resolver directly into the circuit of a d.c. analogue computer, since, like all transformers and synchros, it is strictly an a.c. device. You can, however, modulate an appropriate a.c. voltage with the d.c. computer signal, pass the combined signal through a resolver and other a.c. computing elements, and then demodulate (or rectify) the output signal for use elsewhere in the computer. This procedure is not necessary, of course, if the entire analogue computer operates on a.c.

Polar to Rectangular Co-ordinate Transformation. In a computer it is often necessary to convert a quantity that is specified by a magnitude (R) and an angle (θ) in so-called polar co-ordinates into conventional x and y (rectangular) co-ordinates. Fig. 56 (*a*) illustrates the mathematical relationship between the two co-ordinate systems.

Consider the vector R, representing any quantity at all, which is inclined at an angle θ with respect to the x-axis of the rectangular co-ordinate system.

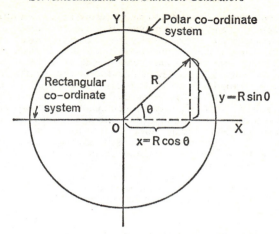

(a) Polar-to-rectangular co-ordinate transformation

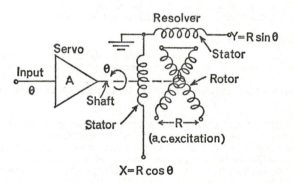

(b) Servo-driven resolver for performing transformation shown in (a) above

Fig. 56 (a) Polar-to-rectangular co-ordinate transformation; (b) servo-driven resolver for performing transformation shown in (a).

By drawing the appropriate horizontal and vertical lines, as shown in Fig. 56 (a), you can 'resolve' this vector into its rectangular components x and y, along the x- and y-axis, respectively. From elementary trigonometry it is evident that the following transformation equations hold:

$$x = r \cos \theta$$
$$\text{and} \quad y = r \sin \theta$$

These transformation equations are easily mechanized in an analogue computer by using a servomechanism in conjunction with either a sine–

cosine pot or a resolver. The set-up using a resolver is shown in Fig. 56 (b). An a.c. voltage of magnitude R is applied to excite the rotor of the resolver. The rotor is positioned to the required angle, θ, by a servo whose input voltage is proportional to θ. The output voltages (x and y) from the stator windings of the resolver are then automatically equal to $R \cos \theta$ and $R \sin \theta$, respectively. It is interesting to note that the inverse rectangular-to-polar co-ordinate transformation is not quite as easily mechanized.

In the present example we have connected the sine–cosine function pot or resolver to the output shaft of a servo, which may be either of the positioning type or of the continuously running (integrating) kind, if rotation for more than 360° is required. Frequently, however, function generators are used at the input of an analogue computer, for 'setting in' quantities that vary in accordance with some function. For a sine function, for example, the wiper of the sine–cosine pot or the rotor of the resolver is directly coupled to a manually operated knob, calibrated in degrees (θ). Other knobs control variable resistors for controlling the magnitude (R) of the 'excitation' voltage. The outputs from the pot or resolver are then fed to other computing elements.

Other Types of Function Generators

Specially shaped cards, sine–cosine pots, and resolvers are all electromechanical function generators, using standard electrical and electromechanical components. There are also purely electronic function generators; you should have passing knowledge of a few of these. Among the most important are diode and photoelectric function generators.

Diode Function Generators

Diode Characteristics. A diode is essentially a voltage-sensitive on–off switch, which has two operating elements, one for emitting a stream of electrons (i.e. a current) and one for collecting it at the other end. If the diode is of the solid-state (crystal) type these elements may be called emitter and collector, respectively; if the diode is a thermionic valve they are called cathode and anode (or plate), respectively. When voltages are placed on these elements the diode will conduct whenever its plate (collector) is positive with respect to the cathode (emitter), and the diode will be an open circuit whenever the anode is negative with respect to the cathode. An 'ideal' diode, therefore, acts as a switch, which can close a circuit whenever a positive applied voltage (E) exceeds a positive cathode bias voltage ($+E_c$), and open the circuit when the applied positive voltage is less than the cathode bias.

As is shown by the dotted line in Fig. 57, practical diodes depart somewhat from this ideal, since their characteristics are not completely linear, especially near the 'on–off' or break points. Moreover, actual diodes have a definite resistance when conducting, rather than zero resistance, as would be the case for a switch. As a result, the slope of the voltage–current characteristic cannot be vertical, as it would be for a switch or an ideal diode. Since the slope has to be adjusted to simulate some function, this is not of great importance.

Diode Limiters. We shall now try to approximate some required mathematical function by a series of connected line segments, using the linear diode characteristics with adjustable 'break points' and slopes. The tools for this are diode circuits known as limiters, usually used in conjunction with an operational amplifier. Let us consider the feedback limiters, basic versions of which are illustrated in Fig. 58.

In each of these circuits a diode with a source of bias voltage (E_c) is connected between the input and output of an operational (feedback) amplifier.

As long as the input, E_i, is less than the bias voltage, E_c, the output voltage, E_o, is simply the product of the input voltage and the amplification; that is, $E_o = AE_i$. When the input voltage exceeds the bias voltage, however, the

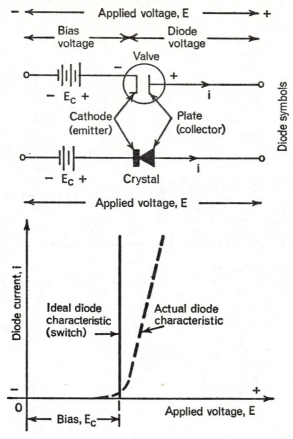

Fig. 57. Characteristics of ideal and actual diodes

diode shorts out the amplifier and only the bias voltage (E_c) appears in the output. Thus, in the positive limiter [Fig. 58 (a)] the maximum positive value of the output voltage is limited to the value of E_c; in the negative limiter [Fig. 58 (b)] the maximum negative value of the output voltage is limited to E_c; and in the dual limiter [Fig. 58 (c)] the output voltage is limited between the positive and negative values, $\pm E_c$. When considering the polarities at the input and output you have to consider, of course, the fact that the amplifier inverts the phase of the input voltage. The slope of the line segment in each of the feedback limiters equals approximately R_f/R_i (the same as the amplification, E_o/E_i). Since the amplification is high, the slope can be adjusted to any desired value.

Multi-segment Function Generator. Finally, let us look at a multi-segment diode function generator, consisting of a number of diode limiters (see Fig. 59). For simplicity, the operational amplifier has been omitted. (The circuit will work reasonably well without it.)

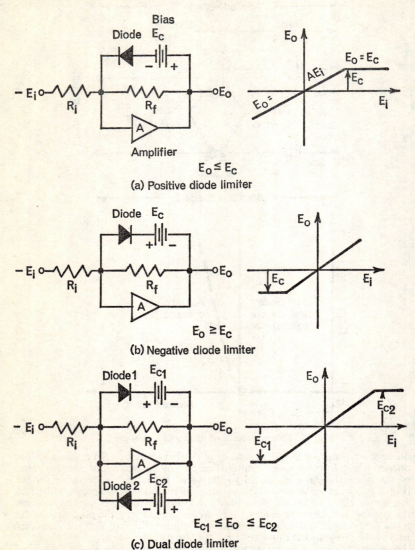

Fig. 58. Diode feedback limiters and output functions: (*a*) positive limiter; (*b*) negative limiter; (*c*) dual limiter.

Servomechanisms and Function Generators 95

The bias voltage for each of the three diodes is obtained through a multiple (adjustable) voltage divider, which is supplied with positive and negative voltages of $+100$ and -100 V with respect to earth. The input voltage, E_i, is fed in through a resistor to a tap on the voltage divider. The output voltage, E_o, is taken across load resistor R_L. A resistor, R, provides the connexion between input and output, when none of the diodes are conducting.

Consider the operation of this arrangement, starting with diode 1. The

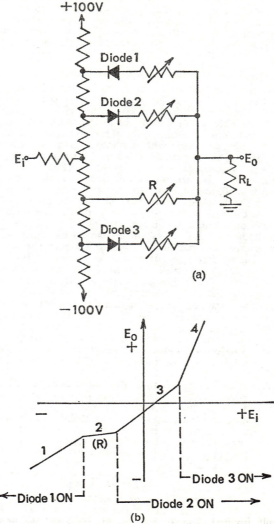

Fig. 59 (*a*) A multi-segment diode function generator and (*b*) output function.

cathode of diode 1 is placed at a high positive potential, so that it can conduct only when the input voltage (E_i) is sufficiently negative to overcome this bias. This is shown by line segment 1 in the input–output curve [see Fig. 59 (b)]. As the input voltage becomes more positive, diode 1 cuts off, but since none of the other diodes are conducting, the current is carried by resistor R, as is shown by line segment 2. As E_i becomes more positive, the anode of diode 2 becomes positive with respect to its cathode, and the diode conducts, as shown by line segment 3. Finally, as the input voltage becomes highly positive, it is able to overcome the negative bias on the anode of diode 3 and the latter conducts. Diode 3 then adds its current to that of diode 2, as is shown by the steep line segment 4. The variable diode series resistors permit adjusting the slopes of the line segments. The 'break points' may be adjusted by changing the taps on the voltage divider.

Photoelectric Mask Function Generator (Photoformer)

Diode function generators are better suited to high-speed computer applications than the servo function generators we have previously considered, since like most electronic devices, they have a very rapid (high-frequency) response, and hence are not limited by the inertia of electromechanical components. Let us look now at another type of all-electronic function generator, the cathode-ray-tube generator with a photoelectrically scanned mask, or photoformer, as it is usually called. This is a 'true' function generator, in contrast to the earlier ones, since it permits the direct generation of any arbitrary function, without the usual approximations. A typical photoformer set-up is shown in Fig. 60, together with the output function generated.

As shown in the block diagram [Fig. 60 (a)], the photoformer consists essentially of an ordinary cathode-ray tube, partially covered by an opaque mask, a photoelectric cell, and two d.c. 'deflection' amplifiers. The electron gun in the cathode-ray tube shoots out a beam of electrons, which make a bright spot when they impinge on the fluorescent screen (similar to that of a television set). The electron beam may be deflected to the left or right and up or down by voltages placed on the horizontal and vertical deflection plates, respectively. As a result, the bright spot on the screen can 'draw' any arbitrary curve, in accordance with the voltages applied to the deflection plates.

The input voltage, x, is applied through a d.c. amplifier to the horizontal deflection plates and guides the electron beam (spot) horizontally across the cathode-ray-tube screen. A feedback loop forces the beam to follow the edge of the opaque mask in the vertical dimension, in accordance with the function to be generated. The vertical deflection voltage required to do this job is a function of the input voltage (x). This vertical deflection voltage is taken as the output voltage, y.

Let us look at the feedback path a little more closely. The bias voltage applied to the vertical d.c. amplifier is of such polarity that it tends to force the beam upward, away from the mask. As soon as the spot on the screen emerges from behind the mask, however, light strikes the photoelectric tube in front of the screen, thus generating an 'error' signal. (You will recall that the error signal in any feedback system corrects the action.) This error voltage, which is of a polarity opposing the bias, is applied to the input of the vertical amplifier and forces the beam downward again, towards the edge of the mask. The upshot is that the feedback loop keeps the spot, as it travels across the screen in accordance with the horizontal deflection or input voltage, x, just below the edge of the mask. The vertical deflection or output voltage (y), hence, is forced to vary as a function of the input voltage (x), or expressed in mathematical form

$$y = f(x)$$

Fig. 60 (b) shows a typical mask representing some function $y = f(x)$. In practice, the opaque cardboard or plastic mask must be designed to represent the function, taking into account the deflection sensitivities of the cathode-

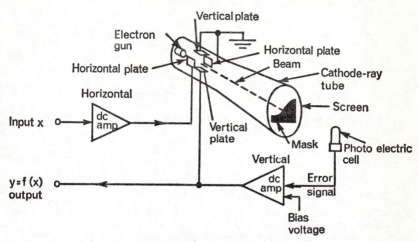

(a) Setup of photoelectric–mask function generator

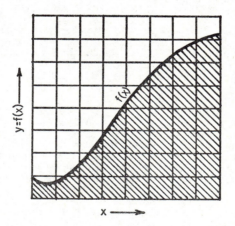

(b) Output function generated (Mask)

Fig. 60. (a) Block diagram set-up of a photoelectric mask function generator; (b) output function generated.

ray tube. The completed mask is then experimentally 'trimmed' for better accuracy. The photocell is usually of the electron multiplier type for high sensitivity to the light spot.

REVIEW AND SUMMARY

A servomechanism is an extension of the feedback (closed-loop) concept.

In an open-loop control system operation is independent of the result (output); correction of the result is obtained manually.

In a close-loop (servo) control system the control operation is a function of the result, or output (see Fig. 42); by the use of a feedback loop the system automatically corrects the result (self-correction).

A positioning servo (closed-loop) control system (Fig. 42) contains the following essential elements:

1. An input signal, or command, to carry out a desired operation.
2. An output, or load, which is positioned in accordance with the input command.
3. A controller (consisting of servo amplifier and motor) for mechanically positioning the load.
4. A comparer, or error detector (differential or sensing device), for comparing the output position with the input. The comparer must generate an output (error) signal that is proportional to the error between the input and output positions.

The controller (amplifier and motor) in a positioning servo drives the load (output shaft) in a direction that decreases the error until the load is positioned exactly in accordance with the input. The error signal is then zero, and the system has 'nulled' itself out.

A synchro transmitter is an electromechanical device (similar to an a.c. generator) that generates a voltage proportional to the angular position of its input shaft; it is one of a family of synchro follow-up devices. It is used in connexion with a synchro receiver (or motor) for giving remote position indication.

A synchro control transformer is a rotatable transformer whose output voltage from the rotatable winding (rotor) is a function of the rotor position and the voltage applied to the fixed (stator) winding; a control transformer can serve as an error detector in a servo.

A computer positioning servomechanism (Fig. 44) usually consists of an input summing network for adding and adjusting the scale of the input signals, a servo amplifier for boosting the input sum signal to a level sufficient to operate the servo motor, reducing gears coupled to the motor shaft, and a servo-driven potentiometer (card) for providing an answer (error) signal proportional to the shaft rotation. The answer (error) signal is fed back to the input network and cancels out part of the input signal. The servo turns in the direction of a decreasing error until the answer voltage equals the input sum voltage (the combined signal is zero), whereupon the servo stops at the desired position. Thus, the servo automatically nulls itself out for zero input signal.

In an alternating-current positioning servo the phase, rather than the polarity, of the input signals must be considered. A voltage that is in phase with a reference voltage is called positive (+), and one that is 180° out of phase with the reference voltage is negative (−). The plus and minus phase a.c. input signals are added by a summing network and strengthened by an a.c. servo amplifier, whose output is fed to the control field winding of a two-phase a.c. (induction) motor; the field winding of the motor is excited by a 90° out-of-phase a.c. voltage. Depending upon the phase (+ or −) of the control field voltage, the motor turns in one or the other direction and drives the answer card in a direction that 'nulls out' the error. The inputs to the answer card are plus and minus phase voltages, applied to the two ends.

Servomechanisms and Function Generators

Hunting is an uncontrolled oscillation of the servo output shaft about its final, zero-signal position. It may be overcome by feeding back an out-of-phase signal to the input of the servo, proportional to the rate (speed) of servo rotation. The feedback signal, acting as a dynamic brake, is generated by an a.c. induction generator (tachometer).

An integrating (velocity) servo (Fig. 48) uses an oppositely phased generator feedback to run at a rate (velocity) exactly proportional to the input voltage. The servo is continuously running and has no answer card. The total amount (revolutions or degrees) of its shaft rotation represents the time integral of the input voltage. An integrating servo will pick up speed until the feedback signal of its generator balances out the input signal, except for the signal required to overcome friction.

A servomultiplier (Fig. 50) consists of a potentiometer card attached to the output shaft of a positioning servo. By automatic adjustment of the setting of the pot in accordance with the servo input voltage (x), the voltage (y) applied to the pot is multiplied by the servo input voltage; the output from the card (pot) wiper, therefore, equals xy. If multiplication by positive as well as negative values is required the potentiometer must be grounded at the centre tap, and positive and negative input voltages ($\pm y$) must be applied to the ends.

In a servodivider the quantity to be divided is applied to the wiper of a servo-driven pot, and the output (quotient) is taken from one end (see Fig. 51). For a servo input voltage, x, and pot input voltage, y, the pot output voltage equals y/x, where $x \geq 1$.

In a servo function generator the output voltage from a servo-driven, non-linear card varies in accordance with a required function of the servo input voltage. The output from the card can be made the desired function by either shaping the width of the card, spacing the turns in a non-uniform manner, or by changing the wire size or composition in accordance with the function (see Fig. 52). Discontinuous functions (e.g. triangular) and trigonometric functions (sine waves, etc.) may be generated by shaping the card and applying the required extreme and inflection values (voltages) of the function directly to fixed taps on the card (see Figs. 53 and 54).

A synchro resolver (Fig. 55) is a rotatable transformer with two fixed stator windings at right angles to each other and one or more (crossed) rotor windings located in the field of the stators. An a.c. voltage (E) applied to the rotor will be resolved into stator output voltages that are sine and cosine functions, respectively, of the angular position (θ) of the rotor ($E \sin \theta$ and $E \cos \theta$); the converse also applies.

Diode function generators make use of the 'on-off' (switch) characteristics of solid-state or thermionic diodes to generate line segments (Fig. 58) or multi-segment functions (Fig. 59) with adjustable slopes and 'break points'. Series or shunt diode limiters, or feedback limiters with operational amplifiers, can be used.

The cathode-ray tube function generator with photoelectrically scanned mask, or photoformer, directly generates an arbitrary function of the input (horizontal deflection) voltage by using a photocell and feedback loop, coupled to the vertical plates, to force the electron beam (spot) to follow the upper edge of the opaque mask, representing the required function (see Fig. 60).

CHAPTER 6

OPERATION OF COMPLETE ANALOGUE COMPUTERS

In the last few chapters we have become acquainted with many of the building blocks that comprise modern analogue computers. There are other components, which fulfil important functions in some computers. These include magnetic amplifiers, transistors, and other semiconductor devices, and various transducers and recording devices, but we do not have sufficient space to describe them here. Some of these devices are also used in digital computers, with which we shall become familiar in the second half of the book. Others you can look up in standard texts on electronics and computers.

System Operation

This chapter is concerned with the analogue computer as an integral system, a sort of self-correcting problem solver. As you may have discovered, the entire analogue computer is really a form of closed-loop (servo) control system that continuously compares the problem with the answer and corrects the latter accordingly. The physical problem may have been written out by a mathematician in the form of differential equations. These are then translated by the operator (or programmer) into corresponding machine equations, with the differential quantities replaced by algebraic 'operators' and the scale of the computer quantities (voltages) properly adjusted to the actual variables. A block diagram of the entire computer system is laid out in schematic form, showing the various computer components required to solve the machine equations. The individual components are then designed to fulfil their proper function in the block diagram layout, the initial conditions of the problem are set in, and there you have it—a complete analogue computing system.

MACHINE EQUATIONS AND SCALE FACTORS

As we have seen, in electronic and electromechanical computers mathematical or physical quantities are usually represented by voltages. The same voltage must always represent the same amount of a particular quantity; that is, we must use a consistent scale to translate from the problem to the computer, and vice versa. For example, if 100 V represents a velocity of 500 mile/h, then 50 V represents 250 mile/h, and 10 V equals 50 mile/h; moreover this scale must remain the same throughout the computer. Finally, since you cannot add apples and oranges, if velocity is an input to a servo computer, then all the other voltages applied to the input of this computer must also be concerned with velocity (either speeding or slowing up the object). Only then can the servo computer turn to a position that corresponds to the sum of all the voltages, representing the total or net velocity. The same goes, of course, for all other quantities represented by voltages, such as forces, angles, positions, accelerations, engine factors, or what have you.

The translation from the physical or mathematical variables to machine variables is made by multiplying the former by the appropriate constant scale factors. Thus, if x, y, and z are the variables of the problem the quantities $X = ax$, $Y = by$, and $Z = cz$ represent the machine variables, with a, b, and c being the constant scale factors. (By convention, machine variables are usually represented by capital letters.) The choice of scale factors is not as completely arbitrary as it might appear from the above. The scale factor must ensure that the machine variable never exceeds the voltage range of the computer, nor overloads any of its components for the entire range of the physical (problem) variable. At the same time the machine variable must be sufficiently large to be relatively unaffected by noise, hum, and stray voltages in the computer, which might impair its accuracy. This generally leads to a scale factor as large as possible without ever exceeding the voltage range of the computer. The voltage range available for any variable, usually called the machine unit, depends, of course, on the particular computer; machine units of 100 V are frequently used for d.c. analogue computers, while 12·5 V a.c. is typical for a.c. machines.

As an example, consider the representation of the forces (thrust and drag) acting on the forward velocity of an aeroplane, by voltages in an a.c. analogue simulator. Assume that the maximum forward force (thrust), x, and the maximum retarding force (drag), y, never exceeds 250,000 pounds (lbf), and that the available machine unit is 12·5 V a.c. The total (net) forward force, z, is of course the algebraic sum of the two forces, or equivalently, their arithmetic difference, since the retarding force (drag) always opposes the thrust, and hence is negative in magnitude. Thus, the simple equation governing the relationship between the forces is

$$\text{Total net force} = \text{Thrust} - \text{Drag}$$

or $z = x - y$. The corresponding machine equations is, then: $Z = ax - by = X - Y$.

From this, the scale factor

$$a = \frac{X}{x} = \frac{12 \cdot 5 \text{ V}}{250{,}000 \text{ lbf}} = 0 \cdot 00005 \text{ V/lbf}$$

and the scale factor

$$b = \frac{Y}{y} = \frac{12 \cdot 5 \text{ V}}{250{,}000 \text{ lbf}} = 0 \cdot 00005 \text{ V/lbf}$$

The final machine equation, therefore, is $Z = 0 \cdot 00005x - 0 \cdot 00005y$, which can easily be set up in a conventional summing network or amplifier.

Time Scale. We have not said anything about the independent variable, namely time, in most physical problems. In many analogue computers the time variable is left alone; that is, the problem is solved in the actual time that it consumes, which is known as real time operation. However, as we shall see later, it is frequently advantageous to either slow down or speed up the time of problem solution with respect to the actual time. By slowing down time, we can study rapidly occurring phenomena in sufficient detail; by speeding up time we can display the solution of a problem visually and immediately observe the effect of changing any of its parameters, such as the initial conditions. In either case, we shall have to choose an appropriate time scale, which relates the actual physical time (t) to machine time (T). The choice of time scale is governed, of course, by the required amount of slowing down or speeding up of the solution, but it must also take into account certain computer characteristics, such as the increasing error incurred with integrators for extended time intervals (you will recall the increasingly non-linear charging curve) and, on the other hand, the increasing unreliability of servos with rapid (high-frequency) operation. Although the machine variables are not affected by a change in the time scale, the rates at which the variables change (either faster or slower than in the physical situation) must be adjusted according to the chosen time scale, of course. This means the appropriate adjustments of integrator and differentiator constants.

For example, if you want to slow down a computer by a ratio of 10:1, with respect to real time, $T = 10t$, and the time scale factor is 10. In contrast, speeding up real time by a factor of 10 would require a scale of $\frac{1}{10}$, so that the machine time is $\frac{1}{10}$ of the real time, or $T = t/10$.

Differential Equations and Differential Operators

While the translation of algebraic problem equations into the corresponding machine equations has proved to be a relatively simple matter, we have yet to demonstrate this for differential equations, which are the 'meat' of any real problem. You can, of course, simply write the differential equation and attempt to 'mechanize' it directly by the use of differentiators and integrators, in conjunction with operational amplifiers, as we have seen to some extent in the preceding chapter. (See the operational-amplifier table, Fig. 41.) You would soon get entangled in unwieldy expressions and computer set-ups, however, if you tried that approach with more difficult problems. Fortunately, certain differential operators and operational methods permit converting many differential equations into corresponding operational equations, which are easily set up and handled by a computer. These operational equations not only look much easier, since all integrals and derivatives are eliminated from them, but they are much easier to work with, being essentially algebraic in nature. Thus, the use of operators is a powerful tool that permits replacing certain differential equations by equivalent algebraic equations, which may be easily solved. We cannot go into these methods here, except to give a few deceptively simple definitions and 'tricks' and hope that you won't misapply them.

The 'p' Operator. In its simplest and most valid form, the differential operator is no more than a symbol or abbreviation for differentiation and integration. Thus, we write shorthand

$$p = \frac{d}{dt}, \quad p^2 = \frac{d^2}{dt^2}, \quad \frac{1}{p} = \int dt, \text{ and so on.}$$

In words, the differential operator p stands for the first time derivative, p^2 for the second time derivative, p^3 for the third time derivative, the reciprocal $(1/p)$ for the time integral, and so on. Essentially, this only helps to write differential equations in simpler form. For example, if you have the messy-looking differential equation

$$a^3 \frac{d^3y}{dt^3} + b^2k^2 \frac{d^2y}{dt^2} + c\frac{dy}{dt} + \int y = E$$

you can write this more simply

$$a^3 p^3 y + b^2 k^2 p^2 y + cpy + \int y = E$$

and that is all you can do at this stage. Of course, a physical equation containing the operator p must be translated into the corresponding machine equation, replacing p with P and $1/p$ with $1/P$. This causes no difficulties if the computer operates in real time. For example, if

$$\frac{dy}{dt} = py = 200, \quad Y = \frac{y}{50} \text{ (or } y = 50Y\text{), and } T = t,$$

then
$$50 \frac{dY}{dT} = \frac{dy}{dt} \text{ or } 50\,PY = py$$

Substituting for py ($= 200$), we obtain

$$50PY = 200, \text{ or } Y = 4(1/P)$$

This last expression, which we have obtained by algebraic means, is, of course, the indicated solution of the machine equation, and it shows that Y is 4 multiplied by the time integral of Y. You can easily check to see that this solution is valid by substituting the original derivative.

If the machine time is not equal to the real time, the machine operator, P, must be divided by the time scale factor. Thus, in the example above, for a slow time, $T = 100t$, $dT = 100\, dt$, and

$$\frac{dy}{dt} = 50\, \frac{dY}{\frac{dT}{100}} = 5000\, \frac{dY}{dT} = 200$$

hence
$$PY = \frac{1}{25} \text{ or } Y = \frac{1}{25P}$$

Similarly, for fast time,

$$T = \tfrac{1}{100}t,\ dT = \tfrac{1}{100}\, dt,\ \text{or } 100\, dT = dt$$

hence,
$$\frac{dy}{dt} = 50\, \frac{dY}{100\, dT} = \frac{1}{2}\, \frac{dY}{dT} = 200$$

and
$$PY = 400, \text{ or } Y = \frac{400}{P}$$

Manipulation of Operators. As we have already seen, certain useful algebraic operations are permitted with the differential operator p, giving it more than just symbolic significance. The operational calculus developed by Oliver Heaviside explores this subject fully, and unless you have studied it, you had better not try anything beyond very simple algebraic transformations, since there are very sharp limits to just what can and cannot be done. However, as an example, we will transform the relatively simple second-order differential equation we developed in Chapter 2 in connexion with the analogous mechanical and electrical oscillation problem. You will recall that to derive the basic equation of the electrical R–L–C circuit (see Fig. 17) we set the applied voltage (E) equal to the sum of the voltage drops across the resistance, capacitance, and inductance of the series circuit. The voltage drop across the resistance (R) is the product of the current and resistance, or iR. The drop across the capacitance (C) we showed to be equal to

$$\frac{1}{C}\int i\, dt, \text{ which in operator language becomes } \frac{i}{pC}.$$

Thus, any time you see a capacitor in *any* circuit, you can write i/pC for the voltage drop across it. Finally, the drop across the inductance (L) was shown to be $L\frac{di}{dt}$, which becomes simply Lpi. Hence, whenever you see an inductor (coil) in a circuit you can write Lpi for the voltage drop across it. Finally, by Kirchhoff's law, the applied voltage (E) must equal the sum of the voltage drops, or

$$E = iR + Lpi + \frac{i}{pC}$$

which is, of course, the same as if you have substituted the operator p directly in the expression

$$E = iR + L\frac{di}{dt} + \frac{1}{C}\int i\, dt$$

To 'solve' the operational expression algebraically for p, we must first clear it of fractions in p, as we would any other algebraic equation. Hence, multiplying both sides by p, we obtain

$$pE = R\,pi + L\,p^2 i + \frac{i}{C}$$

The expression pE on the left side of the equation indicates differentiation of the constant quantity (voltage) E; this is zero, of course, since the rate of change of a constant is zero. With pE equal to zero, we can rewrite the operational equation thus:

$$L\,p^2 i + R\,pi + \frac{i}{C} = 0$$

Since p^2 is the term with the highest order, this indicates a second-order differential equation, and moreover, by substituting d^2/dt^2 for p^2 and d/dt for p, we can rederive the original equation from Chapter 2:

$$L\frac{d^2 i}{dt^2} + R\frac{di}{dt} + \frac{i}{C} = 0$$

Note that we have derived the operational equation without recourse to integration or differentiation, except for replacing pE by 0.

The operational equation can be solved for p like any quadratic algebraic equation, yielding, by application of the well-known formula, two roots, p_1 and p_2, as follows:

$$p_1 = -\frac{R}{2L} + \sqrt{\frac{R^2}{4L^2} - \frac{1}{LC}}$$

and

$$p_2 = -\frac{R}{2L} - \sqrt{\frac{R^2}{4L^2} - \frac{1}{LC}}$$

It is shown in mathematics that the solution of the corresponding differential equation is of the form

$$i = A\varepsilon^{p_1 t} + B\varepsilon^{p_2 t}$$

where p_1 and p_2 are the roots of the operational equation given above. The actual solution depends on the initial values of the charge on C and whether the subtraction of the quantities within the square root comes out positive (real) or negative (imaginary). We shall see later how this equation can be mechanized in an analogue computer.

The Laplace Transforms. While it is always correct to replace the derivatives in a differential equation by the differential operator p, the permitted algebraic operations are limited, and the technique usually consists of manipulating the operational (p) equation into suitable form for the convenient analogue computer set-up and then carrying out the indicated operations (integrations and differentiations) by a corresponding computer components. There is a much more powerful method, with which you should have at least a nodding acquaintance, for converting differential equations into algebraic equations. This is the use of the Laplace transforms, which have been known to mathematicians for over a hundred years, but have been applied to the solution of engineering problems only in the past thirty years or so. Roughly speaking, the Laplace transforms do for the differential equations of engineering what logarithms do for arithmetic: they replace difficult computations with special tables. Logarithms replace the drudgery of multiplication and division by addition and subtraction of logs, and that of taking powers and roots by

Operation of Complete Analogue Computers

multiplication and division of logs, respectively. Similarly, the Laplace transforms replace the lengthy and difficult computations required in the solution of linear differential equations by a table of direct and inverse Laplace transforms. You first look up the direct transforms corresponding to the differential equation of your problem, then solve the resulting algebraic equation for the desired variable (such as the current), and finally look up the inverse transform (just like the antilog), giving the solution of the original differential equation.

We cannot go into the details of the Laplace transform technique here, but essentially the transformation changes a real quantity (such as 1, 2, a, x, y, $f(t)$, etc.) into a function of a complex variable, s, and substitutes complex algebraic terms for derivatives and integrals. (A complex quantity is a combination of a real and an imaginary quantity, the latter containing the square root of -1, such as $\sqrt{-a^2} = a\sqrt{-1}$; $\sqrt{-5} = \sqrt{5}\sqrt{-1}$; $\sqrt{-4x} = 2\sqrt{x}\sqrt{-1}$, etc.) If the original quantities are functions of time $[f(t)]$ the transformed quantities $[F(s)]$ become functions of frequency. Essentially, the Laplace transform is a relation for transforming a real function of time into a complex function of s.

This relation is given by

$$\int_0^\infty \varepsilon^{-st} f(t)\, dt = F(s)$$

You rarely have to compute this forbidding-looking integral, since the required transform usually is listed in the table.

As an example of the Laplace transform, let us compute the current (i) that charges a capacitor (C) through a resistor (R) from a battery (voltage E). The differential equation governing this situation is

$$E = iR + \frac{1}{C}\int i\, dt$$

which you can easily derive for yourself or obtain from the fundamental equation of the R–L–C circuit (given earlier) by leaving out the voltage drop across the inductance, $L\dfrac{di}{dt}$

Looking up the appropriate Laplace transforms, you obtain the simple algebraic equation in terms of s:

$$\frac{E}{s} = iR + \frac{i}{sC}$$

We now can solve for the current, i, and write it in the simple operational form:

$$i = \frac{E}{R}\left(\frac{1}{s + \frac{1}{RC}}\right)$$

Looking up the inverse Laplace transforms for this expression, you obtain the familiar exponential solution for the current

$$i = \frac{E}{R}\varepsilon^{-t/RC}$$

This expression shows that the charging current *initially* (at $t = 0$) rushes into the capacitor limited only by the resistance, in accordance with Ohm's law ($i = E/R$), then builds up exponentially at a rate determined by the time constant, RC, and eventually, after a time equal to several time constants (theoretically $t = \infty$), drops off to zero. (Note that $\varepsilon^{-0} = 1$ and $\varepsilon^{-\infty} = 0$.)

THE COMPUTER BLOCK DIAGRAM

We are now ready to transform the machine equations, obtained by various mathematical techniques, into the corresponding analogue computer set-up by means of the computer block diagram. The initial block diagram, which always precedes the actual computer set-up, relates the machine variables (voltages) to the various computing elements in the same way that the mathematical equation relates the actual problem variables to the various mathematical operations. The machine equation is translated step by step into the corresponding blocks of the computer block diagram, enabling us to 'read' the completed block diagram from the inputs at the left to the outputs at the right, like the equation we started with. Astonishingly, this diagram frequently assumes that the solution of the problem is already known (in implicit form) and then proceeds to add missing portions, similar to working a crossword puzzle backward, until the pattern of problem and solution is complete and all pieces fall into place. Although it may appear disturbing at first, the technique of taking the answer for granted (in the form of an implicit equation) and then working backward is very powerful, and saves a lot of trouble in working out the answer directly.

Implicit Function Technique. As an example of 'the-answer-is-part-of-the-problem' (implicit) technique, consider the analogue computer mechanization of Kepler's astronomical equation for the 'eccentric anomaly':

$$\theta = a + b \sin \theta$$

This is an implicit function, since the unknown—the angle θ in this case—appears on both sides of the equation, and the relation between θ and the known factors, a and b, is not explicitly stated. If you desired an explicit solution you would have to solve the equation for θ in terms of a and b, which is certainly not simple. The solution for θ consists of an infinite series which was first worked out by Lagrange. The first five terms of this series, which may be considered an approximate answer, are as follows:

$$\theta = a + b \sin a + \frac{b^2}{2} \sin 2a + \frac{b^3}{8} (3 \sin 3a - \sin a) + \frac{b^4}{6} (2 \sin 4a - \sin 2a) + \ldots$$

The dots indicate succeeding terms which are even more complicated than those shown. You can imagine the trouble you would have trying to set this series up on an analogue computer so as to obtain an approximate explicit solution for θ.

Now look at Fig. 61, which shows the computer block diagram of the

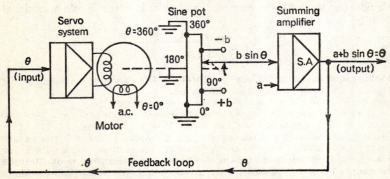

Fig. 61. Analogue-computer block diagram for function $\theta = a + b \sin \theta$.

implicit formula $\theta = a + b \sin \theta$. As you can see, θ appears as an output of the computer at the right, so that—clearly—the equation has been solved in its implicit form. The entire set-up consists of only two components, a positioning servo and a summing amplifier. The unknown, θ, is applied at the input of the positioning servo. (We are not concerned here how this can be done.) A sine function potentiometer with fixed voltage inputs, $+b$ and $-b$, is coupled to the output shaft of the servo, which positions the wiper of the pot to θ. The output voltage from the wiper (brush) of the sine pot, thus, is $b \sin \theta$ for all values of θ between 0° and 360°. This pot output together with the constant voltage a is then applied to the input of a conventional summing amplifier. By the use of an appropriate output transformer, a positive rather than an inverted ($-$) output can be obtained from the summing amplifier, so that the summed output voltage is

$$a + b \sin \theta = \theta$$

which is the desired solution.

The output voltage, representing θ, is fed back to the input of the servo through the feedback loop, which explains how θ was obtained in the first place. The self-correcting ability of the closed-loop servo system assures that the implicit relation will be made true, even if it is initially incorrect. For if θ is not equal to $a + b \sin \theta$, an error will exist between the servo input and the summing amplifier output, and this error signal will cause the servo to rotate in such a direction and to such a position until the error is zero and the relation is true.

Block Diagram Set-up for the Second-Order Differential Equation. As another example, let us try to set up the computer block diagram for the second-order linear differential equation with constant coefficients, which we have encountered several times before in connexion with the mechanical and electrical oscillator problem. As you will recall, the equation for the current in an R–L–C circuit (see Chapter 2) was given by

$$L \frac{d^2 i}{dt^2} + R \frac{di}{dt} + \frac{1}{C} i = 0$$

which may be put into the more general form of a second-order linear differential equation with constant coefficients:

$$\frac{d^2 y}{dt^2} + A \frac{dy}{dt} + By = 0$$

where $A = R/L$, $B = 1/LC$, and $y = i$, in this particular case. After translating this standard differential equation into the corresponding machine equation by introducing the machine variable Y, the machine differential operator P, and the proper scale factors, it will look like this:

$$P^2 Y + APY + BY = 0$$

where
$$P^2 Y = \frac{d^2 Y}{dt} \text{ and } PY = \frac{dY}{dt}$$

(assuming the machine time is the same as the real time, t).

To set up this machine equation on an analogue computer, you must first solve it for the highest derivative of Y (the P^2 term), resulting in

$$P^2 Y = -(APY + BY)$$

where each of the terms on the left and right must be represented by voltages. By inspection of this relation it becomes evident that the two terms at the right of the equation (APY and BY) must be obtained by successive integration of the term on the left-hand side (P^2Y); they must then be added together and multiplied by -1 to obtain the desired equality.

Fig. 62 illustrates one economical analogue computer set-up for solving this differential equation. As before, the equation is solved in its implicit form by applying assumed correct answers to the input of the computer. Thus, the functions Y and PY are applied to the input summing network of a

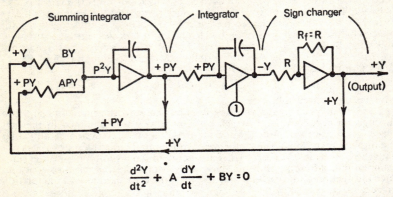

$$\frac{d^2Y}{dt^2} + A\frac{dY}{dt} + BY = 0$$

Fig. 62. Functional computer set-up for solving a second-order linear differential equation.

summing integrator, as shown. The input resistors and the integrating capacitor are chosen so that Y is multiplied by the constant scale factor B and PY is multiplied by the constant A. Their sum, therefore, by the defining relation above, is:

$$APY + BY = -P^2Y$$

This function ($-P^2Y$) is integrated and inverted in sign by the first integrator, so that its output voltage is $\frac{1}{P}(P^2Y) = +PY$. This voltage is applied to the input of a second, straight integrator, whose constants are adjusted for a scale factor of unity. The output of this integrator, therefore, is the integrated and inverted function

$$-\frac{1}{P}(PY) = -Y$$

The voltage $-Y$, finally, is inverted in sign to $+Y$ through a sign-changing operational amplifier, whose input resistance equals its feedback resistance ($R = R_f$). The output voltage, $+Y$, represents the solution of the differential equation. This voltage ($+Y$) and the output of the first integrator ($+PY$) are fed back to the input of the computer, which results in the closed-loop system that automatically enforces the differential equation.

Initial Conditions. Any actual physical problem starts from a given situation. For example, in the mechanical oscillator problem described in Chapter 2 we stated that the weight (m) was initially displaced by a certain amount by extending the spring before releasing it. Thus, the initial displacement (x_o) at the beginning of the problem ($t = 0$) is the initial condition of the problem. The behaviour of the mechanical oscillator at any time thereafter is, of course, described by the differential equation we have developed. Similarly, in the electrical oscillator problem the capacitor was first charged from a battery voltage (E) to an initial charge (q_o) before it was allowed to discharge through the resistor and inductor (coil). Thus, the initial condition in this case is the initial capacitor charge ($q_o = CE$) at the beginning, or zero time ($t = 0$).

When the differential equation representing a physical system is solved by successive integrations a constant of integration appears together with the integrated function during each integration. The values of these integration constants must be determined by the given initial conditions of the problem. Thus, the same differential equation may represent a problem for a variety of initial conditions, differing only by constants. Since a constant appears for each integration, a second-order differential equation (containing a second derivative) requires two integrations and therefore has two constants to be specified by initial conditions. For example, in addition to the initial charge, you may have to specify the initial charging current (i_o) at zero time. (For a capacitor charging through a resistance R, the initial current $i_o = E/R$ at $t = 0$.)

Since an analogue computer set-up must represent a given problem exactly, the correct initial conditions of the problem must be set into it before the start of a computer run. This resetting is accomplished by applying the proper initial-condition voltage to each of the integrators in the set-up. The various integrating capacitors are then charged up (or the integrating servos turn to the positions) corresponding to the initial conditions of the problem. When the computer runs through the problem it will start from these initial conditions (integrator outputs), but the behaviour of the machine variables thereafter is governed by the differential equation. In the computer block diagram, the initial-condition inputs are shown by encircled quantities next to each integrator. Thus, the initial condition for the second integrator in Fig. 62 is $Y_o = 1$, shown by ①, which means that the integrator output starts with '1'. No initial condition is shown for the first integrator (in Fig. 62); this signifies that the initial output voltage of the integrator is zero [i.e. $(PY)_o = 0$].

Operational Amplifier Solutions of Differential Equations

Although most differential equations must be 'synthesized' block by block on the computer functional diagram, it is sometimes possible to use the versatile operational amplifier directly to solve rather difficult equations. The technique consists of inserting the appropriate combinations of resistors and capacitors into the input and feedback networks of the amplifier, such as we have seen already in simple form in the operational amplifier table (see Fig. 41). The R–C combinations form complex impedances (having resistive and reactive components), which result in output functions that contain combinations of integrals and derivatives of the input voltage and the impedances.

Fig. 63 is an expansion of the operational amplifier table (Fig. 41), showing a few of the many possible combinations of complex impedances and the resulting output functions. In each case the type of differential equation solved by the operational amplifier circuit is shown at the left, while the column at the right shows the resulting output function in the form of an operational equation. As before, the differential operator p stands for the derivative, d/dt, while its reciprocal, $1/p$, stands for the integral $\int dt$.

The top row in Fig. 63 shows the solution of a simple linear differential equation of the first order, obtained by the insertion of a parallel resistor-capacitor combination into the feedback loop of the amplifier. This is also known as a 'lag function' because the output voltage (E_o) lags the input by the time constant $R_f C_f$. The second diagram, with the series R–C combination in the feedback loop, again illustrates the solution of a first-order linear differential equation, by means of an integration and scale change. The third diagram shows the rather complex output function resulting from the insertion of a series R–C combination in the input network and a parallel R–C combination into the feedback loop. This set-up will solve a second-order differential

Description of operation and equation	Circuit diagram	Operational form of output function $p = \frac{d}{dt}; \frac{1}{p} = \int dt$
Solution of first order linear differential equation: $\frac{dE_o}{dt} + AE_o = E_i$	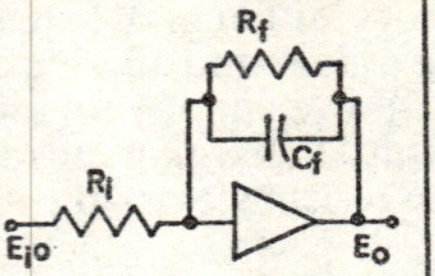	$E_o = -\frac{R_f}{R_i}\left[\dfrac{1}{1 + R_f C_f p}\right] E_i$
Solution of first order linear differential equation: $AE_i + B\int E_i dt = E_o$	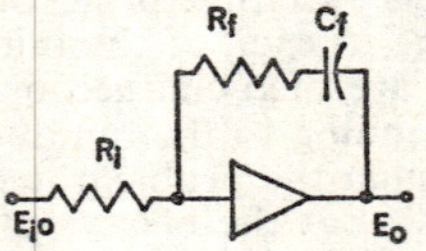	$E_o = -\left[\dfrac{R_f C_f p + 1}{R_i C_f p}\right] E_i$
Solution of second order linear differential equation: $A\frac{dE_o}{dt} + (B+C)E_o + C\int E_o dt = E_i$	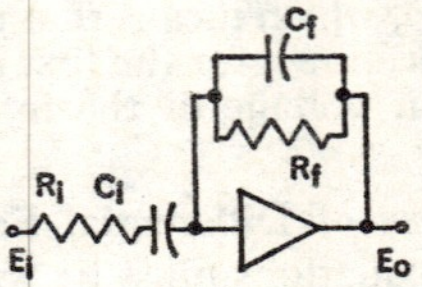	$E_o = -\left[\dfrac{R_f C_i p}{(R_i C_i p + 1)(R_f C_f p + 1)}\right] E_i$
Solution of second order linear differential equation: $A\frac{dE_i}{dt} + (B+C)E_i + C\int E_i dt = E_o$	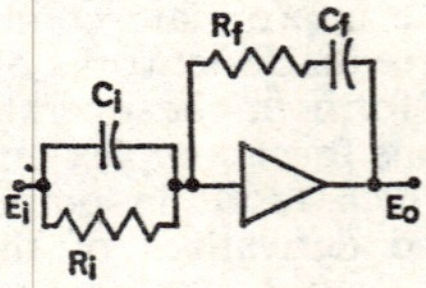	$E_o = -\left[\dfrac{(R_i C_i p + 1)(R_f C_f p + 1)}{R_i C_f p}\right] E_i$

Fig. 63. Solution of differential equations by operational amplifiers with complex impedances.

equation with constant coefficients, A, B, and C. (The second-order form becomes apparent by differentiation of the equation shown at the left.) Finally, in the last diagram the input and feedback R–C networks are interchanged, resulting in the solution of the same type of second-order differential equation with the variables E_i and E_o interchanged.

A TYPICAL ANALOGUE COMPUTER APPLICATION: AIRCRAFT FLIGHT SIMULATION

We have studied in some detail all the ingredients that go into the make-up of a modern analogue computer. Let us now look at a typical application,

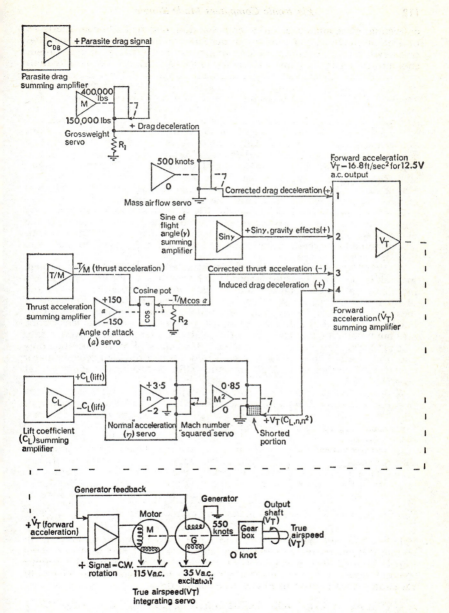

Fig. 64. True air-speed simulator system of a jet aeroplane. (After Curtiss-Wright Corporation flight simulator.)

consisting of the analogue computer simulation of some flight characteristics of a jet aircraft. Fig. 64 shows a considerably simplified block diagram of the true air-speed simulator system for a modern jet aircraft. The complete flight simulator, built by the Curtiss-Wright Corporation, comprises literally hundreds of these interrelated systems, which take up the volume of a large room. The various computer components are assembled around an actual aircraft cockpit, which contains all the instruments and controls of the aircraft. The trainee pilot sits in his normal position in the cockpit and operates the controls of the simulator in exactly the same way as if he were flying the real aircraft. The controls and instruments of the simulator respond to his actions and give exactly the same indications as those of the real aircraft, including engine vibrations, gusts of air, control counterpressures, thunderstorms, etc. (The external conditions are simulated by the instructor.) If the trainee should stall the aircraft or actually 'crash' it the cockpit instruments will soon tell him of this embarrassing situation, and though he saves his life (the purpose of the simulator), he 'sweats' as if he had crashed the real thing.

True Air Speed. The block diagram (Fig. 64) isolates a few of the many flight-simulator components required for simulating the true air speed of the plane. Many of the inputs shown at the left of the diagram consist of quantities computed elsewhere in the computer, and it is not possible to show here how they were obtained.

Air speed is simply the speed of the aircraft relative to the mass of air through which it is moving. The true air speed can be computed by considering all the factors that affect the total thrust and total drag, and hence the acceleration, of the aircraft. True air speed must be distinguished from the indicated or measured air speed, displayed by the air-speed indicator on the pilot's instrument panel. The indicated air speed must be corrected for instrument errors, air compressibility, and changes in the relative air density to convert it into true air speed. The part of the air-speed system shown here is concerned only with the direct simulation of true air speed, though the complete simulator contains, of course, additional components for displaying the indicated air speed.

Forward Acceleration. The computer obtains the true air speed (v_T) by integrating the plane's forward acceleration, which is the rate of change of true air speed (dv_T/dt which is usually symbolized by $\dot{v}_T$). Now any kind of acceleration, according to one of Newton's laws of motion, is the ratio of force to mass ($a = f/m$). Hence, if we know the mass (or weight) of the aircraft and the various forces that tend to speed up or retard the plane we can compute its forward acceleration, and from this we can obtain (by integration) the true air speed. The forward acceleration is the difference between the acceleration due to the total thrust of the jet engines and the deceleration (slowing up) due to the total drag acting on the aircraft. The total drag is made up of a number of factors, the chief ones being the parasitic drag on the wings and fuselage due to air friction and the induced drag caused by a rearward component of the lift. For this reason, all factors that affect the lift of the aircraft also affect the induced drag, and hence the total drag. We must therefore sum up all the factors that contribute to acceleration due to thrust (called thrust acceleration) and all the factors that contribute to deceleration due to drag, compute the difference between the two (or the algebraic sum), and then we have the forward acceleration of the aircraft. This job is done by the forward acceleration summing amplifier (see Fig. 64).

Forward Acceleration Summing Amplifier. To take into account the various factors described above, the forward acceleration summing amplifier has a number of input signals obtained from other sources. Input (3) of the summing amplifier is a negative (—) voltage representing the thrust acceleration. (The signal is negative to take into account the fact that the amplifier will invert its phase, making it + in the output.) The negative thrust acceleration (−thrust/mass, $-T/m$) signal, originally obtained from the output of the thrust acceleration summing amplifier, is fed through a cosine function card on the angle-of-attack (α) servo to obtain the component of the thrust acceleration that is parallel to the flight path. This cosine component of the thrust acceleration $\left(\dfrac{T}{m}\cos\alpha\right)$ is then fed to input (3) of the forward acceleration summing amplifier.

Parasite Drag. Input (1) of the forward acceleration summing amplifier represents the deceleration due to parasite drag. Note that this signal is positive (+), or opposite in phase to the thrust acceleration signal. The positive output of the parasite drag summing amplifier is fed through a divider card on the gross weight servo in order to obtain the ratio of parasite drag to aircraft mass, which is the negative acceleration (deceleration) produced by the drag. This drag signal is then corrected for the relative air density in relation to the air speed by feeding it through a multiplier card on the mass-air-flow servo. (Mass-air-flow = Relative air density × True air speed.) The corrected negative drag deceleration signal is applied to input (1) of the forward acceleration summing amplifier.

Induced Drag. The part of the drag that is affected by the lift of the aircraft (i.e. the induced drag) is simulated by positive and negative signals from the lift coefficient (C_L) summing amplifier. These signals are applied to a multiplier card on the normal acceleration servo to obtain the acceleration component 'normal' (perpendicular) to the lift, which is the induced drag acceleration. This signal is then further multiplied by a card on the 'Mach number squared' servo to correct it for the effects of air compressibility when the plane is approaching the speed of sound. Since this effect is noticeable only at high air speeds, the card is shorted to ground (shown by the shaded portion) for low values of the air speed. The corrected induced drag deceleration signal is fed to input (4) of the forward acceleration summing amplifier. Note that this signal, too, is positive (+), or opposite in phase to the thrust acceleration signal.

Gravity Effects. Finally, gravity contributes to the total drag on the aircraft. The sine component of the flight path angle (γ) represents the downward force (gravity) of the weight of the aircraft. Accordingly, a positive output signal from the sine-of-flight-path angle ($\sin\gamma$) summing amplifier is applied to input (2) of the forward acceleration summing amplifier.

Summation. The positive drag deceleration and negative thrust acceleration signals are summed up in the input network of the forward acceleration summing amplifier. The sum, which is a negative (−) signal, is inverted in phase by the amplifier, so that the positive output of the amplifier is proportional to the total forward acceleration of the aircraft, or equivalently, to the rate of change of true air speed + $\dot{v}_T$.

True-air-speed Integrating Servo. Since forward acceleration is the rate of change of true air speed, all that remains to be done is to integrate the output of the forward acceleration summing amplifier in order to obtain the true air speed. As we have seen in the preceding chapter, this can be done by making the total rotation of a servo output shaft proportional to the product of the changing rate of speed and the time elapsed. Accordingly, the positive output signal ($+\dot{v}_T$) from the forward acceleration summing amplifier is applied to the input of the true-air-speed integrating servo, resulting in clockwise rotation of the servo output shaft proportional to true air speed. (Note that the servo is of the generator feedback type we have described.) By attaching the appropriate cards (potentiometers) to the servo shaft, the true-air-speed output can be made available to other portions of the computer. In the actual simulator, relays simulate the effect of the brakes after landing by feeding in a negative air-speed ($-\dot{v}_T$) signal in place of the forward acceleration signal.

Repetitive Computers

The portion of a flight simulator we have just described operates in real time, that is, in a 1 : 1 time relation to the actual aircraft. This is, of course, necessary for realistic simulation. The simulator as well as all other analogue computers we have studied to this point are classed as 'slow', since they operate either on the same time scale as the problem or even slower. However, the term 'slow' should not be taken in a disparaging sense, since, because of the parallel operation of many computer elements, most problems are actually solved extremely rapidly. These computers are classed as 'slow' to distinguish them from the 'fast' or 'repetitive' analogue computers. A repetitive computer operates on a time scale that is considerably speeded up compared with the actual problem time. Repetitive computers solve problems in the same way as the slow computers we have studied, but they differ from the latter in that they repeat automatically the computer solution (or 'run') at a rapid rate somewhere in the range of from 10 to 100 times per second (10 − 100 c/s). During each cycle the machine variables conform to the prescribed differential equations, and at the end of the run they are reset to their initial values to begin another cycle.

The repetitive computer has the outstanding advantage that its rapidly repeated solutions can be displayed on the screen of a cathode-ray oscilloscope (similar to a television screen), and hence variations in the problem set-up can be observed immediately. Particularly, you can see immediately the result of varying initial conditions or the coefficient settings of the multiplier pots without having to wait for the end of a run. Because of their rapid operation, repetitive computers usually contain all-electronic components rather than electromechanical devices.

Future Outlook

The analogue computer has recently been in sharp competition with the digital types, which have enjoyed increasing popularity. There is no doubt that the digital computer is particularly well suited in medium and large-sized installations for complex applications, where high precision and wide versatility are required. More recently, desk-sized digital computers have also become available for general-purpose applications. It is possible that in this category the digital type will replace many of the previously popular small analogue installations. The analogue computer, however, is enjoying a renaissance of its own in the area of special, fixed-purpose applications where, because of its low cost, small size, light weight, and simple maintenance, it has no peer. Whenever it is required to solve a not-too-complex particular problem with medium (engineering) precision, minimum maintenance, and minimum investment, the analogue computer would appear to be a natural choice. Moreover, the analogue computer can do the job as rapidly as is required for most applications. The rapidity of the digital computer is something of a 'white elephant'. The digital type may perform individual computations in microseconds or less, but it may have to carry out thousands of these computations before approaching even a partial result. Moreover, the devices that print out the answers are usually considerably slower than the electronic portions, thus slowing up the entire system.

Combined Computers. The controversial choice of analogue computers *versus* digital computers is becoming somewhat academic, since many present-day computers combine some of the features of both techniques. This trend will increase as improved analogue–digital converters (see pages 289–96) become available and components are further miniaturized.

New Developments. The computer field in general and analogue computers in particular are in a stage of rapid development. Almost daily commercial companies introduce computers which demonstrate some new features and techniques. The present-day trend is towards low-cost computers of 'modular' construction, which permits later expansion by 'add-on' units. By packaging all important computer elements, such as operational amplifiers, servos, function generators, and coefficient-setting potentiometers, into small units (modules), additional computer modules of the required type can be purchased later, as needed, and interconnected with the general computer scheme. Thus, a basic desk-top analogue computer may sell for about £600 and contain a 'pegboard' or 'patchcord' for interconnexions, a few amplifiers, multipliers, and function generators. By plugging in additional modules and adding entire computer cabinets, such a unit can be expanded into a large, floor-sized computer in the £25,000 price bracket.

Another apparent trend is the ever-increasing use of solid-state devices, such as crystal diodes, transistors, and controlled rectifiers, for amplifying and switching, as well as the use of magnetic and memory devices. We shall become acquainted with some of these devices in the digital computer portion of this volume, where they appear as essential elements; the general purpose of these new components is to achieve drastic reductions in size, weight, power consumption, and heat dissipation, while at the same time taking advantage of their increased reliability.

The combination of digital and analogue techniques is proceeding at a rapid pace. Repetitive analogue computers are being equipped with digital memories to store intermediate results of lengthy or iterative (trial-and-error) problems. Digital computers are being programmed to set up and connect together the components of an analogue computer to solve automatically certain classes of differential equations. In process control, digital techniques in combination with analogue inputs and outputs are being increasingly utilized for improved computation accuracy.

REVIEW AND SUMMARY

An analogue computer is a form of closed-loop (servo) control system that continuously compares the problem with the answer and corrects the latter accordingly.

The problem (physical) variables are translated into corresponding machine variables by multiplying them by constant scale factors. The scale factor is generally made as large as possible without exceeding the voltage range of the computer or overloading any of its components.

The time scale of a computer is either the same as that of the actual problem (1:1), in which case the computer operates in 'real time', or it is slower or faster than the actual problem time. To 'slow down' a computer, the actual time must be multiplied by the required scale factor; to 'speed up' the computer, the time must be divided by the required scale factor.

Differential operators and operational methods are used to translate differential equations into corresponding operational equations, which are algebraic in nature and are easily manipulated and set up on the computer. The operator p stands for the first derivative, d/dt, while its reciprocal, $1/p$, stands for the integral $\int dt$.

$$(p^2 = \frac{d^2}{dt^2}; p^3 = \frac{d^3}{dt^3}, \text{etc.})$$

The machine operator, P, must be corrected for the time-scale factor, if the machine time is not equal to the real (problem) time.

Laplace transforms convert real functions of time [$f(t)$] into complex (real + imaginary) functions of frequency [$F(s)$]. The relation between the functions is given by:

$$\int_0^\infty \varepsilon^{-st} f(t)\, dt = F(s)$$

The Laplace transforms for a given differential equation may be looked up in a table of direct transforms; the resulting algebraic equation is then solved for the desired variable, and the corresponding solution of the differential equation is obtained from a table of inverse transforms.

The computer block diagram relates the machine variables (voltages) to the computing elements in the same way as the mathematical equation relates the problem variables to the required mathematical operations. Like the equation, the computer block diagram is 'read' from the inputs at the left to the outputs (answers) at the right.

If a given equation is not easily solved in explicit form (e.g. $x = k/y$), it may be solved by the computer as an implicit equation ($xy = k$). In the implicit-function technique the unknown variable (answer) is part of the input to the computer; the computer elements are set up to perform the required mathematical operations as if the answer were known. The answers (outputs) or partial answers are then fed back to the input, resulting in a closed-loop (feedback) system that automatically enforces the mathematical relations of the original equation (see Figs. 61 and 62).

The values of the physical variables and their derivatives in a differential equation, at the time the problem is first considered ($t = 0$), are known as the initial conditions of the problem. When the equation is solved the constants of integration must be replaced by the given initial conditions; similarly, in an analogue computer each integrator must be set to the voltages corresponding to the initial conditions before each computer 'run'.

Some differential equations can be solved directly by inserting complex impedances (parallel and series combinations of resistors and capacitors) into the input and feedback networks of an operational amplifier (see Fig. 63).

A repetitive computer operates on a fast time scale, from 10 to 100 times the 'real' (problem) time, which permits continuous display of the solution on the screen of a cathode-ray oscilloscope.

CHAPTER 7

INTRODUCTION TO THE DIGITAL COMPUTER

We now return to our starting-point—counting. Our exploration of the analogue computer has led us into some highly sophisticated mathematical concepts. By the use of analogue methods we were able to set up a direct model or mathematical analogy of some problem, observe its dynamic behaviour, and obtain an 'answer' by solving its underlying differential equations. We now leave the world of physical phenomena occurring in 'real time' and enter the abstract realm of pure number—the quantitative, discrete, numerical data that comprise arithmetic. The digital computers, whose operation shall concern us for the remainder of this book, break down all problems presented to them into a 'program' of numerical (and sometimes alphabetical) data and a series of instructions as to what to do with the data. After processing the data the digital computer prints out the answers, again in discrete numerical (or letter) form. Given this basic frame of digital computer operation, we must now study the types of number systems (there are many) that can best be handled by a machine, the most logical ways of performing the required arithmetical operations and controlling their sequence, and the possible means of 'communicating' with the computer; that is, the manner in which to present the problem in the required form of data and instructions (input) and how to translate the answers (output) into understandable terms. As we did with the analogue computer, we shall 'synthesize' the complete digital computer from a series of easily comprehensible building blocks.

EARLY DIGITAL COMPUTERS

The early history of digital computing devices was covered briefly in Chapter 1. You will recall the ancient abacus and the variety of digital adding machines and calculators which were introduced in the seventeenth century by Pascal and Leibniz. The earliest programmed digital machine probably goes back to the punched-card loom invented in 1801 by Jacquard. Operating somewhat in the fashion of a mechanical player piano, the Jacquard machine attained automatic digital process control of woven figured fabrics through a loom controlled by punched cards. You will recall that the great pioneer of automatic computers, Charles Babbage, applied the punched-card idea to program his 'analytical engine' (in 1833), which contained all the concepts of a truly automatic computer, but was never completed because of the insufficiently developed machinist's art (Babbage's 'folly'). It was not until 1886 that Hollerith at the United States Bureau of the Census developed a successful punched-card machine for sorting and tabulating census data.

The First Automatic Digital Computer

The first large-scale fully automatic digital computer was almost certainly the IBM (International Business Machines Corporation) Automatic Sequence-Controlled Calculator. This machine, which later became known as Mark I, was the brainchild of Prof. Howard Aiken of Harvard. Its development began in 1937 and went on until 1944.

The Harvard–IBM Mark I calculator has all the important functional components of an automatic digital computer—input, memory, arithmetic (processing) unit, control, and output—except that its actual computing (arithmetic) section is not separate, as in later types of computers, but is closely allied to the memory operations. The input to the machine, consisting of 23-digit decimal numbers and operation instructions, is fed in either by regular IBM punch cards, punched tape, or by hand-set dial switches. Depending upon the coded instructions, the machine can perform automatically any desired sequence of operations, such as adding, subtracting, multiplying, dividing, and transferring or clearing numbers, as well as calculating logarithms, exponentials, sine functions, etc. However, by present-day standards, the Mark I computer is slow. To add or subtract numbers takes about $\frac{1}{3}$ second, multiplication approximately 5 seconds, division up to 16 seconds, and to compute a logarithm or exponential to 23 decimal places may take up to 90 seconds. This must be compared with the fantastic speeds of recent electronic computers, which can perform similar mathematical operations within a few millionths of a second (microseconds) or less.

In Britain, meanwhile, work was being concentrated on the development of pulse techniques and circuitry for war-time radar. These techniques were to become, as we shall see, the basis of every electronic digital computer.

Although more advanced electromechanical computers were built in the 1940s, the major industrial efforts were soon turned upon the infinitely more rapid and more reliable electronic digital computers. The first one, called ENIAC (Electronic Numerical Integrator and Calculator), was constructed between 1942 and 1945 at the Moore School of Electrical Engineering of the University of Pennsylvania.

The First Electronic Digital Computer

The Moore School's ENIAC contains 18,000 thermionic valves (a feat never again duplicated). Its appearance immediately made all electromechanical (relay) computers obsolete because of its capability of performing 5000 additions per second, compared to the top speed of 5–10 additions per second of a relay computer. ENIAC was intended primarily for the calculation of ballistic trajectories compiled in firing tables; this is an extremely time-consuming job if done by hand or desk calculators. Except for its small memory capacity and relative slowness of input-output and some arithmetic operations, the ENIAC is essentially similar to a large number of more efficient electronic digital computers that followed it. The machine was constructed at a cost of about £170,000 and was moved in 1947 to the Ballistic Research Laboratories of the United States Army at Aberdeen, Maryland.

CLASSIFICATION OF DIGITAL COMPUTERS

The digital-computer family classification shown in Fig. 65 displays a small cross-section of the maze of digital computers that arose in the past 20 years as refinements of the Mark I and the ENIAC prototypes. As with analogue computers, we may divide the digital types into the two broad categories of general-purpose and special-purpose computers. However, in contrast to analogue computers, these categories are not rigidly defined for digital computers, since the numerical (arithmetic) operation of the latter is essentially always 'general-purpose' and, with minor modifications, may be applied to any problem. Within the general and special-purpose categories, digital computers may be further subdivided into mechanical, electromechanical, and electrical or electronic types, the latter two being the only important surviving types because of their inherently high speed.

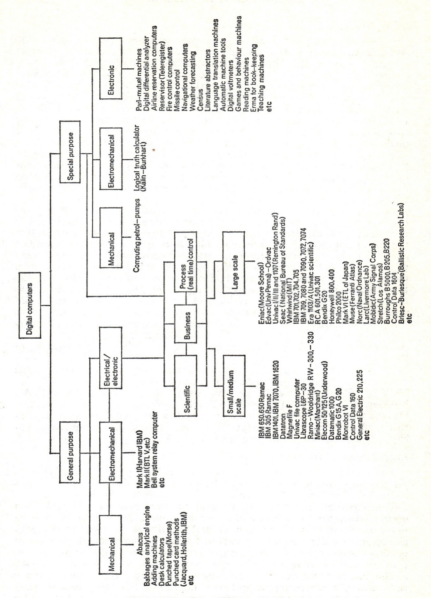

Fig. 65. Digital computer classification.

General-purpose Digital Computers

Historically, the mechanical digital computers came first, of course. You will recall the abacus (and soroban), the early adding machines and calculators (Pascal and Leibniz), Babbage's ingenious but never completed 'analytical engine', and the still-surviving punched-tape (Morse) and punched-card machines (Hollerith, IBM, etc.). The next great advance, as we have pointed out, was the development of digital computers based upon more refined electromechanical devices (relays, stepping switches, etc.), of which the Mark I and the succeeding Mark II were the prototypes. In this category also belong two of Bell Telephone Laboratories' digital relay computers, which were made available to military research agencies in 1946. These two computers were far more versatile and reliable than Mark I and Mark II and signify the highest point of development reached by the 'mechanical brains'. Except for telephone-system applications, these remarkable machines, too, were soon made obsolete by the rapid development of the electronic digital computers in the early post-war years.

Electronic Computers. Modern electronic digital computers serve three major areas of application: scientific, business data processing, and industrial process control; the latter, since it involves real problems happening in actual time, is also known as real-time control. The computers themselves may be considered either general-purpose or special-purpose types, depending upon whether they fulfil a broad scientific, business, or industrial function, or a more narrowly specialized purpose. (Hence the overlap in the chart, Fig. 65.) Either large-, medium-, or small-scale machines may serve any of these application areas, although the larger machines are generally used in business and scientific applications, while smaller special-purpose machines usually suffice for process control. However, distinctions of size and type are gradually beginning to fade, since modern solid-state design and modular construction permit expansion of existing computers to any required size and flexibility of use.

Large-scale electronic digital computers date their development from the early university- and government-sponsored computers, such as ENIAC and EDVAC (Electronic Discrete Variable Automatic Calculator). The EDVAC was developed between 1946 and 1952 at the University of Pennsylvania, and was the first computer to contain an internally stored program, a serial arithmetic calculator, and cyclic delay line storage. (We shall explore the significance of these developments in a later chapter.)

Special-purpose Digital Computers

The special-purpose digital computers are primarily offshoots of general-purpose machines and overlap with them in scientific, business, and process-control applications, as has been previously mentioned. Again a subdivision into mechanical, electromechanical, and electrical or electronic computers appears convenient (see Fig. 65). Any sort of numerical counter or register may be considered a special-purpose mechanical digital computer. The mileage counter on a speedometer, and petrol pumps that compute the number of gallons dispensed and the total price, are familiar examples of mechanical computers that have been with us for a long time.

The electromechanical category consists primarily of specialized relay computers. An interesting example is the Logical Truth Calculator, devised in 1947 by two Harvard University undergraduates. Bored by the drudgery of working out lengthy 'truth' tables during a course on mathematical logic, they decided to build a machine that would solve these logic problems automatically. They succeeded in building an electromechanical calculator, consisting of about £50 worth of simple parts (mostly switches, lights, and relays), which would determine the 'truth value' of logical problems fed into it. We shall deal with mathematical (symbolic) logic in considerable detail in a later chapter, but it is of interest to note at this time that the automatic calculation of logical truth is built into all present-day digital computers.

As shown in the chart (Fig. 65), the number of special-purpose computers in the electrical/electronic category is legion. Whenever a large number of related data must be handled quickly, a special-purpose digital computer can do the job. Applications are of a widely diverse range: the pari-mutuel machines used at the race tracks and airline reservation status computers; fire control, missile, and navigational computers; weather and business forcasting; automatic machine tools; language translators; reading machines (in development); machines that can learn and teach; machines to keep track of cheques; machines that can play games and profit by experience; machines that can simulate human adaptive behaviour; etc.

CHAPTER 8

SURVEY OF NUMBER SYSTEMS

Since digital computers deal with numbers, we turn now to a central question in digital design, namely, what kind of numbers can be handled most easily. Although we are accustomed to working primarily with the decimal system, there are many other systems for numerical calculations, some of which are far better suited to the capabilities of digital machines.

THE UNITARY SYSTEM

The simplest system of counting is not the decimal system, but a one-to-one comparison between the objects to be counted and the count, or tally. Thus, a rancher makes a single mark each time one of a herd of cattle enters an enclosure and keeps account of the herd by tallying up the total number of marks. He might make his task easier by breaking the long series of tallymarks into groups of five, possibly by making a diagonal stroke through each set of four marks (like this: ⊮), a system of tallying which is still very popular for keeping track of things. You add or subtract in the unitary system by making more groups of single marks or by crossing some out, as required. Although simple, this system obviously wastes writing time and space, and is clumsy for handling any calculations except a simple tally.

THE DECIMAL SYSTEM

Based upon the ten fingers, the decimal system has been used from time immemorial. It was natural that Pascal and Leibniz used decimal counting wheels in their early calculators, since these counters were just as easy to make as any others. Even in the pioneer electronic digital computers, such as the ENIAC, decimal counting elements were used, though this was wasteful of both equipment and time, as we shall see presently. The decimal system is not directly used any longer in digital computer practice, but you should be aware of how it operates in order to understand other numerical systems.

When you write down the number 487, for example, you are actually using shorthand for the more complete expression $4 \times 100 + 8 \times 10 + 7 \times 1$, which may be written in powers of ten: $4 \times 10^2 + 8 \times 10^1 + 7 \times 10^0$ (any number raised to the zero power equals 1). It is apparent from the example that in the decimal system any number can be represented in ascending powers of 10. A number raised to a power is called a base; hence, 10 is the base of the decimal system. (The base used is also known as the radix of the number system.) Thus, the first right-hand digit of any decimal number represents the number of times 10 is taken to the zero power ($10^0 = 1$) or units. This number is expressed by discrete characters, or coefficients, running from 0 to 9. The first digit to the left of the units represents, of course, the number of times 10 is taken to the first power ($10^1 = 10$), or the tens. The next digit to the left represents the number of times 10 is taken to the second power ($10^2 = 100$), or the hundreds, and so on. You can see that the unit column at the far right is the least significant digit of any number, since it cannot change its value by more than 9. The next more significant digit is the tens column to the left of the units, since it can change the value of the number by as much as 9×10 or 90. Each column to the left becomes successively more significant, with the extreme left column (the hundreds column in the example) being the most significant.

Rules of Counting. The positional significance of the digits in any number system becomes evident through the way the digits assemble when counting. Thus, in the decimal system the count begins with 0 and proceeds through each discrete digit in turn, until the complete set of coefficients, 0 to 9, has been exhausted (i.e. 0, 1, 2, 3, 4, 5, 6, 7, 8, 9). If you want to continue counting, you must again return to 0, but you carry a 1 to the next more significant column to the left (the tens column), to indicate that one cycle of 0 to 9 has been completed. The number 10, hence, indicates the completion of one cycle of units (coefficients) from 0 to 9, with no or 0 additional units. You continue counting by passing again through the complete set of 0 to 9, with 1 in the tens column, or 10, 11, 12, 13, 14, 15, 16, 17, 18, 19. Again you have used up all discrete coefficients, and with the next number, 20, you carry a 2 to the left (tens) column, to indicate that two cycles have been completed. This counting process may be continued through nine complete cycles, until you reach the number 99, at which time you have run out of coefficients in the second column. You therefore carry a 1 to the third or hundreds column, making it 100, to indicate the completion of the tens column cycles, and so on.

Using the experience gained from counting in the decimal system, we can formulate three counting rules which hold for any system of numbers.

1. The base, or radix, of a counting system equals the number of available discrete characters, or coefficients.

Example: In the decimal system there are ten discrete characters (coefficients), 0 to 9, in any column. Hence, the base, or radix, of the system is 10.

2. Whenever a column holding the highest coefficient receives another count it goes back to 0 and shifts a carry count to the next more significant column to the left.

Examples: In the decimal system, $9 + 1 = 10$; $19 + 1 = 20$; $99 + 1 = 100$; $999 + 1 = 1000$.

3. The column at the extreme right, or least significant column, counts units. Each count in the second column equals the base, or radix, of the system. Multiply the value of any column by the base to get the value of the next more significant column to the left.

Examples: In the decimal system the right-most column counts from 0 to 9; each count in the second column equals the base, 10. The value of the second column (10) multiplied by the base (10), or $10 \times 10 = 100$, which is the value of the third column (hundreds). The value of the fourth column to the left is $100 \times 10 = 1000$, and so forth.

Using these three simple rules, we can build up number systems with any radix and any number of coefficients, as we shall see presently.

THE BINARY NUMBER SYSTEM

Let us now try out another number system, known as the binary, because it has only the two values (bi = 2) 0 and 1. Having two discrete coefficients, the radix (or base) of the binary number system by rule (1) above is 2. To understand why the binary system is so popular and desirable for digital computers, you need only reflect that most electrical and electronic components have only two stable states, namely on and off, current or no current. This bistable nature of electrical devices can easily represent the two characters of the binary system, by associating zero with off and one with on. A few examples of how this works out are shown in Fig. 66. A switch in 'off' (open) position represents binary 0; in 'on' (closed) position, it represents 1. A relay that is de-energized (open contacts) may stand for 0, one that is energized (closed contacts) represents 1. A valve or transistor may be 'off' (non-conducting), representing 0, or 'on' (conducting) representing 1. Finally, a magnet magnetized to saturation in one direction can stand for binary 0, while representing 1 when magnetized in the opposite direction.

The presence or absence of a certain signal also may signify either binary value. Thus, the presence or absence of a hole punched in a card may signify 1 or 0, respectively. The presence of an electrical pulse may stand for 1; its absence for 0. A high voltage level may signify 1, a low level 0; and so on.

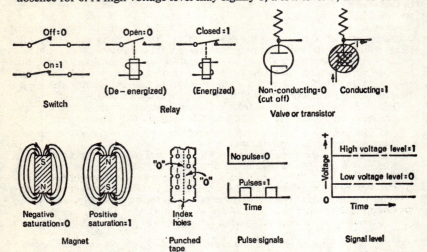

Fig. 66. Binary values 0 and 1 represented by bistable components and signals.

Binary Counting. Counting in the binary system is very simple, following the general rules we have developed for the decimal system. We start off counting 'zero, one . . .', whereupon all the binary digits have been used up. Hence, by rule (2) above, the count starts again with 0 and a 1 'carry' is placed in the column to the left, giving 10 (read 'one, zero'). Decimal number 2, therefore, corresponds to binary 10. Continuing the count gives 11, corresponding to decimal 3. Now once again all the 'bits' (binary digits) for the second column are used up, the count reverts to 0, and a 1 carry is placed in the third column to the left, giving 100 ('one, zero, zero') for decimal number 4. Continuing the count by placing 1 in the first column gives 101 ('one, zero, one'), equivalent to decimal 5. Since the bits in the first column are used up again, the column reverts to 0 in the next count, with 1 placed in the second column, giving 110 for decimal 6. The next count gives 111 ('one, one, one'), which stands for decimal 7. The table opposite shows the binary count to 15, together with its decimal equivalents.

As you can see, four binary columns are required to express decimals up to 15. Though we have not shown it in the table, usually four digits are used for any binary number (up to decimal 15) by placing zeros in the unused columns. (This does not change the value.) Thus, in the four-column system, decimal 3 is expressed as 0011, and decimal 4 as 0100, rather than 11 and 100, respectively.

Survey of Number Systems

By applying general rule (3) for counting, you can compute the value of each binary column. The extreme right-hand, or least significant, column counts units (0, 1), of course. Each count in the second column equals the radix (base), or 2. Each count in the third column equals twice the base, or $2 \times 2 = 4$. Similarly, each count in the fourth column equals $4 \times 2 = 8$, each count in the fifth column, $8 \times 2 = 16$, and so on. It is apparent from

Decimal Notation	Binary Notation
0	0
1	1
2	10
3	11
4	100
5	101
6	110
7	111
8	1000
9	1001
10	1010
11	1011
12	1100
13	1101
14	1110
15	1111

this that in the binary system the column value increases by powers of two, the first column being the zero power of 2 (i.e. 2^0), or units. This relation is useful to keep in mind for computing the decimal equivalent of a binary number. For example, the binary number 10001 is shorthand for

$$(1 \times 2^4) + (0 \times 2^3) + (0 \times 2^2) + (0 \times 2^1) + (1 \times 2^0)$$
$$= 1 \times 16 + 0 + 0 + 0 + 1$$

which adds up to decimal number 17, as you can see. As another example, consider the conversion of the binary number 101011:

```
1 0 1 0 1 1
          └──── 1 × 2⁰ = 1 ×  1 =  1
        └────── 1 × 2¹ = 1 ×  2 =  2
      └──────── 0 × 2² = 0 ×  4 =  0
    └────────── 1 × 2³ = 1 ×  8 =  8
  └──────────── 0 × 2⁴ = 0 × 16 =  0
└────────────── 1 × 2⁵ = 1 × 32 = 32
                                 ───
1 0 1 0 1 1 (binary)           =  43
                                  (decimal)
```

To convert a decimal number into the equivalent binary, you must carry out the reverse procedure. You first find the highest power of 2 that 'goes into' the decimal, thus obtaining the most significant (extreme left-hand) 1 digit of the binary. You then subtract the value of this power of 2 from the original number, and try the next highest power of 2 that will go into the remainder. If it goes, you write a binary 1, if it doesn't, a binary 0. Again subtract the value of the power of 2 from the remainder, and try the next-highest power of 2 into the second remainder, thus obtaining the third binary digit (from the left). Continue this process until there is no remainder. As an example, let us find the binary equivalent for decimal number 53:

$$\begin{array}{rl} \text{Decimal } 53 = & 1 \times 2^5 + \ldots \text{ (write binary 1)} \\ (2^5) = & -32 \\ \hline 21 = & 1 \times 2^4 + \ldots \text{ (write 1)} \\ (2^4) = & -16 \\ \hline 5 = & 0 \times 2^3 + 1 \times 2^2 + \text{ (write 01)} \\ (2^2) = & -4 \\ \hline 1 = & 0 \times 2^1 + 1 \times 2^0 \text{ (write 01)} \\ (2^0) = & -1 \\ \hline & 0 \end{array}$$

Collecting the binary digits (from left to right), we obtain the answer 110101, which is equivalent to decimal 53.

Binary Fractions. The value of binary fractions can be calculated in a similar manner. When we write the decimal fraction 0·35076, for example, we mean this to be shorthand for

$$(3 \times 10^{-1}) + (5 \times 10^{-2}) + (0 \times 10^{-3}) + (7 \times 10^{-4}) + (6 \times 10^{-5})$$
$$= 3/10 + 5/100 + 0/1000 + 7/10,000 + 6/100,0000$$
$$= 35,076/100,000.$$

Similarly in the binary number system. The first column to the right of the binary point is the most significant and represents the first negative power of two (2^{-1}), which is one-half ($\frac{1}{2}^1$). The second column to the right of the binary point represents $2^{-2} = \frac{1}{2}^2$ or $\frac{1}{4}$; the third column is $2^{-3} = \frac{1}{8}$, the fourth, $2^{-4} = \frac{1}{16}$, and so on. Evidently, the column value to the right of the binary decreases by powers of two, the extreme right-hand column being the least significant. For example, the binary fraction 0·10001 stands for

$$(1 \times 2^{-1}) + (0 \times 2^{-2}) + (0 \times 2^{-3}) + (0 \times 2^{-4}) + (1 \times 2^{-5})$$
$$= \tfrac{1}{2} + \tfrac{0}{4} + \tfrac{0}{8} + \tfrac{0}{16} + \tfrac{1}{32}$$
$$= \tfrac{1}{2} + \tfrac{1}{32}$$
$$= \tfrac{16}{32} + \tfrac{1}{32}$$
$$= \tfrac{17}{32} \text{ (decimal system equivalent)}$$

Similarly, we can compute the equivalent of the binary fraction 0·101011, as follows:

$$\begin{array}{l} 1 \times 2^{-1} = \tfrac{1}{2} = \tfrac{32}{64} \\ 0 \times 2^{-2} = \phantom{\tfrac{1}{2}} = \tfrac{0}{64} \\ 1 \times 2^{-3} = \tfrac{1}{8} = \tfrac{8}{64} \\ 0 \times 2^{-4} = \tfrac{0}{16} = \tfrac{0}{64} \\ 1 \times 2^{-5} = \tfrac{1}{32} = \tfrac{2}{64} \\ 1 \times 2^{-6} = \tfrac{1}{64} = \tfrac{1}{64} \end{array} \qquad 1\ 0\ 1\ 0\ 1\ 1$$

$$\text{(decimal) } \tfrac{43}{64} = \cdot 1\ 0\ 1\ 0\ 1\ 1 \text{ (binary)}$$

Binary Addition. Adding binary numbers is simple. Since there are only two characters, 0 and 1, there are only four possible combinations when adding two numbers. As shown below in tabular form, adding two 0s, results in 0, of course; 0 and 1, or 1 and 0, gives a sum of 1; and adding two 1s results in 0, with a 1 to carry:

ADDEND	0	0	1	1
AUGEND	+0	+1	+0	+1
SUM	0	1	1	0
CARRY	0	0	0	1

These four combinations may be summarized even more briefly in the binary addition table shown below, in which the intersections of the row and column digits give the results of addition.

Binary Addition Table

+	0	1
0	0	1
1	1	10

To add two numbers, you can either count two digits in sequence or you can simply 'remember' the results of adding the four possible combinations of digits, as given in the addition table above. A digital computer does not bother to count digits during addition, but relies on the addition table stored away in its 'memory'.

For example, let us add 'longhand' the binary equivalents of the two numbers 43 and 51 (= 94):

```
ADDEND:      43 =   1 0 1 0 1 1
AUGEND:     +51 =  +1 1 0 0 1 1
                  ─────────────
SUM:         94 = 1 0 1 1 1 1 0
                      ↙   ↙ ↙
CARRY:              1 0 0 0 1 1
```

As indicated by the arrows, the 'carry' from each column is shifted over to the next column to the left and is added to the sum. To check whether the result obtained above is correct, let us convert the binary sum (1011110) into an equivalent decimal:

$1011110 = (1 \times 2^6) + (0 \times 2^5) + (1 \times 2^4) + (1 \times 2^3) + (1 \times 2^2) + (1 \times 2^1) + (0 \times 2^0) = 64 + 0 + 16 + 8 + 4 + 2 + 0 = 94$ (Ans.)

The answer, 94, checks correctly, since $43 + 51 = 94$.

The addition of binary fractions is carried out in exactly the same manner as the addition of binary integers. To verify this, you can place a (binary) point in front of the two binaries in the example above and add. The result will be the same as before. However, conversion of the two numbers and sum into decimal system fractions shows that you have added $\frac{43}{128} + \frac{51}{128} = \frac{94}{128}$ (instead of $43 + 51$).

Binary Subtraction. The rules of subtraction are the same in the binary as in the decimal system. You can obtain the binary difference of two numbers by applying the binary addition table in reverse, with the proviso that you cannot subtract a larger number from a smaller one. Thus, if according to the table $1 + 1 = 10$, it follows that $10 - 1 = 1$. Similarly, if $1 + 0 = 1$, then $1 - 1 = 0$ and $1 - 0 = 1$; moreover, if $0 + 0 = 0$, then $0 - 0 = 0$. However, what meaning is to be given to $0 - 1$? Well, just as in decimal subtraction, a digit must be 'borrowed' from the column to the left when a larger digit is to be subtracted from a smaller one. In the case of binary subtraction, a 1 is 'borrowed' from the column to the left, and together with the 0 makes it a complete cycle, 10. As an example, let us subtract binary 110 (= 6) from 10011 (= 19) by this method:

```
                          1
BORROW:                  1̷0 10
MINUEND:       19 =   1  0̷  0̷  1  1
SUBTRAHEND:     6 = −        1  1  0
                    ─────────────────
DIFFERENCE:    13 =       1  1  0  1
```

As you can see, subtraction by borrowing is none too simple, since you have to go several columns to the left to 'borrow' the necessary 1, changing all the minuend digits in the process. An alternative, which is used for decimal subtraction by many people, consists of adjusting the minuend as required and then adding a 'carry' to the lower (subtrahend) digit in the next column to the left. Thus, you can subtract 3,568 from 40,043, for example, as follows:

		10	10	14	13	
MINUEND:	4	0̸	0̸	4	3̸	
SUBTRAHEND:	—		3	5	6	8
CARRY:		1	1	1		
DIFFERENCE:		3	6,	4	7	5 (Ans.)

The method is actually much simpler than it looks, since the adjustments and 'carries' are all done in the head. Thus, you might mentally 'verbalize' the example above as follows: '8 and 5 is 13, put down 5 and carry 1; 1 plus 6 is 7, and 7 is 14, put down 7 and carry 1; 1 plus 5 is 6 and 4 is 10, put down 4 and carry 1; 1 plus 3 is 4 and 6 is 10, put down 6 and carry 1; 1 and 3 is 4, put down 3.' The answer you write down is 36,475.

Applying this 'carry' method to the previous binary example, we write

	10	10					
	1	0̸	0̸	1	1	=	19
	—		1	1	0	=	− 6
CARRY:	1	1					
		1	1	0	1		13

and you say (mentally): '0 and 1 is 1, put down 1; 1 and 0 is 1, put down 0; 1 and 1 is 10, put down 1 and carry 1; 1 and 1 is 10, put down 1 and carry 1; 1 and 0 is 1, put down 0 or nothing at all, since this is not a significant digit'. The answer is 1101.

As another example, let us subtract binary 11101 (=29) from 110011 (=51) by the 'carry' method:

		11	10	10				
MINUEND:	1	1̸	0̸	0̸	1	1	=	51
SUBTRAHEND:	—	1	1	1	0	1	=	−29
CARRY:	1	1	1					
DIFFERENCE:		1	0	1	1	0	=	22

Again you might verbalize: '1 and 0 is 1, put down 0; 0 and 1 is 1, put down 1; 1 and 1 is 10, put down 1 and carry 1; 1 plus 1 is 10 and 0 is 10, put down 0 and carry 1; 1 plus 1 is 10 and 1 is 11, put down 1 and carry 1; 1 and 0 is 1, put down 0 or nothing.'

Subtraction by Complementing. Subtraction may also be accomplished by adding the complement of the number to be subtracted, as you may remember from arithmetic. In the decimal system the nines complement is used, which requires each of the digits of the number to be subtracted from 9. For example, the complement of 54,673 is 45,326, as shown in the calculation below.

	99,999
NUMBER:	−54,673
COMPLEMENT:	45,326

You can, of course, determine the complement directly by inspection, and do not need to subtract the number from 99,999. If 54,673 is to be subtracted from some number, and instead the complement (45,326) is added to it, the answer will be 99,999 too large. This situation can be corrected by adding an additional 1, making it 100,000, and ignoring the initial 1 in the answer, a process that may be called end-around carry. For example, suppose you want to subtract 54,673 from 73,825 by the nines-complement method. The calculation is shown below together with conventional subtraction for comparison:

Complementing

```
       73,825
      +45,326   (Complement) of 54,673
      ───────
      ①19,151
     +    ↘→ 1  (End-around Carry)
      ───────
      =19,152   (Ans.)
```

Subtracting

```
       73,825
      −54,673   (Subtrahend)
      ───────
       19,152   (Ans.)
```

Although the complement method looks more complicated, it is actually very easy to perform, and is much simpler in the case of binary subtraction, as we shall see presently.

The equivalent of the nines complement in the binary system is the ones complement. The ones complement of a binary number is simply the difference between 1 and each of the digits of the number. Since $1 - 0 = 1$ and $1 - 1 = 0$, complementing in the binary system simply consists of putting down a 1 for a 0, and a 0 for a 1. Thus the complement of 10010 is 01101, and the complement of 11101 is 00010. To subtract by complementing in the binary system, you add the ones complement plus 1, and ignore the initial 1 (end-around carry), just as in the decimal system. Let us try this method on the previous example of 110011 (= 51) minus 11101 (= 29):

```
        1 1 0 0 1 1
     +  0 0 0 1 0     (Complement of 11101)
       ─────────────
     ①  1 0 1 0 1
     +         ↘────→ 1  (End-around Carry)
       ─────────────
        1 0 1 1 0     (Ans. = 22)
```

The answer, 10110 (= 22), checks with that previously obtained by conventional binary subtraction. As another example, consider the following problem:

```
   Subtracting                Complementing
  1 0 0 1 0 1 = 37           1 0 0 1 0 1
− 1 1 0 0 1 = 25           + 0 0 1 1 0
  ─────────────              ─────────────
  0 0 1 1 0 0 = 12          ①  0 1 0 1 1
                           +         ↘────→ 1
                              ─────────────
                     (Ans.)  0 1 1 0 0
```

The extra zeros in front of the binary answers are, of course, of no consequence.

Most digital computers perform binary subtraction by adding the ones complement. Since addition is carried out by reference to the binary addition table stored in the memory, and complementing consists of inspection (also a memory function), all real arithmetic is avoided in this way.

Binary Multiplication. In any number system multiplication consists of adding a number to itself as many times as is specified by the multiplier. Some computers actually perform multiplication in this crude manner. However, most computers—as well as people—refer to a multiplication table stored in their memory, thus avoiding the time-consuming process of repetitive addition. As shown below, the binary multiplication table consists of only four entries, of which three are 0, since any number multiplied by 0 equals 0. The four possible combinations are $0 \times 0 = 0$, $0 \times 1 = 0$, $1 \times 0 = 0$, and $1 \times 1 = 1$.

Binary Multiplication Table

×	0	1
0	0	0
1	0	1

In actual practice binary multiplication reduces to copying the multiplicand whenever the multiplier digit is 1, and not copying it (or writing zeros) whenever the multiplier digit is 0. As in decimal multiplication, you must, of course, also shift one place to the left after obtaining each partial product, and in the end add up all the partial products to obtain the answer. The following example illustrates the simple procedure:

```
MULTIPLICAND:       1 1 0 1 = 13
MULTIPLIER:      ×  1 0 1 = × 5
                   ─────────
                    1 1 0 1   (First Partial Product)
                  0 0 0 0     (Second Partial Product)
                1 1 0 1       (Third Partial Product)
                ─────────────
PRODUCT:        1 0 0 0 0 1 = 65   (Ans.)
```

We could, of course, have omitted the second partial product of zeros by simply shifting over to the left one additional place. Thus, binary multiplication can be reduced to the process of copying the multiplicand whenever the multiplier digit is 1, then shifting over one place to the left, if the next multiplier digit is a 1, or shifting over one additional place for each 0 in the multiplier. The partial products are then added to obtain the answer. As an illustration consider the following problem:

```
MULTIPLICAND:      1 1 0 1 1 0 =   54
MULTIPLIER:    ×   1 1 0 0 1 1 = × 51

                     1 1 0 1 1 0  (Copy)
                   1 1 0 1 1 0    (Shift and Copy)
               1 1 0 1 1 0        (Shift, Shift, Shift, and
                                   Copy)
             1 1 0 1 1 0          (Shift and Copy)
             ──────────────────────
PRODUCT:   1 0 1 0 1 1 0 0 0 0 1 0 = 2,754 (Ans.)
```

Survey of Number Systems

After adding all the partial products, you obtain the binary number 101011000010, which after conversion into a decimal turns out to be 2754. You can easily verify that the product of 54 × 51 is 2754.

You may have noted in the example above that it is rather awkward to add a number of partial products in binary notation. The method used in most computers is to add the partial products together as soon as they appear, with proper regard to shifting, of course. An example using this method follows:

```
MULTIPLICAND:            1 1 1 1 =   15
MULTIPLIER:         ×    1 1 0 1 = × 13

PRODUCT 1:               1 1 1 1
PRODUCT 2:         1 1 1 1

SUM:               1 0 0 1 0 1 1
PRODUCT 3:         1 1 1 1

FINAL SUM:     1 1 0 0 0 0 1 1 =   195  (Ans.)
```

The final answer, when converted to a decimal, checks with conventional multiplication of 15 × 13 = 195.

Binary Division. Division in any system is the inverse of multiplication. It is the process of determining how many times one number (the divisor) can be subtracted from another number (the dividend), while still leaving a positive remainder. This is exactly what we mean when we ask ourselves 'how many times does one number (divisor) go into another (dividend)?' By definition, therefore, division can always be carried out by subtracting the divisor from the dividend a number of times until one more subtraction would leave a negative remainder. The number of times this can be done is the result, or quotient. Many computers actually divide by repeated subtraction, though in the paper-and-pencil method we avoid this time-consuming process with the aid of multiplication (long division). Either method is perfectly legitimate, as the following example will illustrate:

Problem: Divide 478 by 94.   Answer: 5, remainder 8
Conventional Method (Long Division)   *Repeated Subtraction*
```
         5  (Quotient)                           478
     94)478                                    −  94
        470                                     ───
        ───                                      384  one (subtraction)
          8 (Remainder)                           94
                                                ───
                                                 290  two
                                                  94
                                                ───
                                                 196  three
                                                  94
                                                ───
                                                 102  four
                                                  94
                                                ───
                          (Remainder)   8  *five* (Ans.)
```

Similarly, binary division can be accomplished by either conventional long division or by repeated subtraction. For example, dividing binary 110111 (= 55) by 101 (= 5), yields binary 1011 (= 11), using long division.

Long Binary Division

```
                    1 0 1 1   Quotient
Divisor 1 0 1 ) 1 1 0 1 1 1   Dividend
                1 0 1
                -----
                  1 1 1
                  1 0 1
                  -----
                    1 0 1
                    1 0 1
                    -----
                    0 0 0   (Remainder)
```

Note the comparative ease of performing binary 'long' division compared with the decimal system. In binary division we never need to try multiples of the divisor to find the largest one that will 'go into' the dividend. We are either able to subtract the divisor, in which case the quotient digit is 1, or we are not able to subtract the divisor, yielding a quotient digit of 0. And, of course, we must not forget to shift to the right after each subtraction, as in decimal division. The problem below illustrates that this method is far simpler than repeated subtraction.

Problem: Divide binary 10000010 (= 130) by 101011 (= 43). Answer: 3, remainder 1.

```
       Long Division                    Repeated Subtraction
         011 (= 3) (Ans.)              10000010
101011 ) 10000010                    −   101011  one (subtraction)
         101011                        ----------
         ------                         1010111
         0101100                         101011  two
          101011                        ----------
          ------                          101100
          000001 (remainder = 1)          101011  three (Ans.: 3 =
                                         ----------        binary 11)
                                          000001 (remainder = 1)
```

Imagine that the quotient is decimal 20, instead of 3; you would have to subtract the divisor 20 times to obtain the answer.

THE BIQUINARY OR 'TWO-FIVE' SYSTEM

Let us briefly touch on the biquinary system of numerical notation, so-called because it uses both twos and fives. The biquinary system was used in early digital (relay) computers to reduce the number of counters from ten (in the decimal system) to seven. As implied by its name, the biquinary system uses two groups of digits, one group of two (bi-) and another group of five (quinary). The quinary part keeps counting the five digits, 0, 1, 2, 3, 4, and then repeats the sequence again. The bi- or two part keeps track of the number of times the quinary part has run through its sequence. This is indicated by changing the position of a 1 digit. Thus, decimal zero is expressed as 01 (bi-) 00001 (quinary); that is, both the two and five parts have the 1 in the extreme right position. Decimal 1 is expressed by 01 00010; decimal 2 by 01 00100; 3 by 01 01000; 4 by 01 10000; 5 by 10 00001, and so on.

Note that the position of the 1 in the bi-part changes to the left after completing the first cycle of the quinary part. A comparison of decimal and biquinary notation is shown below.

Decimal Notation	Biquinary Notation
	Bi-Quinary
0	01 00001
1	01 00010
2	01 00100
3	01 01000
4	01 10000
5	10 00001
6	10 00010
7	10 00100
8	10 01000
9	10 10000

By comparing the table above with the illustration of the abacus in Chapter 1 (Fig. 1), you can see that the positions of the beads in groups of 2 and 5 in the abacus corresponds exactly to the positional significance of the digits in the biquinary system. The abacus, thus, is the first computing device to employ the biquinary number system.

OCTAL NUMBER NOTATION

The octal number system came into being because of the difficulty of dealing with long strings of binary 0s and 1s and converting them into decimals. When testing a computer, sample problems must be hand-fed into it, which may involve binary numbers with perhaps 40–60 digits each. These are extremely awkward to read or handle. Moreover, to find out what these binaries stand for in terms of ordinary (decimal) numbers, you have to go through the lengthy conversion process using powers of 2, since—as we have seen—direct substitution of decimals is not possible.

The octal system overcomes both these disadvantages. It is essentially a shorthand method for replacing groups of three binary digits by a single octal digit, running from 0 to 7. Moreover, having eight coefficients, the radix (base) of the octal system is 8 (octo- = 8). Hence, each octal digit has 8 times the weight of the next less significant digit to the right, or equivalently, octal digits increase by powers of eight. This radix has been deliberately chosen because it stands in direct proportion to the binary system, whose digits increase by powers of two, as you will recall. Groups of three binary digits, therefore, increase by powers of eight, just as do single digits in the octal system; thus, conversion from one system to the other becomes a simple matter. (The reason why the conversion of binary numbers into decimals is so awkward is the absence of a simple relation between powers of two and powers of ten.)

The table overleaf shows the manner in which octal numbers are formed. A cycle consists of eight numerals, 0, 1, 2, 3, 4, 5, 6, and 7, and then starts over again with 10 (read 'one, zero', not 'ten'), 11, 12, etc. To obtain the proper equivalence, the binary numbers are written in groups of three, with all the non-significant zeros filled in.

COMPARISON OF DECIMAL, OCTAL, AND BINARY NOTATION

It is apparent from the table that the first, or least significant, column of the octal system represents eight to the zero power (8^0) or units; the second (next most significant) column represents eight to the first power (8^1), or eights, and so on. Thus, the octal number '15' does not mean decimal fifteen, but rather $8^1 + 5 \times 8^0 = 8 + 5 = 13$ (decimal).

Binary-Octal Conversion. Conversion from binary to octal numbers, or vice versa, is extremely simple. You just divide the binary into groups of three bits each, starting at the right and filling in zeros to the left of the significant

Decimal	Octal	Binary
0	0	000
1	1	001
2	2	010
3	3	011
4	4	100
5	5	101
6	6	110
7	7	111
8	10	001 000
9	11	001 001
10	12	001 010
11	13	001 011
12	14	001 100
13	15	001 101
14	16	001 110
15	17	001 111
16	20	010 000

digit whenever required to make a complete group. (This has no effect on the value of the number.) You then consult the table or your memory for the equivalent octal value of each 3-bit binary group. You need only remember the values from 0 to 7. For example, the binary number 11101111 is converted into an equivalent octal number as follows:

$$11101111 = 011 \quad 101 \quad 111 \quad \text{(binary)}$$
$$= \quad 3 \quad\quad 5 \quad\quad 7 \quad \text{(octal)}$$

You can easily check to see that octal 357 actually represents binary 11101111 by converting both to the equivalent decimal. Thus, octal

$$357 = (3 \times 8^2) + (5 \times 8^1) + (7 \times 8^0)$$
$$= (3 \times 64) + (5 \times 8) + (7 \times 1)$$
$$= \quad 192 \quad + \quad 40 \quad + \quad 7 = 239 \text{ (Ans.).}$$

Similarly, binary

$11101111 = (1 \times 2^7) + (1 \times 2^6) + (1 \times 2^5) + (0 \times 2^4) + (1 \times 2^3) + (1 \times 2^2) + (1 \times 2^1) + (1 \times 2^0) = 128 + 64 + 32 + 0 + 8 + 4 + 2 + 1 = 239$ (Ans.).

To convert an octal to a binary number, you simply write the binary triple-bits under each of the octal digits, and ignore the zeros to the left of the left-most (most significant) 1. For example, octal number

$$167 = 1 \quad 6 \quad 7$$
$$= 001 \quad 110 \quad 111 = 1110111 \text{ binary (Ans.)}$$

Inspection shows that this is correct, since—starting from the right—binary $111 = 7 \times 8^0$, binary $110 = 32 + 16 = 48 = 6 \times 8^1$, and binary 001 (last to the left) $= 1 \times 2^6 = 64 = 1 \times 8^2$. Thus, 167 is the octal equivalent of binary 1110111.

Octal Addition. You add octal numbers just like decimals, keeping in mind, however, that the next digit after 7 is 0, with 1 to carry. For example, $3 + 5 = 10$, and $7 + 5 = 14$ (see table on page 134). Hence, when adding columns of numbers, you must carry 1 as soon as the addition exceeds 7. The following two examples illustrate the procedure:

Octal Addition

```
octal      647  =    423 (decimal)
         + 275  =    189 (decimal)
         ─────
          1144  =    612 (decimal)

octal      777  =    511 (decimal)
         + 777  =  + 511
         ─────
          1776  =   1022 (decimal)
```

By converting each of the octal numbers to decimals, as was done above, you can check the results of octal addition. Octal subtraction may be carried out in a similar manner, or by adding the sevens complement of the subtrahend and the 'end-around carry'. The sevens complement is obtained by inspection of the difference between 7 and each of the digits in the octal subtrahend, in a manner analogous to that explained for binary subtraction by means of the ones' complement.

Octal Multiplication. The rules for multiplying in the octal system are the same as those explained for the decimal and binary systems. You can determine the partial products from the octal multiplication table given below and then add the partial products and carry digits in accordance with the adding procedure above.

Octal Multiplication Table

×	0	1	2	3	4	5	6	7
0	0	0	0	0	0	0	0	0
1	0	1	2	3	4	5	6	7
2	0	2	4	6	10	12	14	16
3	0	3	6	11	14	17	22	25
4	0	4	10	14	20	24	30	34
5	0	5	12	17	24	31	36	43
6	0	6	14	22	30	36	44	52
7	0	7	16	25	34	43	52	61

Consider, for example, the multiplication of octal 725 by octal 34, shown overleaf. Multiplying in the usual manner, you find from the table above that octal $4 \times 5 = 24$; hence, put down 4 and carry 2. The carry is combined with the product of $4 \times 2 = 10$, to yield 12. Hence, put down 2 and carry 1. This 1 is again combined with the product of $4 \times 7 = 34$, to yield $34 + 1 = 35$. Thus, the first partial product is 3524, as shown. Similarly, the second partial product is 2577 and is shifted one place to the left. Octally adding the two partial products yields the answer, octal 31,514, which is equivalent to decimal 13,132. You can verify the answer by converting octal 725 to decimal 469 and octal 34 to decimal 28. Multiplying, decimal $469 \times 28 = 13,132$, which checks with the above.

Octal Multiplication

```
octal   725              469 (decimal)
      ×  34            ×  28 (decimal)
      ------           ------
        3524             3752
        2577              938
      ------           ------
       31,514 (Ans.)   13,132 (decimal Ans.)
```

You can see that the octal system, like any other, is completely self-contained. You can multiply by finding the product of any two digits, and by shifting the partial product to the left; conversion to decimals is entirely unnecessary, except for checking final results. A machine that can do these basic things can multiply in any assigned number system. As an exercise, if you wish, work out the procedures for octal division, either by repeated subtraction or by 'long division', as was shown for the binary system.

HEXADECIMAL NOTATION

To represent the ten decimal characters, 0 to 9, requires four binary digits (since $8 = 1000$ and $9 = 1001$). However, four binary digits can be arranged into 16 different combinations, as is evident from the table of binary notation on page 125. Hence, the use of a four-bit code to represent decimals from 0 to 9 wastes six out of these possible 16 combinations. In order not to waste the six unused combinations, notation in the scale of 16, or hexadecimal notation, has been used in some computers. The hexadecimal system is a combination of the ten numbers, 0 to 9, and a choice of six letters of the alphabet, which are also treated as numbers. The letters representing digits 10 to 15 may be the alphabetic sequence u to z, or more descriptively, t for *t*en, e for *e*leven, d for

Comparison of Decimal, Hexadecimal, and Binary Notation

Decimal	Hexadecimal	Binary
0	0	0000
1	1	0001
2	2	0010
3	3	0011
4	4	0100
5	5	0101
6	6	0110
7	7	0111
8	8	1000
9	9	1001
10	u (or t)	1010
11	v (or e)	1011
12	w (or d)	1100
13	x (or h)	1101
14	y (or f)	1110
15	z (or i)	1111
16	10	
17	11	
18	12	
19	13	
20	14	

twelve (*d*ozen), *h* for *t*hirteen, *f* for *f*ourteen, and *i* for *f*ifteen. Any other six marks could be used equally well, as long as the total adds up to 16 symbols to represent the 16 possible combinations of 4 bits. Thus, each of the symbols in the hexadecimal system exactly replaces one four-digit binary combination, permitting direct conversion.

The table opposite gives a comparison between the decimal, hexadecimal, and binary number stems. Note that counting in the hexadecimal system runs from 0 to 9 and then from *u* to *z* (or *t* to *i*). The cycle then starts over again with 10 (read 'one, zero' representing decimal 16) and runs to 19, and then from 1*u* to 1*z* (representing decimal 31). The basic rules for arithmetic in the hexadecimal system are the same as those discussed for the decimal, binary, and octal systems, with the letters being treated as ordinary numerals. A computer system that employs both numbers and letters, such as the hexadecimal, is called alphanumeric.

BINARY-CODED DECIMAL NOTATION

The binary number system is the simplest and the best suited for digital computers. The decimal system, on the other hand, is the most convenient and the most familiar throughout the world. Hence, if computers must work in the binary system there must be a simple method available for converting from binaries to decimals, and vice versa. The standard method using powers of 2 is awkward, and though computers can be instructed to perform the conversion, the human operators who must test the computers find it very time-consuming to convert long strings of binaries into decimals. We have seen that the octal system is a shorthand way of writing binaries, but it is of no help in converting them to decimals. To overcome this difficulty various binary codes have been devised to translate each decimal digit separately into an equivalent 4-bit binary combination, and vice versa, thus saving computation time. The hexadecimal system is one such code, using all 16 possible 4-bit combinations. Other codes exist, with varying advantages and disadvantages; some of these codes (for numbers) are summarized below. The numbers on top of the columns give the relative 'weight' of each column. The 8–4–2–1 code is a straight binary count.

Binary-coded Decimal Notations

Decimal Digit	Excess-3 Code	2–4–2–1 Code	8–4–2–1 (straight binary) Code	7–4–2–1 Code	5–4–2–1 Code	8–4–2–1 Army Fieldata Code	
0	0011	0000	0000	0000	0000	11	0000
1	0100	0001	0001	0001	0001	11	0001
2	0101	0010	0010	0010	0010	11	0010
3	0110	0011	0011	0011	0011	11	0011
4	0111	0100	0100	0100	0100	11	0100
5	1000	1011	0101	0101	1000	11	0101
6	1001	1100	0110	0110	1001	11	0110
7	1010	1101	0111	1000	1010	11	0111
8	1011	1110	1000	1001	1011	11	1000
9	1100	1111	1001	1010	1100	11	1001

Note that you do not need to perform any computations to convert a decimal number into one of the binary codes shown in the table. You simply substitute, digit by digit, the corresponding code combination for each of the decimal digits. For example, the decimal number 715 looks like this in the various codes:

Decimal Number 715:	7	1	5
In excess-3 code:	1010	0100	1000
In 2–4–2–1 code:	1101	0001	1011
In 8–4–2–1 code:	0111	0001	0101
In 7–4–2–1 code:	1000	0001	0101
In 5–4–2–1 code:	1010	0001	1000

The same process of direct table substitution is used for reconverting one of the coded binaries into the corresponding decimal.

Of the various binary decimal codes, the excess-3 code has enjoyed considerable popularity in digital computer practice. As you will note from the table, this is simply the binary number system shifted up by three places so that binary 3 (0011) becomes zero in excess-3 code, binary 4 (0100) becomes one, and so on. The first three and the last three of the 16 possible four-bit binary combinations are not used, but the unused combinations are symmetrically distributed, which has certain advantages. For one thing, the nines complement (used in decimal subtraction) can be formed in excess-3 code, simply by inverting the digits, that is, by writing 0 for 1, and 1 for 0. For example, the nines complement of 715 in the decimal system is 999 − 715 = 284. Inverting the excess-3 numeral for 715, given above, yields 0101 1011 0111, which equals decimal 284, as you can verify from the table. Another advantage is that the addition of two numbers whose sum is greater than 10, produces a simultaneous carry in both the decimal and excess-3 systems. Codes for commercial computers are beginning to be standardized for use with artificial computer languages, described in Chapter 13.

REVIEW AND SUMMARY

The simplest system of counting, the unitary system, consists of a one-to-one comparison between the objects to be counted and the tally. However, the system is awkward and wasteful.

The decimal system, consisting of ten coefficients, 0 to 9, and the radix (base) 10, is used in mechanical digital computers and in early electronic types. In the decimal system a number is represented in ascending powers of ten, the weight of each digit being 10 times that of the adjacent digit at right. The extreme right (least significant) column thus represents ones, the next column to the left represents tens, the next hundreds, and so on.

Rules of Counting: (1) The radix (base) of a counting system equals the number of available discrete characters, or coefficients.

(2) Whenever a column holding the highest coefficient receives an additional count it cycles back to 0 and shifts a carry count to the next more significant column at the left.

(3) The farthest right, or least significant, column counts units. Each count in the second column equals the radix. The value (weight) of any column multiplied by the radix gives the weight of the next more significant column at left.

The binary number system, having the two coefficients 0 and 1 and the radix 2, is well adapted to the bistable (on–off) nature of many electrical and electronic computing devices. A four *bi*nary-digi*t* (*bit*) counting system permits representing all decimals from 0 to 9. (0 = 0000, 1 = 0001, 2 = 0010, 3 = 0011, 4 = 0100, etc.)

Binary digits to the left of the binary point (integers) increase in ascending powers of 2, each column having twice the weight of the adjacent (less significant) column at right; binary digits to the right of the binary point (fractions) decrease by negative powers of 2, that is, each column has half the weight of the adjacent (more significant) column at left. To convert a binary numeral into a decimal, express the digits (from left to right) in descending powers of 2 and add up the value of all digits. To convert a decimal into a binary, find the highest power of 2 that 'goes into' the decimal, which represents the most significant binary digit, subtract this value from the decimal, and find the next highest power of 2 that goes into the remainder, forming the next most significant binary digit, and so on.

The rules of arithmetic (addition, subtraction, multiplication, and division) in any positional number system, such as the binary, are the same as those in the decimal system. The binary addition and multiplication tables consist of four entries, each, as follows:

+	0	1
0	0	1
1	1	10

(Addition)

×	0	1
0	0	0
1	0	1

(Multiplication)

The biquinary number system, used in the abacus and some digital computers, consists of two groups of digits, one of 2 (bi-) and one of 5 (quinary). A half-cycle consists of 5 position changes of the quinary digit, whereupon the binary digit changes position, and the quinary digit counts over again.

The octal number system, with 8 coefficients and the radix 8, replaces groups of 3 binary digits by single octal digits, running from 0 to 7 (octal 0 = binary 000, 1 = 001, 2 = 010, etc.). Conversion from binary to octal, or vice versa, consists of direct substitution of a 3-bit binary group by the equivalent octal digit, or the reverse.

The hexadecimal notation is a coded decimal system consisting of 16 numerical and alphabetic (alphanumerical) coefficients, thus using up all 16 combinations of a 4-bit binary code. A few other systems of coding decimals by 4-bit binaries are the excess-3 code, the 8–4–2–1, the 7–4–2–1, the 5–4–2–1, and 2–4–2–1 codes; the numerals indicating the weight of each column (see text).

CHAPTER 9

BUILDING BLOCKS OF DIGITAL COMPUTERS—I: COMPUTER LOGIC (BOOLEAN ALGEBRA)

Socrates: 'What Plato is about to say is false.'
Plato: 'Socrates has just spoken the truth.'
This is a logical paradox dating back to Aristotle and is one of many which had not been resolved until modern times. It is a paradox, for if Socrates spoke the truth, then Plato's statement must be false; but, if Plato's statement is false, then Socrates did not speak the truth and, hence, what Plato said must have been true. If Plato spoke the truth, then Socrates also spoke the truth and, hence, what Plato said is false; and so on *ad infinitum*. A similar puzzle goes like this:
'A barber shaved all persons in his home town who did not shave themselves. Did the barber shave himself?' We leave it to you to work out the circular consequences of this paradox, but the two examples illustrate the sort of logical games to which Aristotle's formal logic has led. It was not until about 1910 that Russell and Whitehead, in their celebrated *Principia Mathematica*, resolved these and other difficulties of Aristotle's formal logic.
What have the paradoxes and syllogisms of formal logic in common with sophisticated electronic computers? Very little, one would think. As a matter of fact, the subject of logic was and is taught in the philosophy department of all universities, a department that concerns itself with the eternally recurring problems and arguments first formulated by the ancients. What was it, then, that happened in logic that suddenly catapulted it to a matter of utmost importance for digital-computer design?
Mathematical Logic. Two great English mathematicians and logicians, Augustus De Morgan (1806–71) and George Boole (1815–64), first systematized Aristotle's formal logic into the powerful techniques of mathematical logic required for calculating, rather than reasoning out, problems of logical truth. The evolution of mathematical logic to its present form may be credited to Boole's masterful contribution, entitled *An Investigation of the Laws of Thought on Which Are Founded the Mathematical Theories of Logic and Probabilities*, which was published in 1854. In this long-neglected work Boole worked out the mathematical rules for combining statements (propositions) that would yield logically valid conclusions. As we shall see later on, Boole transformed the statements into abstract symbols, valid for all cases, and derived the rules for manipulating the symbols correctly by methods that have become known as Boolean algebra. Sixty years later another pair of famous English mathematicians and philosophers, Alfred North Whitehead and Bertrand Russell, recognized Boole's great contribution in their *Principia Mathematica* (1910–13), in which they succeeded in proving the essential equivalence of mathematics and logic. A few years later, Hilbert and Ackermann in Germany published their *Mathematical Logic* (1928), a text that has remained the classic on the subject.
Switching Logic. The new 'symbolic' logic became in time a useful tool for the analysis of language and its meanings (semantics) and continued to be taught in philosophy departments without anyone having the slightest idea of the revolutionary influence it was to have on computer technology. In 1937 a

young research assistant at Massachusetts Institute of Technology, Claude E. Shannon, was studying for his Master of Science degree in electrical engineering. His thesis was concerned with switching circuits, in particular with the most effective methods for obtaining a desired result with a minimum number of elements. The thesis posed the question as to whether there were mathematical methods for calculating the effects resulting from various possible switch and relay combinations, and it proved successfully that the algebra of logic (i.e. Boolean algebra) was precisely the means for solving these problems. On the basis of his thesis, Shannon, in 1938, published a paper entitled 'A Symbolic Analysis of Relay and Switching Circuits'. This epochal paper marks the entry of abstract mathematical (symbolic) logic into the practical affairs of engineering and computer design.

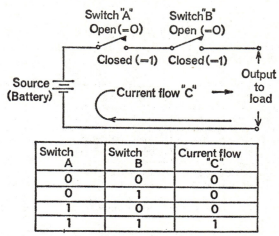

Fig. 67. Switches in a series illustrating logical AND.

As an example of Shannon's application of symbolic logic to switching circuits, consider two switches (A and B) that are connected in series with a current source (battery) and some load (output device) (see Fig. 67). We would like to know for which switch positions current will flow through the output terminals to energize the load. In this simple case you will intuitively come to the immediate conclusion, 'only when both switches are closed'. This is, of course, correct, but for the sake of clarity and logical procedure let us make a table listing all possible switch combinations and their results.

Switch 'A' closed	Switch 'B' closed	Current flows to load
No	No	No
No	Yes	No
Yes	No	No
Yes	Yes	Yes

The table confirms our intuition that current flows to the load for only one possible switch combination; namely, if both switch A and switch B are simultaneously closed. We can simplify the table and put it in mathematical-appearing form by writing 0 for 'No' and 1 for 'Yes'. More generally, let 0 be the symbol whenever some condition is not true, or is false (that is, switch is not closed, current does not flow, etc.), and let 1 be the symbol whenever some condition is true (that is, switch is closed, current does flow, etc.). Using this shorthand form, we obtain

(Switch) 'A'	(Switch) 'B'	(Current Flow) 'C'
0	0	0
0	1	0
1	0	0
1	1	1

0 = False
1 = True

In this general form, the table represents the simple statement: 'C (current flow) is true only when both A and B are true simultaneously.' This table is known in symbolic logic as the 'truth table' for the logical *AND* (also called conjunction or logical product).

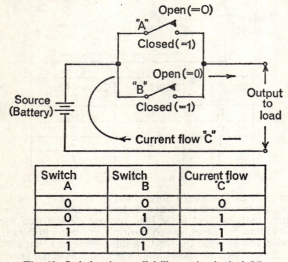

Switch A	Switch B	Current flow "C"
0	0	0
0	1	1
1	0	1
1	1	1

Fig. 68. Switches in parallel illustrating logical OR.

As a second example, consider what happens if the same two switches (A and B) are connected in parallel between the source and the load (output) (see Fig. 68). Let us again make a table to find out for which switch combinations current will flow:

Switch 'A' closed	Switch 'B' closed	Current flows to load
No	No	No
No	Yes	Yes
Yes	No	Yes
Yes	Yes	Yes

Computer Logic (Boolean Algebra)

It is evident from the table that current flows when either one or both of the switches are closed. In other words, for the parallel connexion, three out of four possible combinations are successful. Again substituting 0 for 'No' or 'False', and 1 for 'Yes' or 'True', we obtain the following:

(Switch) 'A'	(Switch) 'B'	(Current Flow) 'C'	
0	0	0	0 = False
0	1	1	1 = True
1	0	1	
1	1	1	

This truth table asserts generally that C is true whenever A or B, or both, are true. It is known in symbolic logic as the truth table for the inclusive OR (also called alternation or logical sum). The reason it is called 'inclusive' is because it is used in the sense of 'and/or'. (There is also an 'exclusive' OR, in the sense of 'either/or'.

FUNDAMENTALS OF BOOLEAN ALGEBRA

The algebra of logic is an abstract structure using letter symbols that may have any meaning whatsoever, as with the conventional algebra taught in school. However, there is one big difference that is immediately apparent: the letter symbols representing dependent or independent variables (x, y, z, etc.) in conventional algebra may have any value whatsoever, while the variables of logical algebra (A, B, C, p, q, r, etc.) can have only two values; they are either true or false, one (1) or zero (0), pulse or no pulse, closed switch or open switch, etc., depending upon the form in which the variables are represented. Logical variables, thus, are two-valued (also called dyadic or binary), as are the propositions of Aristotle's formal logic, which are either true or false and never in-between. Multiple-value systems of logic have been constructed, using various degrees of probability (a three-value system employs true, false, or probable), but for computer purposes the two-value system has so far proved most advantageous because its logical variables can be represented directly by corresponding binary variables (binary digits or bits), which—as you will recall—also can have only two values, zero (0) or one (1). We will therefore construct our elementary algebra of logic in terms of the two values, true ($= 1$) and false ($= 0$), and to keep it from becoming too abstract we shall relate it to elementary electrical (switching) circuits whenever possible. In this way we shall be able to interpret the significance of the various logical operations in terms of digital-computer building blocks (to be studied in later chapters).

Combinations of Binary Variables

Since logical variables are binary (i.e. have two values), a certain number of binary variables taken together yield a finite number of possible combinations. For example, if two binary variables are represented by two switches, each of which may be either ON or OFF, both switches taken together yield four possible combinations of states. These are, as we saw in the earlier examples, OFF–OFF (0, 0), OFF–ON (0, 1), ON–OFF (1, 0), and ON–ON (1, 1). If we used three switches to represent three binary variables, each of which can be either 1 (ON) or 0 (OFF), we would get eight possible combinations of states, as shown in the table below. Four variables or switches would result in 16 combinations, and so on. In general, the number of possible combinations of binary variables (switches) is 2^n, where $n=$ number of variables (or switches). (If $n = 2$, $2^n = 2^2 = 4$; if $n = 3$, $2^n = 2^3 = 8$; if $n = 4$, $2^n = 2^4 = 16$, etc.)

Combinations of Three Binary Variables (or Switches)

Combination #	Binary variables (or switches) A B C	
1	0 0 0	
2	0 0 1	$0 = $ OFF
3	0 1 0	$1 = $ ON
4	0 1 1	
5	1 0 0	
6	1 0 1	
7	1 1 0	
8	1 1 1	

If you check back in the previous chapter you can see that the table of combinations above also represents the eight binary digits 0 to 7; hence, writing the combinations in binary order is an easy way of keeping track of them.

FUNCTIONS OF BINARY VARIABLES

In conventional algebra a dependent variable (y) may be some function of an independent variable (x), written $y = f(x)$, or possibly of several independent variables, $y = f(u, v, w, x$, etc.). In physical language this means that the result, or output (dependent variable), of some process is dependent on the manner in which the inputs (independent variables) vary. Since the independent variables may have any value at all, the number of possible (output) functions is unlimited, or infinite.

Similarly, in Boolean algebra a dependent binary variable (C) may be a function of one or several independent binary variables (A, B, etc.), each of which may have a value of 0 or 1. This is written $C = f(A, B$, etc.). In terms of the switch analogy, you can interpret this physically by reasoning that the output or current flow, C (dependent variable), is a function of the inputs or settings of switches, $A, B, \ldots$ (independent variables). Since the number of input or switch settings is limited, the number of possible output functions is also limited. In contrast to conventional algebra, therefore, the number of possible functions of binary variables is finite. However, again in terms of the switch analogy, the number of output functions is not limited to the number of input combinations (switch settings), since combinations of these settings are possible. Each combination of possible settings (i.e. combination of a combination) represents an output function. Thus, a single input (independent binary) variable can have four possible output (dependent) functions, two variables can have 16 output functions, three variables can have 256 functions, and so on. In general, the number of possible functions (combinations of combinations) of binary variables is 2^{2^n}, where n equals the number of binary variables. (If $n = 2$, $2^{2^2} = 2^4 = 16$; if $n = 3$, $2^{2^3} = 2^8 = 256$, etc.)

Function Tables. The possible functions of a number of binary variables may be listed in an orderly manner in a function table, which is analogous to the truth table listing the possible combinations of binary variables. As with the truth table, the function table may be interpreted as describing the outputs

Computer Logic (Boolean Algebra)

(functions) resulting from all possible combinations of input (switch) settings in an electrical circuit (that is, from the variation of the independent binary variables). Whenever a function in the table appears of interest in a computer application, a corresponding logic circuit can be built to represent that function. No single circuit, however, can represent all possible logical functions of binary variables. The examples that follow will illustrate the relation of the function table to possible circuit representations.

Functions of One Binary Variable

One binary variable, A, can have only two values, 0 or 1, but there are four different combinations, or functions, of these values, as shown in the table of functions of one variable, $B = f(A)$. Electrically, this may be interpreted as four different possible outputs resulting from the settings of a single bistable device, such as a switch. A separate circuit could be built to represent each of the four functions listed in the following table.

Functions of One Binary Variable

(Input) A	Output Functions $B = f(A)$			
	f_0	f_1	f_2	f_3
0	0	1	0	1
1	0	0	1	1

The first function, f_0, states that $B = 0$, which means (in electrical circuit terms) that the output is always zero, or false, regardless of the input (switch setting). The second function, f_1, is the reverse, negation, or complement of the input ($B = $ not A); the third function, f_2, is identical to the input ($B = A$); and the fourth function, f_3, states that the output is always one or true ($B = 1$), regardless of the input. We can easily see that three of these functions

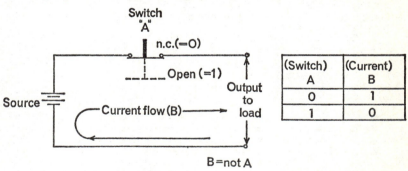

Fig. 69. Logical negation represented by the truth table and a normally closed switch.

are of no interest to computer design. An output function that is always false (f_0) or always true (f_3) is not responsive to the input, and hence is of no value. Similarly, we do not need to build an electrical circuit to duplicate the input; hence, the identity function f_2 ($B = A$) also can be discarded. This leaves the negation function f_1 ($B = $ not A) as the only one of value in computer design.

Negation. The logical negation (complement) of a variable, $B = $ not A, is variously symbolized by $B = \bar{A}$, $B = A^1$, or $B = \sim A$; in the present volume we shall always use a dash ($\bar{A}$) over the variable to be negated. In words, the negation equation $B = \bar{A}$ asserts that B is true whenever A is false, and B is false whenever A is true. This is summarized by the truth table consisting of the A-column (input) and the f_1 function column in the complete function table (see Fig. 69). In a circuit logical negation is represented by a switch that is normally closed (n.c.), as is shown in Fig. 69. In this case the unactuated, normally closed switch A represents the 'zero' or false condition, while actuating (opening) the switch sets it to the 'one' or true condition. Output current flow ($B = 1$) results when the switch is left in its normally closed ($A = 0$) condition.

Functions of Two Binary Variables

Two binary variables, A and B, can have four different combinations of truth values (n = 2; $2^n = 2^2 = 4$) and 16 possible combinations of these combinations or functions (since $2^{2^2} = 16$). The table of functions of two variables, $C = f(A,B)$, is constructed by writing the four different combinations of variables A and B (i.e. 0—0; 0—1; 1—0; and 1—1) as reference columns at the left, and then writing all their possible combinations as function columns to the right of the reference columns. This may be done in an orderly manner by writing the 16 binary (four-digit) numbers, 0 to 15, each representing a single function. In the function table below, this sequence has been altered for the purposes of explanation, though all binary digits from 0 to 15 are present. Electrically, the function table may be interpreted as representing 16 separate circuits, each consisting of two bistable (switch) inputs, A and B, and an output (C) corresponding to one of the function columns in the table.

Each of the 16 function columns in the table below, together with the two reference (input) columns at the left, comprises a truth table that represents

Functions of Two Binary Variables

(Inputs) A B	Output Functions $C = f(A,B)$															
	f_0	f_1	f_2	f_3	f_4	f_5	f_6	f_7	f_8	f_9	f_{10}	f_{11}	f_{12}	f_{13}	f_{14}	f_{15}
0 0	0	0	1	0	1	0	0	0	1	0	1	1	1	0	1	1
0 1	0	0	1	1	0	0	0	1	0	1	1	0	1	1	0	1
1 0	0	1	0	0	1	0	1	0	0	1	0	1	1	1	0	1
1 1	0	1	0	1	0	1	0	0	0	1	1	1	0	0	1	1

a specific logical operation, $C = f(A,B)$, which may be realized by an electrical circuit. We shall now proceed to analyse the logical meaning of these 16 function columns together with their electrical circuit equivalents. To simplify our task, we shall use some graphical aids consisting of truth table charts and Venn diagrams. As shown in Fig. 70, the truth table charts are simply rectangular-co-ordinate presentations of truth tables; the rows and columns represent the independent logical variables (inputs) A and B respectively, while the intersections of the rows and columns represent the truth value of the dependent (output) variable, C. If the truth value of the output variable is 1 (true) the intersecting square is shaded; if it is 0 (false) it is left blank. Thus, in the chart representation of the logical AND, the output variable, C, is 1 (true) only if inputs A and B are both 1; hence, only the lower right square, at the intersection of $A = 1$ and $B = 1$, is shaded, while all other squares (representing 0) are left unshaded. In contrast, the truth table chart for the logical OR is shaded in three squares, wherever either one of the input

Computer Logic (Boolean Algebra)

variables, *A* or *B*, or both are equal to 1, and it is left blank only in the square representing both *A* and *B* equal to zero (false).

Venn Diagrams. The Venn diagrams, originated by the nineteenth-century mathematician John Venn, are another graphic equivalent of truth tables. Some people visualize logical operations more easily in the form of Venn

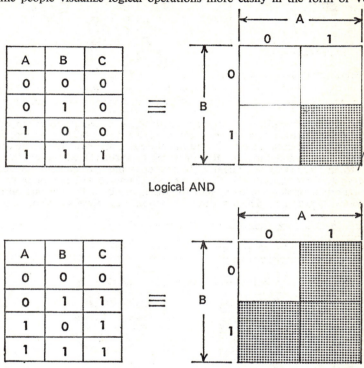

Fig. 70. Truth table chart representations of logical AND and OR.

diagrams than in the form of truth tables or charts. As illustrated in Fig. 71, a Venn diagram consists of two overlapping circles inside a rectangle. The circles marked '*A*' and '*B*' represent the two independent (input) variables *A* and *B*, respectively. Whenever the truth value of the dependent (output) variable *C* is 1 (true), the area of the independent (input) variables is shaded. The following simple rules assist in interpreting Venn diagrams:

1. Everything inside the complete circle '*A*' represents the input variable *A*; everything outside this circle (but inside the rectangle) is not-*A*, or $\bar{A}$.

2. Everything inside the complete circle '*B*' represents the input variable *B*; everything outside this circle is not-*B*, or $\bar{B}$.

3. The overlapping area of the two circles, marked '*AB*', is common to both variables, *A* and *B*.

4. The shaded area represents the true condition of the input variables for which the output variable $C = 1$ (true).

Fig. 71 illustrates some examples of typical Venn diagrams. The first diagram, completely unshaded, represents a truth table in which the output variable C is not true for any combination of the input variables A and B, and hence C is always false, or $C = 0$. The second diagram illustrates the

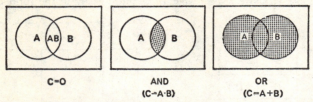

Fig. 71. Venn diagram representations for $C = 0$, AND, and OR.

logical AND, for which the output variable, C, is 1 (true) only when both A and B are 1 (true). The only area belonging to both A and B is the common overlap (AB) between the circles, and therefore this area is shaded. The third diagram illustrates the logical OR, for which C is true (1), whenever A or B, or both, are true (1). Consequently, the area covering both circles is shaded.

Always-false (Contradictory) Function f_0. Refer now to the function table for two binary variables (on page 146) for the analysis of the various output functions, $C = f(A,B)$. The first output function, f_0, is always false (0, 0, 0, 0)

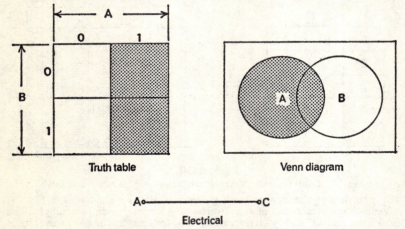

Fig. 72. Representations of function $C = A$.

no matter what the input, and is known in logic as contravalid or contradictory. (Only a contradictory proposition could always lead to a false conclusion.) The truth table chart for this function consists of four blank squares (indicating 0, 0, 0, 0) and the Venn diagram consists of the unshaded circles within the bare rectangular frame, as we have seen. Although the contravalid function is not of interest in computer design, it could be represented electrically by an earthed terminal, which shorts out any input signal (pulse). Alternatively, a negative voltage derived from a battery could also represent an output that is always false.

Function f_1. The second output function in the table, f_1, is 0, 0, 1, 1, and is equal to the input variable A, as you can verify by comparing it to the A (extreme left) column. Thus, we can write the simple logical equation $C = A$ for this function to indicate the equality. As shown in Fig. 72, the truth table chart for this function consists of the two shaded squares in the right column. The Venn diagram has circle A shaded to indicate that $C = 1$ whenever $A = 1$, which is equivalent to stating that $C = A$. The function has no significance in computers, since the output equals one of the inputs and no useful new condition results. Electrically, the function $C = A$ can be represented by a simple connexion (wire) from input A to output C.

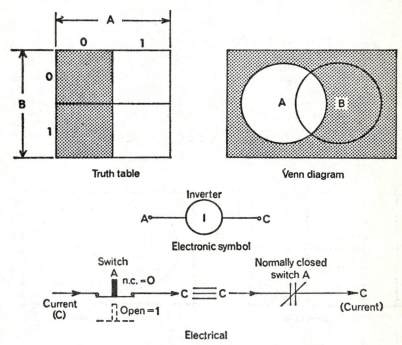

Fig. 73. Representations of NOT function $C = \bar{A}$.

NOT Function f_2. The function f_2 (1, 1, 0, 0) is the denial or negation of f_1, since it is true (1) whenever the latter is false (0), and vice versa. Since f_1 equals the input variable A (i.e. $C = A$), f_2 represents the negation of A, as expressed by the equation $C = \bar{A}$ (read 'C equals not-A, or the complement of A'). Fig. 73 illustrates the truth table and Venn diagram representations for this function. Note that the Venn diagram is shaded everywhere except inside the A circle, in accordance with the rule that everything outside this circle is not-A, or $\bar{A}$. Electrically, this function is of some significance, since it negates, or inverts, one of the inputs. Symbolically this is shown by connecting the output C to the input A through an inverter (I).

Inverter circuits reverse the polarity or phase of an input signal, so that a positive signal comes out negative or zero, and vice versa. We shall study such circuits in a later chapter. In its simplest form the NOT function can be represented by a normally closed switch, A, as shown in Fig. 73. When the switch is in its normally closed (0) position, output current (C) flows, representing a 1 (one); setting the switch to its open (1) position interrupts the output current, which represents 0. Hence, C is 1 whenever A is 0, and C is 0 when A is 1, or simply $C = \bar{A}$. Note (in Fig. 73) that a normally closed switch or relay contact is usually symbolized by two parallel vertical lines crossed by a diagonal line, while for a normally open switch the diagonal is omitted. (The representation in Fig. 69 is, of course, also acceptable.)

Functions f_3 and f_4. The next two functions in the table represent nothing new. The output function f_3 (0, 1, 0, 1) equals the input variable B, as you can verify by a comparison of the two columns. Hence, we can write the logical equation $C = B$ for this function. In the Venn diagram for this function the B circle would be shaded to indicate that $C = 1$ whenever $B = 1$. Electrically, the function is represented by a simple wire connexion linking the input B to the output C. Function f_4 (1, 0, 1, 0) is the negation, or complement, of f_3, or, equivalently, the negation (complement) of the input variable B (0, 1, 0, 1), and therefore is represented by equation $C = \bar{B}$. The function is represented by a Venn diagram in which everything is shaded except the B circle, and this is shown symbolically by connecting output C to input B through an inverter (I). Again, the simplest electrical representation is a normally closed switch (B) that interrupts the output current (C). No illustration is given for functions f_3 ($C = B$) and f_4 ($C = \bar{B}$), since they are analogous to functions f_1 ($C = A$) and f_2 ($C = \bar{A}$), illustrated in Figs. 72 and 73, respectively.

Logical Product (AND) Function f_5. We recognize this function (0, 0, 0, 1) immediately as the logical AND, since C is true (1) only if both A and B are simultaneously true. This function is known as the logical product of the variables (inputs) A and B, and it is written like a product, $A \cdot B$ or, sometimes $A \times B$ (read 'A and B'). When it is evident that the logical product is meant, the product sign may be omitted altogether, giving the logical equation $C = AB$ for 'C is the logical product of A and B'.

Fig. 74 illustrates various representations of the logical product. As we have already seen (in Fig. 71), the Venn diagram is shaded only for the common overlap (AB) between circles A and B. Electrically, the AND function can be represented by the series connexion of two normally open switch contacts, A and B. An output current, C, results only when both switches are closed. In actual computer practice, electronic 'gates', rather than switches, are used to implement the logical product. (We shall study such gates in Chapter 10.) Accordingly, the electronic symbol for the logical product is the AND gate, consisting of a number of inputs (A and B in this case) and a single output (C), as shown in Fig. 74. The letter 'a' (for 'and') or the product symbol ($\cdot$) may be placed inside the circular segment to indicate an AND gate, or else an ordinary block symbol may be used. The design of an AND gate is such that it emits a true ($= 1$) output signal (C) only if all its input signals (A and B) are simultaneously true (1).

Logical Product Function f_6. Function f_6 (0, 0, 1, 0), another logical product, asserts that 'C is true (1) whenever A and not-B ($\bar{B}$) are simultaneously true'. By writing the complement (negation) of the B input column, $\bar{B} = 1, 0, 1, 0$, you can verify that A and $\bar{B}$ are both 1 only in the third row, and hence f_6 is 1 only in this row of the function table. The corresponding logical equation is $C = A \cdot \bar{B}$, or simply, $C = A\bar{B}$.

As shown in Fig. 75, the Venn diagram presentation consists of shading circle A, except where it overlaps with B, since 'not-B' excludes the overlap from being true (1). Electrically, the equation $C = A\bar{B}$ can be represented simply by the series connexion of a normally open switch, A, and a normally

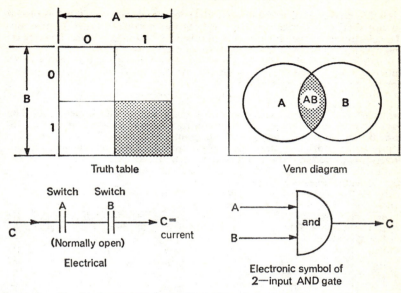

Fig. 74. Representations of logical product $C = A \cdot B$ (AND)

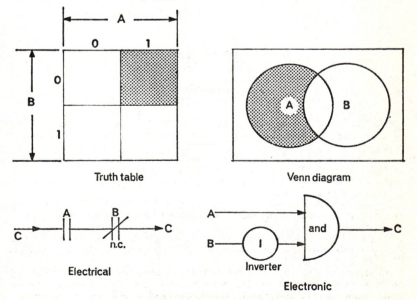

Fig. 75. Representations of logical product $C = A \cdot \bar{B}$.

closed (n.c.) one, B. Current flow ($C = 1$) occurs whenever switch A is placed in closed (1) position and switch B is left in its normally closed (0) position. The electronic symbol for this function is an AND gate with an inverter (I) inserted into the B input connexion, to indicate not-B, or $\bar{B}$.

Logical Product Function f_7. This function (0, 1, 0, 0) is analogous to f_6, differing only in that A is being denied, rather than B. The logical equation for this function, thus, is

$$C = \bar{A} \cdot B, \text{ or simply } C = \bar{A}B$$

which is read 'C is true whenever not-A and B are simultaneously true'. The graphic representations of this function (not shown) are similar to Fig. 75, except that the B circle in the Venn diagram is shaded (excluding the overlap with A), the normally closed and open switches are interchanged, and an inverter is placed in the A input of the logical AND gate symbol.

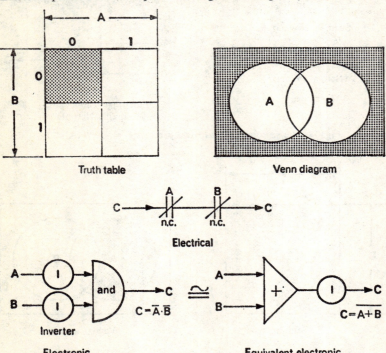

Fig. 76. Representations of joint denial (AND NOT, NOR).

Logical Product (NOR) Function f_8. As you can see, the next function, f_8 (1, 0, 0, 0), states that 'C is true (1) only if both A and B are false (0)', since C is 1 only in the first row of the function table, where both A and B are 0. This is known in logic as the joint denial (AND NOT) function and it is expressed by the logical equation

$$C = \bar{A} \cdot \bar{B} \ (C \text{ equals not-}A \text{ and not-}B)$$

As we shall see a little later, by the rules of logic (De Morgan's law)

Computer Logic (Boolean Algebra)

function f_8 is also the complement (negation) of the logical OR, known as the NOT OR, or simply NOR function. A common-sense reflection shows that this must be so: if C is true only if both A and B are false, we may express this equivalently by 'neither A nor B are true'.

The Venn diagram presentation of the joint denial or NOR function (Fig. 76) consists of shading the rectangle, except the two circles A and B, since both A and B are false. Electrically, two normally closed switches, A and B, represent joint denial, while the electronic implementation consists of inverters placed in both inputs of an AND gate, so that an output signal (1) results only when inputs A and B are both false (0). Equivalently, the function can be implemented by inverting the output of an OR gate, as Fig. 76 also shows.

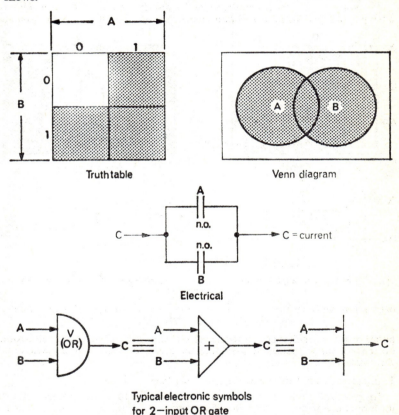

Fig. 77. Representations of OR function, logical sum $C = A + B$.

Logical Sum (OR) Function f_9. The function f_9 (0, 1, 1, 1), which we have already met, is known as the logical sum or inclusive OR function, since it asserts that 'C is true (1) when either A or B, or both, are true (1)'. The logical symbol for OR is the ordinary plus (+) sign, or sometimes the symbol $\vee$ (for disjunction). (We shall use the + for OR.) The logical equation for function f_9, thus, is

$$C = A + B \text{ (read 'A or B')}$$

As shown in Fig. 77, the truth table chart for the logical sum is shaded for three out of four combinations of input variables A and B, and is left blank (indicating the absence of a true output) only when both A and B are false. The circles in the Venn diagram are both shaded, indicating that the logical sum (C) is true when either or both inputs A and B are true. As we have already seen (Fig. 73), the OR function may be represented electrically by two parallel connected switches, A and B. An output current ($C = 1$) flows when

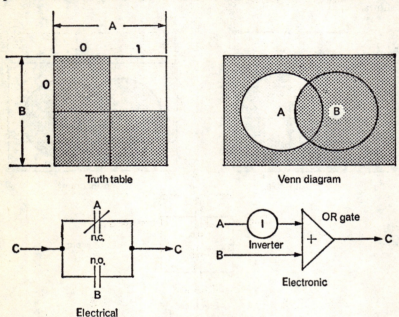

Fig. 78. Representations of logical sum $C = \bar{A} + B$.

either or both switches are set to the closed (1) position. Finally, the electronic symbol for the logical sum is the OR gate, consisting of several inputs (A and B in this case) and a single output. The logic symbol for OR (+ or $\sqrt{}$) or the word 'OR' may be used to identify an OR gate, or else a triangular or block shape may be employed to distinguish it from the AND gate. Any OR gate is designed to provide a true (1) output during the time that one or more of its inputs are true (1) and provide a false (0) output whenever none of its inputs are true. We shall study OR gates in the next chapter.

Logical Sum Function f_{10}. Logical function f_{10} (1, 1, 0, 1) is the logical sum of $\bar{A}$ (not-A) and B, since C is true (1) whenever either A is false (not-A is true) or B is true, or both A is false and B is true. C is false (0) only when both A is true ($\bar{A}$ is false) and B is false. This may be expressed by the logical equation $C = \bar{A} + B$ (read 'not-A or B').

In the Venn diagram (Fig. 78) the function is represented by shading the entire rectangle outside circle A, since everything outside is not-A. Furthermore, since B is true simultaneously, the entire circle B (including the overlap with A) is also shaded. Electrically, the function can be realized by two parallel connected switches, one (A) being normally closed and the other (B) normally open. Output current ($C = 1$) flows when switch A is left in its closed (0) position or switch B is placed in the closed (1) position, or if both switches are placed in the described positions. Electronically, the function is fulfilled by inserting an inverter into the A input lead of an OR gate, thus inverting that input to not-A.

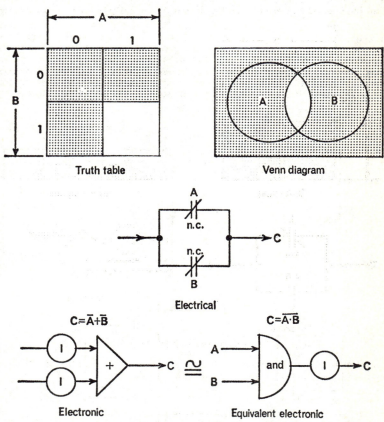

Fig. 79. Representations of NAND function $C = \bar{A} + \bar{B}$ (or $C = \overline{AB}$).

Logical Sum Function f_{11}. In logical function f_{11} (1, 0, 1, 1) the input variable B, rather than A, is denied, so that the output variable C is true (1) whenever either A is true or B is false (not-B is true), or both these conditions are fulfilled simultaneously. Hence, we can write for this function the logical sum $C = A + \bar{B}$ (read 'A or not-B'). The graphic representations of this function are the same as those shown for function f_{10} (see Fig. 78), except that the A and B symbols are interchanged everywhere.

Logical Sum (NAND) Function f_{12}. The next function, f_{12} (1, 1, 1, 0), is the negation or complement of the logical product (AND) function f_5 (0, 0, 0, 1), as you can verify immediately by comparing the two output (C) columns. For this reason it is known as the NOT AND, or more briefly, as the 'NAND' function (symbolized $C = \overline{AB}$). In logic it is also called the alternate denial, since it asserts that 'C is true (1) if either A is false (not-A is true) or B is false (not-B is true), or both are false (0); C is false, if both A and B are true'. This is summarized by the logical sum equation $C = \bar{A} + \bar{B}$ (read 'not-A or not-B') or the equivalent product equation $C = \overline{AB}$ (read 'A NAND B). (The equivalence arises from De Morgan's laws, to be described shortly.)

The Venn diagram for the alternate denial or NAND function (see Fig. 79) is exactly the inverse of the AND diagram shown in Fig. 74; that is, everything is shaded except the overlap between the circles (since not-A is everything outside circle A, and not-B is everything outside circle B, leaving only the common overlap). Electrically, the NAND function can be portrayed by two parallel connected, normally closed switches, A and B, as shown in Fig. 79. Output current ($C = 1$) flows if either or both switches are left in their normally closed (0) positions (i.e. either or both inputs are false). No current

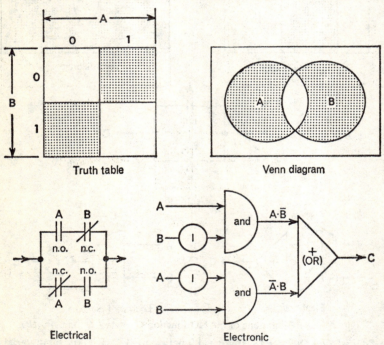

Fig. 80. Representation of circle-sum (exclusive OR) function $C = A\bar{B} + \bar{A}B$ (or $C = A \oplus B$).

flows ($C = 0$) when both switches are placed in the open (1) position (i.e. both inputs are true). The electronic circuit that carries out the NAND function may consist of an OR gate with inverters placed in both input leads to inhibit (negate) input signals A and B; equivalently, an AND gate can be used, whose output is passed through an inverter.

Exclusive OR Function f_{13}. By comparing the A and B input columns in the function table with the output (C) column f_{13} (0, 1, 1, 0), you can see that C is true (1) whenever the truth values of the input variables A and B are different (i.e. when A is not equal to B), and C is false (0) whenever the truth values of A and B are alike (i.e. A and B are both zero or both one). You might shorten this definition by saying that 'C is true if A is true and B is false or if A is false and B is true'. This is known in logic as the exclusive OR. The translation of this statement into a logical equation takes the form $C = (A \cdot \bar{B} + \bar{A} \cdot B)$, or more simply $C = A\bar{B} + \bar{A}B$. This equation, involving the logical functions AND, NOT, and OR, is sometimes shortened to a briefer symbolic form, called the circle-sum: $C = A \oplus B$ (read 'A circle-sum B'), where the encircled plus symbol $\oplus$ means exactly the same as the more complete logical statement given above.

The Venn diagram for the circle-sum (Fig. 80) shows both circles shaded with the exception of the overlap, indicating that either A or B may be true (or false), but that not both can be simultaneously true (or false). Electrically, the circle-sum function can be simulated by two sets of series-connected switches (A and B), both sets being connected in parallel. The first set has switch B normally closed (for not-B), while the second set has switch A normally closed (for not-A). This arrangement portrays the 'exclusive OR', since it will yield an output current ($C = 1$) if either switch A is set to closed (1) position and switch B is left in its normally closed (0) position, or if A (of the second set) is left in closed (0) position and B is set to closed (1) position. The same result can be obtained electronically by connecting the outputs of two AND gates, each with inputs A and B, as inputs of an OR gate. One of the AND gates has an inverter inserted into the B input lead (for not-B), while the other has an inverter inserted into the A input lead (for not-A). You can literally translate this gate arrangement, shown in Fig. 80, as 'either A AND not-B, OR not-A AND B', which is, of course, the equation of the circle-sum.

The Logical Equivalence Function f_{14}. The next output function, f_{14} (1, 0, 0, 1), asserts that 'C is true (1) whenever A is equal to B (that is, both are either true or false), and C is false (0) whenever A is different from B (i.e. A is not equal to B)'. This is known as the logical equivalence function. Another way of stating this equivalence is: 'C is true if either A and B are both true (1), or if A and B are both false.' This is easily translated into the logical equation $C = (A \cdot B + \bar{A} \cdot \bar{B})$, or simply, $C = AB + \bar{A}\bar{B}$.

Note the Venn diagram (Fig. 81) for this case; everything is shaded, including the overlap (for A and B), but not the circles A and B (to indicate not-A and not-B). Electrically, the equivalence is portrayed by hooking two sets of series-connected switches in parallel, one set (A and B) having both switch contacts normally open (for $A \cdot B$), while the other set has both contacts normally closed (for $\bar{A} \cdot \bar{B}$). Electronically, we again have the combination of AND gates inserted into the input leads of an OR gate, but in this case both inverters are placed in the inputs (A and B) of the same AND gate, to obtain the double negation $\bar{A} \cdot \bar{B}$.

Always True (Tautological) Function f_{15}. As you can see for the last function, f_{15} (1, 1, 1, 1), C is always true ($C = 1$) regardless of the values of the inputs A and B. Propositions that are always true are called valid or tautological in logic (all the principles of logic are tautological) and they have no particular computer significance. Both the truth table and the Venn diagram of this function are completely shaded (see Fig. 82) to indicate that C is always true, or $C = 1$. Electrically, the function can be represented by a positive voltage connected to an output terminal (C), to indicate that $C = 1$ at all times.

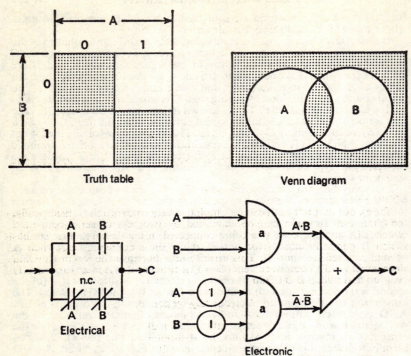

Fig. 81. Representations of the logical equivalence function $C = AB + \bar{A}\bar{B}$.

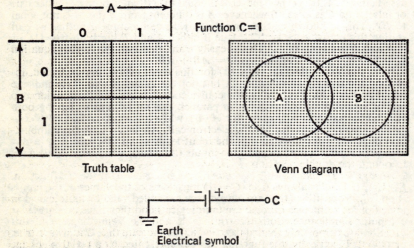

Fig. 82. Representations of 'always true' (tautological) function $C = 1$.

FUNCTIONS OF THREE OR MORE VARIABLES

You will recall that three binary variables, A, B, and C, can have eight different (input) combinations of truth values ($2^3 = 8$), and hence may have 256 (2^{2^3}) possible output functions, $D = f(A, B, C)$. We shall not bother to construct a function table for 256 output functions, but the principles are exactly the same as for the table of two functions. The definitions we have constructed for the AND, NOT, OR, NAND, NOR functions, etc., of two variables apply also to three or more input variables. Wherever the two-input definitions demand agreement of two 0s or two 1s in the truth table, multiple-input tables require simultaneous agreement of all input columns. You can, therefore, construct the appropriate multiple-input truth table for yourself whenever a particular logical statement involving three or more variables must be tested or defined.

Some Useful Logical Relations

By means of the definitions of the logical sum (OR), the logical product (AND), and the negation (NOT), we can establish some useful logical relations. The validity of a particular expression is always tested by substitution in the appropriate truth table, a procedure that is known as case analysis. Let us first turn to a few relations based upon the logical sum (OR function). For purposes of quick reference the truth table of the logical sum function (f_9) is reproduced again below.

Logical Sum (OR)

A	B	$A + B$
0	0	0
0	1	1
1	0	1
1	1	1

The first relation states that 'A or 0 equals A', or logically:

1. $A + 0 = A$. Although the arithmetic form makes this statement look obvious, a logical statement is not arithmetic and consequently cannot be taken for granted. We test the statement by case analysis; that is, by letting A have its two possible values, 0 or 1, in the truth table above. When $A = 0$, the first row of the table shows that $0 + 0 = 0$. When $A = 1$, the third row gives $1 + 0 = 1$. Hence, the statement is proven for both cases, since the output ($A + B$) always equals the truth value of A.

The second relation states that 'A or 1 equals 1', which means that the statement is always true (1). In logical equation form:

2. $A + 1 = 1$. Again substituting in the truth table, $A = 0$ gives $0 + 1 = 1$ (second row of table), and $A = 1$ gives $1 + 1 = 1$ (fourth row); hence, the statement is proved.

The next statement is 'A or A equals A', or in logical form:

3. $A + A = A$. If $A = 0$ we obtain $0 + 0 = 0$ (first row), and if $A = 1$ the fourth row gives $1 + 1 = 1$, thus proving the statement.

Finally, we have the relation 'A or not-A is always true':

4. $A + \bar{A} = 1$. $A = 0$ results in $\bar{A}$ (not-A) becoming 1, and row 2 of the truth table shows that $0 + 1 = 1$. If $A = 1$, then $\bar{A} = 0$, and $1 + 0 = 1$ (third row). Hence, the statement is proved.

Turning to the logical product (AND) function (f_5 in the function table), we can establish the following relations by means of the truth table reproduced below:

Logical Product (AND)

A	B	$A \cdot B$
0	0	0
0	1	0
1	0	0
1	1	1

5. $A \cdot 0 = 0$. If $A = 0$, then $0 \cdot 0 = 0$ (first row), and if $A = 1$, then $1 \cdot 0 = 0$ (third row); hence, the statement is proved. Similarly, we can show:
6. $A \cdot 1 = A$.
7. $A \cdot A = A$.
8. $A \cdot \bar{A} = 0$.

By recalling that $0 \cdot 0 = 0$ and $1 \cdot 1 = 1$ (see table above) and grouping the following factors into pairs, you can also show that:
9. $A \cdot A \cdot A \cdot A \ldots \cdot A = A$
Similarly 10. $A + A + A + A \ldots + A = A$

Testing for Logical Equivalence

The truth table provides a simple means for testing the logical equivalence of two logical expressions, a matter that is of considerable concern to computer designers, who must use the minimum number of logical components to accomplish certain results. Two logical expressions are equivalent if their dependent (output) variables have the same truth value whenever their independent (input) variables have the same truth values. The significance of logical equivalence to computer design is this: if several input signals are applied to either of two 'black boxes', which may differ in design and logical operation, the boxes are equivalent, provided their outputs are the same for all possible combinations (1 or 0) of the input signals.

To test the logical equivalence of two expressions, a truth table must be constructed for both expressions. If the two expressions have the same truth value for each case in the truth table, then the expressions are equivalent and can be substituted for each other. (The number of cases or rows in the truth table depends upon the number of combinations of the input variables; it is 2^n, where n = number of variables.) The following examples will clarify the procedure.

Example 1: Prove that $A + (A \cdot B) = A$.

A	B	$A \cdot B$	$A + (A \cdot B)$	$A + (AB) = A$
0	0	0	0	1
0	1	0	0	1
1	0	0	1	1
1	1	1	1	1

The first three columns in the table above represent the truth table for the logical product. The fourth column is constructed in accordance with the definition of the logical sum (OR) and consequently a 1 is placed in the column when either the A column or the $A \cdot B$ column, or both, have a truth value of 1, and a 0 is placed in the column when the truth value of both columns is 0. Finally, the fifth column is a comparison of the first column (A) and the fourth column ($A + AB$) and a 1 is placed in it whenever the truth value of the two columns is the same. Since the final column results in four 1s, the statement is proved equivalent for all four cases.

Example 2: Prove that $A \cdot (A + B) = A$.

A	B	$A + B$	$A \cdot (A + B)$	$A = A \cdot (A + B)$
0	0	0	0	1
0	1	1	0	1
1	0	1	1	1
1	1	1	1	1

The first three columns are the truth table for the logical sum. The fourth column represents the logical product of the first column (A) and the third column ($A + B$), in that a 1 is entered only when both columns have a truth value of 1. The fifth column at the right is the comparison of the first column (A) and the fourth column, $A \cdot (A + B)$; a 1 is entered whenever both columns, representing the two statements to be compared, are in agreement. Since the comparison is true (1) for all four cases, the two statements are proved logically equivalent.

Example 3: Prove that $A + (\bar{A} \cdot B) = A + B$.

As the table below shows, this is done in the same way as in the previous two examples. The truth table proves the equivalence.

A	B	$\bar{A} \cdot B$	$A + \bar{A} \cdot B$	$A + B$	$A + B = A + \bar{A} \cdot B$
0	0	0	0	0	1
0	1	1	1	1	1
1	0	0	1	1	1
1	1	0	1	1	1

Example 4: Prove that $A \cdot (\bar{A} + B) = A \cdot B$.

As we shall see presently, this final example can be proved equivalent to Example 3 by the use of De Morgan's laws, and hence does not require a separate proof. However, as an additional exercise, let us work out the detailed truth table for the last statement, again using the definitions of AND, OR, and NOT.

A	B	$(A \cdot B)$	$\bar{A}$	$(\bar{A} + B)$	$A \cdot (\bar{A} + B)$	$A \cdot B = A(\bar{A} + B)$
0	0	0	1	1	0	1
0	1	0	1	1	0	1
1	0	0	0	0	0	1
1	1	1	0	1	1	1

Though the examples given are relatively simple, the truth table method permits testing the logical equivalence of far more complex expressions. The substitution of a simple logical expression for a more complex one that has been proven equivalent is by no means a trivial game, but on the contrary, is of great significance to logical designers. For instance, Example 3, page 161, states that the combination of an OR gate and an AND gate with an inverter in one input may be replaced by a simple OR gate. Similarly, Example 4,

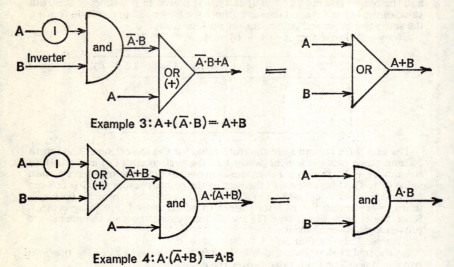

Fig. 83. Logical computer equivalents of Examples 3 and 4 in text.

page 161, states that a combination of an AND gate and an OR gate with an inverter in one input can be replaced by a simple AND gate. The two computer configurations of Examples 3 and 4, together with their logical equivalents, are shown in Fig. 83.

DE MORGAN'S RULES

There are various logical methods, rules, and tricks which the designer uses to minimize the number of computer components required to implement a particular logical function. In general, any binary logical function may be represented electronically by a group of AND gates (conjunctions) whose outputs are applied to an OR gate (disjunction). Conversely, it is also possible to implement the same function by a group of OR gates (often called buffers) whose outputs feed into an AND gate. To accomplish these alternate configurations in the simplest way, logical statements must frequently be converted from logical sums (disjunctions) to logical products (conjunctions), and vice versa. This conversion is easily accomplished by means of two rules first given by De Morgan.

1. *Negation (Complement) of Logical Sum*

The first of De Morgan's rules concerns the negation, or complement, of a logical sum. It states:

The negation (complement) of a logical sum is equivalent to the logical product of the negations (complements) of the variables making up the logical sum. This may be expressed by the following logical equation:

$$\overline{A + B} = \bar{A} \cdot \bar{B} \quad \ldots \ldots \ldots \quad (1)$$

Computer Logic (Boolean Algebra)

This logical equivalence explains why the joint denial (not-A AND not-B) was stated to be the same as the negation of the OR function (i.e. NOT OR, or NOR) in our analysis of the functions of two binary variables. (Also see Fig. 76.)

To prove De Morgan's first rule, we again use the truth table test.

De Morgan's First Rule: Negation of Logical Sum

A	B	$\bar{A}$	$\bar{B}$	$\bar{A} \cdot \bar{B}$	$A + B$	$\overline{(A + B)}$	$\overline{(A + B)} = \bar{A} \cdot \bar{B}$
0	0	1	1	1	0	1	1
0	1	1	0	0	1	0	1
1	0	0	1	0	1	0	1
1	1	0	0	0	1	0	1

The first four columns in the table above, A, B, $\bar{A}$, and $\bar{B}$, are self-explanatory. The fifth column, $\bar{A} \cdot \bar{B}$, represents a logical product that is true (1) only when both $\bar{A}$ and $\bar{B}$ are 1, which is the case only in the first row. The next column, $A + B$, is the logical sum, as previously defined, and the following column, $\overline{A + B}$, is the negation of the sum, obtained by writing 1 for 0, and 0 for 1. Finally, the last column at the right is the comparison of $\overline{A + B}$ and $\bar{A} \cdot \bar{B}$ columns. They prove to be the same in all four cases (represented by 1, 1, 1, 1) and, consequently De Morgan's first rule is proved.

2. *Negation (Complement) of Logical Product*

The second of De Morgan's rules states:

The negation of a logical product is equivalent to the logical sum of the negations (complements) of the variables comprising the logical product. This may be expressed by the logical equation:

$$\overline{A \cdot B} = \bar{A} + \bar{B} \quad \ldots \ldots \ldots \quad (2)$$

De Morgan's second rule explains why the alternate denial ($\bar{A} + \bar{B}$) was stated to be equivalent to the negation of the AND function (i.e. NOT AND or NAND = $\overline{AB}$) in the analysis of the two-variable function table (see Fig. 79).

The truth table proving De Morgan's second rule is given below.

De Morgan's Second Rule: Negation of Logical Product

A	B	$\bar{A}$	$\bar{B}$	$A \cdot B$	$\overline{A \cdot B}$	$\bar{A} + \bar{B}$	$\overline{A \cdot B} = \bar{A} + \bar{B}$
0	0	1	1	0	1	1	1
0	1	1	0	0	1	1	1
1	0	0	1	0	1	1	1
1	1	0	0	1	0	0	1

The derivation of the table above is self-explanatory.

Applications and Examples. For practical application the two De Morgan rules can be summarized as follows:

Any logical (binary) expression equals the negation (complement) of the expression obtained by changing all ANDs (logical products or conjunctions) to ORs (logical sums or disjunctions) and vice versa, and by replacing all variables with their negations (complements). This may be generalized in logical equation form for any number of binary variables:

$$\overline{(A + B + C + D + \ldots + K)} = \bar{A}\,\bar{B}\,\bar{C}\bar{D}\ldots\bar{K} \quad \ldots \ldots \quad (3)$$
and $(\overline{ABCD\ldots K}) = \bar{A} + \bar{B} + \bar{C} + \bar{D} + \ldots \bar{K} \quad \ldots \ldots \quad (4)$
Fig. 84 illustrates the computer equivalents of De Morgan's rules.

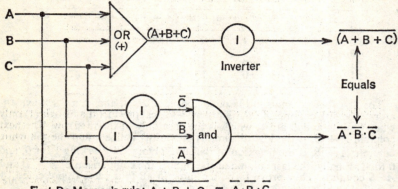

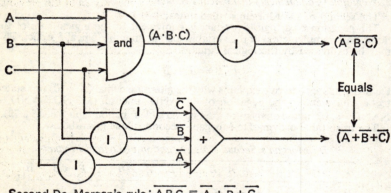

Fig. 84. Computer logical set-ups illustrating De Morgan's rules.

LAWS OF REARRANGEMENT

Ordinary algebra has a number of rules and laws which permit rearranging and simplifying complex expressions. They are known as the commutative, associative, and distributive laws. Boolean algebra has essentially these same laws, except that they are applied somewhat differently, as we shall presently see.

Computer Logic (Boolean Algebra)

1. *Commutative Laws.* The commutative laws simply state that it is immaterial which of several quantities is taken first during addition and multiplication. For two logical variables, this can be expressed by

$$A + B = B + A \quad \ldots \ldots \quad (1)$$
and
$$A \cdot B = B \cdot A \quad \ldots \ldots \quad (2)$$

The two commutative laws apply to any number of variables, of course, and they can be strictly proved (as can all the other basic laws) by case analysis with the appropriate truth table. You can try this as an exercise.

2. *Associative Laws.* Applying equally to logical products (AND) and logical sums (OR), the associative laws assert that it is immaterial how the various terms of a logical expression are grouped together, or associated. This can be expressed in logical form for three variables:

$$A \cdot (B \cdot C) = (A \cdot B) \cdot C = A \cdot B \cdot C \quad \ldots \quad (3)$$
and
$$A + (B + C) = (A + B) + C = A + B + C \quad \ldots \quad (4)$$

Again, these two rules apply to any number of logical variables. Although they may appear obvious, the associative laws are not trivial in computer design, as shown by the illustration of logically equivalent ways of performing the AND and OR operations (Fig. 85). In the case of the AND circuit, equivalent logical operations may be performed by either three, two, or a single AND gate. Use of the simplest expression results in a single gate, at a considerable saving in electronic components and cost. Similarly, the OR operation can be performed either by two equivalent or gate connexions or, most economically, by a single OR gate with three inputs.

3. *Distributive Laws.* Conventional algebra has a distributive law for addition, which requires that a multiplier of a sum operates on each term of the sum. This rule applies equally well to Boolean algebra, and may be stated in logical equation form as follows:

$$A \cdot (B + C) = AB + AC \quad \ldots \ldots \quad (5)$$

Conversely, the rule implies, of course, that common multipliers may be factored out of sums, as in conventional algebra. This operation frequently permits considerable circuit simplification in logical design. Consider, for example, the logical equation $E = AC + AD + BC + BD$.

By factoring out common multipliers A and B, we obtain

$$E = A(C + D) + B(C + D)$$

and factoring again the common logical sum $(C + D)$, there results

$$E = (A + B) \cdot (C + D)$$

Fig. 86 illustrates how these equivalent expressions, obtained by factoring, result in a reduction from four AND gates and one OR gate (required for mechanizing the first expression above) to a simple combination of two OR gates and one AND gate (required for mechanizing the last expression).

Unlike ordinary algebra, Boolean algebra also has a distributive law for multiplication. This states that when adding a term to a logical product, the added term must be combined with, or distributed to, each factor of the product. This may be expressed by the unfamiliar-looking logical equation:

$$A + B \cdot C = (A + B) \cdot (A + C) \quad \ldots \quad (6)$$

Since this does not have an equivalent in conventional algebra, let us try to prove its validity by the use of previously derived relations. First, let us multiply out the right side of equation (6):

$$(A + B)(A + C) = AA + AB + CA + CB$$
$$= A + AB + CA + CB$$

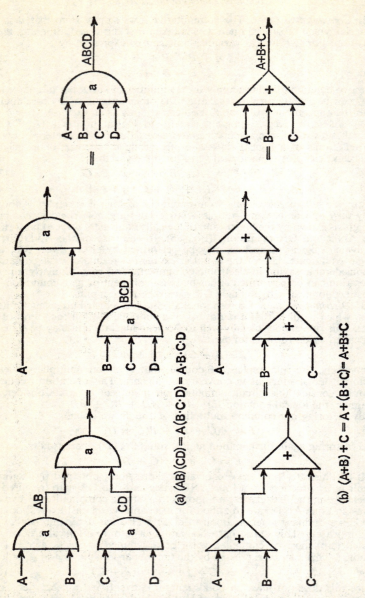

Fig. 85. Logically equivalent ways of performing (*a*) AND operations; (*b*) OR operations.

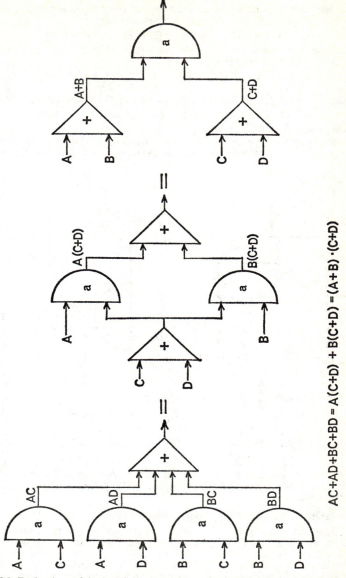

Fig. 86. Reduction of logical interconnexions obtained by factoring of common terms.

since we have previously established (useful relation 7, on page 160) that $A \cdot A = A$. Factoring out 'A' from the first two terms
$$(A + B)(A + C) = A(1 + B) + CA + CB$$
But we have previously shown (relation 2, page 159) that $1 + A = 1$; hence, $1 + B = 1$, and the equation becomes
$$(A + B)(A + C) = A + CA + CB$$
$$= A(1 + C) + CB$$
Again, $1 + C = 1$, and inverting the factors in $C \cdot B$, we obtain
$$(A + B)(A + C) = A + BC$$
which proves rule (6) given above.

The converse of the distributive law for multiplication (rule 6), which we have just derived, again proves more useful in logical design than the original

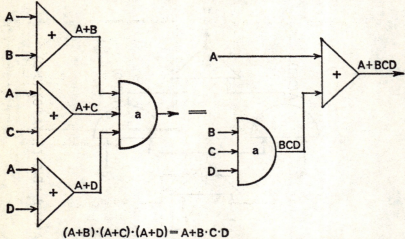

Fig. 87. Simplified logic circuit obtained by application of distributive law for multiplication.

expression, as is shown in Fig. 87. Here the equivalent logical expressions
$$(A + B)(A + C)(A + D) = A + BCD$$
permit a reduction from three OR gates and one AND gate (for the expression on the left side) to one OR gate and one AND gate (for the expression on the right side).

Translating Truth Tables into Logical Equations

The logical operations to be performed by a computer are frequently put into schematic form as truth tables or charts. They must be translated from the tablular form into Boolean algebra so that the appropriate circuits for performing the indicated operations can be developed. In many cases the circuit designer can carry out this translation by inspection of the truth table, as we shall see, or by employing simple graphical means. After obtaining the complete logical function in algebraic form it must be reduced to its simplest form in order to minimize the required number of computer components. In the following pages we shall give some illustrations of translating truth tables into corresponding Boolean expressions and also a few of the methods employed for reducing (minimizing) the resulting expressions.

Example 1: Write the logical equation for the following truth table and reduce it to the simplest logical form.

Inputs		Output	
A	B	C	
0	0	0	
0	1	1	(Condition 1)
1	0	1	(Condition 2)
1	1	1	(Condition 3)

Solution: By inspection you immediately recognize the truth table as the OR function of logical sum $(A + B)$. Thus, in this simple case you could immediately write down the result that the output (C) is true, if either input A or input B, or both, are true; i.e.

$$C = A + B$$

However, let us assume you do not recognize the truth table as the OR function; how would you go about translating it into a Boolean equation? The general method is as follows. Any truth table describes the conditions of the independent (input) variables for which the output function (dependent variable) is true (1). These conditions are logical alternatives, expressed by OR (+), which taken together completely describe the output (dependent) function. In the present example three conditions exist for which the output C is true (1).

Condition 1: C is true, if A is false (0) and B is true (1), or C is 1, if A is 0 and B is 1. If A is 0 (false), not-A ($\bar{A}$) must be 1 (true). Hence, condition 1 finally may be expressed: C is 1, if $\bar{A}$ is 1 and B is 1, which becomes $C = \bar{A} \cdot B$, since the 'and' indicates a logical product.

Condition 2: C is 1, if A is 1 and B is 0 (or $\bar{B}$ is 1), which becomes

$$C = A \cdot \bar{B}$$

Condition 3: C is 1, if A is 1 and B is 1, which becomes $C = A \cdot B$.

Now the truth table states that C is true for either condition 1 OR condition 2 OR condition 3. Hence, we can form the logical sum of the three conditions (products), and obtain:

$$C = \bar{A} \cdot B + A \cdot \bar{B} + A \cdot B$$

This equation completely describes the truth table above, and you can check its correctness by substituting the truth values (0 or 1) for A and B for any of the rows (conditions) in the table. For instance, inserting the values $A = 0$ and $B = 1$ given in the second row of the table (condition 1), results in

$$C = 1 \times 1 + 0 \times 0 + 0 \times 1 = 1$$

Reducing to Simplest Terms (*Minimizing*). The equation directly obtained from the truth table is in its complete or canonical form. For practical computer use the expression must be reduced to its simplest term, known as its minimal form. There are various ways of going about this, but no specific method exists, except for some complicated chart procedures. Experience in recognizing familiar forms and the clever manipulation of De Morgan's rules, the basic laws (commutative, associative, and distributive), and the useful relations we have developed earlier, will assist in minimizing the canonical (complete) equations. Conventional algebraic techniques, such as factoring, may be employed with caution, except that cancellation of terms and removal by subtraction from both sides is not allowed. The completion of the solution for Example 1 by two alternate methods will illustrate possible procedures.

Method 1: The expression
$$C = \bar{A}B + A\bar{B} + AB$$
may be factored, resulting in
$$C = \bar{A}B + A(\bar{B} + B)$$
But we have previously established (by useful relation 4) that the addition of a proposition and its complement (negation) is always true (1); hence,
$$B + \bar{B} = 1$$
and
$$C = \bar{A}B + A = A + \bar{A}B$$
A previous example proved that the latter expression is equivalent to $(A + B)$, but we can show this again by using the distributive law for multiplication, distributing A to both factors of the product. Thus,
$$C = A + \bar{A}B = (A + \bar{A}) \cdot (A + B)$$
However, $(A + \bar{A}) = 1$, as before, and we obtain as the final answer
$$C = 1 \cdot (A + B) = A + B \qquad \text{(Ans.)}$$
which is, of course, the irreducible equation of a logical sum.

Method 2: Useful relation (3) states that the addition of an identical (redundant) term to an OR expression does not change it; that is $A + A = A$, and $A + A + A + \ldots + A = A$. Hence, let us add the redundant term AB to the existing AB term in the complete expression for the truth table. Thus,
$$C = \bar{A}B + A\bar{B} + AB = \bar{A}B + AB + A\bar{B} + AB$$
Factoring out B from the first two terms and A from the last two terms:
$$C = B(\bar{A} + A) + A(\bar{B} + B)$$
But, as we have seen (useful relation 4), $\bar{A} + A = 1$ and $\bar{B} + B = 1$. Hence $C = B \cdot 1 + A \cdot 1 = A + B$ (Ans.), which checks with the result obtained by method 1 above.

Example 2: Derive the Boolean expression for the truth table below and reduce it to minimal form.

A	B	C	D (output)	
0	0	0	0	
0	0	1	1	(condition 1)
0	1	0	1	(condition 2)
0	1	1	1	(condition 3)
1	0	0	0	
1	0	1	0	
1	1	0	1	(condition 4)
1	1	1	0	

Solution: As shown in the table, there are four conditions for which the output function, D, is true (1).

Condition 1: D is 1, if A is 0, ($\bar{A}$ is 1) and B is 0 and C is 1,
$$\text{or } D = \bar{A} \cdot \bar{B} \cdot C$$
Condition 2: D is 1, if A is 0 and B is 1 and C is 0,
$$\text{or } D = \bar{A} \cdot B \cdot \bar{C}$$
Condition 3: D is 1, if A is 0 and B is 1 and C is 1,
$$\text{or } D = \bar{A} \cdot B \cdot C$$

Condition 4: D is 1, if A is 1 and B is 1 and C is 0,
$$\text{or } D = A \cdot B \cdot \bar{C}$$
Finally, the truth table shows that D is true (1) for either condition 1 OR condition 2 OR condition 3 OR condition 4. Hence, we can write:
$$D = \bar{A}\bar{B}C + \bar{A}B\bar{C} + \bar{A}BC + AB\bar{C} = \bar{A}\bar{B}C + \bar{A}BC + \bar{A}B\bar{C} + AB\bar{C}$$
(rearranging)

Factoring out $\bar{A}C$ from the first two terms and $B\bar{C}$ from the last two terms, we obtain
$$D = \bar{A}C(\bar{B} + B) + B\bar{C}(\bar{A} + A)$$
But, as previously established, the sum of a proposition and its complement is 1. Hence, $\bar{B} + B = 1$, and $\bar{A} + A = 1$, and the expression reduces to:
$$D = \bar{A}C \cdot 1 + B\bar{C} \cdot 1 = \bar{A}C + B\bar{C} \quad \text{(Ans.)}$$
Since none of the symbols are common to both terms, this last expression is in its minimal form and cannot be further reduced. You can check its correctness by substituting values for any conditions given in the truth table. Condition 1, for example, lists $A = 0$, $B = 0$, and $C = 1$. Hence, $\bar{A} = 1$, $\bar{C} = 0$, and by substituting we obtain
$$D = \bar{A}C + B\bar{C} = 1 \times 1 + 0 \times 0 = 1$$
which proves the answer obtained above.

In general, the minimizing of a Boolean expression is more complicated than is indicated by the examples above and requires elaborate chart methods.

REVIEW AND SUMMARY

Logical variables are two-value (binary), being either true ($= 1$) or false ($= 0$) (represented electrically by 'pulse–no pulse', ON–OFF, etc.).

The number of possible truth value combinations of n binary variables is 2^n (2 variables $= 4$ combinations; 4 variables $= 16$ combinations, etc.).

The number of possible functions (combinations of combinations) of n binary variables is 2^{2^n} (i.e. 2 variables $= 16$ functions; 3 variables $= 256$ functions).

A truth table lists the truth values (1 or 0) of a dependent binary variable (conclusion or output) as a function of all possible truth value combinations of the independent binary variables (propositions or inputs).

A function table lists all possible (output) functions (combinations of combinations) of a dependent variable as a function of all possible combinations of the independent (input) variables. The table may be interpreted electrically as describing the outputs (currents) resulting from all possible combinations of input (switch) settings. A circuit can be built to represent any logical function.

One binary variable, A, has two truth values, 0 for 1, and four output functions, $B = f(A)$. These are: (1) $B = 0$ (always false); (2) $B = 1$ (always true); (3) $B = A$ (identity); and (4) $B = \bar{A}$ (B equals not-A, called the negation or complement).

Two binary (input) variables, A and B, have four combinations of truth values (0–0, 0–1, 1–0, and 1–1) and 16 output functions, $C = f(A, B)$. These are: (1) $C = 0$ (always false or contradiction); (2) $C = A$; (3) $C = \bar{A}$ (C equals not-A; negation or complement); (4) $C = B$; (5) $C = \bar{B}$ (negation); (6) $C = A \cdot B$ (read 'A AND B', logical product); (7) $C = A \cdot \bar{B}$; (8) $C = \bar{A} \cdot B$; (9) $C = \bar{A} \cdot \bar{B}$ (read 'NOT A AND NOT B', called joint denial; also equal to $\overline{A + B}$, read 'neither A NOR B'); (10) $C = A + B$ (logical sum, read 'A OR B'); (11) $C = \bar{A} + B$; (12) $C = A + \bar{B}$; (13) $C = \bar{A} + \bar{B} = \overline{A \cdot B}$

(alternate denial, NOT AND or NAND); (14) $C = A\bar{B} + \bar{A}B = A \oplus B$ (exclusive OR or 'circle-sum'); (15) $C = AB + \bar{A}\bar{B}$ (logical equivalence); and (16) $C = 1$ (always true or tautology). Each of these functions can be 'mechanized' by a circuit.

Venn diagrams, consisting of two overlapping circles (A and B) inside a rectangle, are the graphic equivalent of truth tables for two binary variables. The following rules apply: (a) Everything inside complete circles 'A' and 'B' represents the input variables A and B, respectively; (b) everything outside circles 'A' or 'B' represents $\bar{A}$ or $\bar{B}$, respectively; (c) the overlapping area of the circles is common to both A and B and hence represents the logical product $A \cdot B$; (d) the shaded area of the diagram represents the truth value combination of the input variables for which the output variable C is true ($C = 1$).

Logical equivalence, such as $C = A$, is represented electrically by a simple connection (wire) from terminal A to terminal C.

The NOT function (negation or complement), such as $C = \bar{A}$, is represented electrically by a normally closed (= 0) switch; and electronically by an inverter circuit that inverts the phase or polarity of an input signal (A to $\bar{A}$).

The AND function (logical product $C = A \cdot B$) is represented electrically by two normally open switches in series, and electronically by an AND gate, which emits a true (= 1) output signal (C), if all its input signals (A and B) are simultaneously true (1).

The OR function (logical sum $C = A + B$) is represented electrically by two normally open switches in parallel, and electronically by an OR gate that provides a true output signal (C) whenever one or more of its inputs (A, B) are true (= 1).

The following are useful logical relations for reducing (minimizing) logical (Boolean) expressions:

Logical Sums	*Logical Products*
$A + 0 = A$	$A \cdot 0 = 0$
$A + A = A$	$A \cdot A = A$
$A + A + A + A + \ldots A = A$	$A \cdot A \cdot A \cdot A \cdot \ldots A = A$
$A + \bar{A} = 1$	$A \cdot \bar{A} = 0$
$A + 1 = 1$	$A \cdot 1 = A$
$A + A \cdot B = A$	$A(A + B) = A$
$A + \bar{A} \cdot B = A + B$	$A(\bar{A} + B) = A \cdot B$

De Morgan's Rules: Any logical expression is equal to the complement (negation) of the expression obtained by changing all ANDs (logical products) to ORs (logical sums), and vice versa, and by replacing all variables with their complements (negations).

Thus, $\overline{A + B} = \bar{A} \cdot \bar{B}$, or $\overline{A + B + C \ldots + K} = \bar{A}\bar{B}\bar{C}\ldots\bar{K}$
and $\overline{A \cdot B} = \bar{A} + \bar{B}$, or $\overline{ABCD \ldots K} = \bar{A} + \bar{B} + \bar{C} + \bar{D} + \ldots + \bar{K}$.

Commutative Laws: $A + B = B + A$ and $A \cdot B = B \cdot A$
Associative Laws: $A + (B + C) = (A + B) + C = A + B + C$
and $A \cdot (B \cdot C) = (A \cdot B) \cdot C = ABC$
Distributive Laws: $A \cdot (B + C + D) = AB + AC + AD$
and $A + B \cdot C = (A + B) \cdot (A + C)$
or $(A + B)(A + C)(A + D) = A + BCD$

A truth table can be translated into the equivalent logical (Boolean) expression by taking the logical sum of all the conditions (expressed as logical products of the input variables) for which the output (dependent) function is true (1). The resulting canonical (complete) expression may then be reduced to minimal form by algebraic manipulation using the basic laws, De Morgan's rules, and other relations.

CHAPTER 10

BUILDING BLOCKS OF DIGITAL COMPUTERS—II: ELECTRONIC DEVICES

In the preceding chapters we have looked into the mathematical and logical fundamentals that underlie the operation of automatic digital computers. We must now turn to the devices and circuits which are used to mechanize the 'logico-mathematical' operations performed by computers. We use a combined word for the two types of operations, since it develops that the logical decisions required for carrying out the instructions of the computer program are very much the same as the mathematical operations needed for performing the computer arithmetic. The rules of binary (Boolean) logic control binary arithmetic, and consequently the arithmetic portion of a digital computer consists of a variety of logic circuits hooked together in 'logical chains'.

To appreciate the equivalence of binary arithmetic and logical operations, consider the simple 'truth table' for the addition of two binary digits, A and B.

Truth Table for Addition of Two Binary Digits

Digit A	Digit B	Sum	Carry 'C'
0	0	0	0
0	1	1	0
1	0	1	0
1	1	0	1

The first two columns of the addition table represent all the possible combinations of the two binary digits A and B. They might also represent the logical combinations of two independent binary variables, or propositions (A and B), in a truth table. The third column of the table shows the arithmetic sum ($A + B$) of the possible combinations of the two binary digits, while the fourth column shows the 'carry' digit. The 'carry' is 0 for the three combinations $0 + 0 = 0$, $0 + 1 = 1$, and $1 + 0 = 1$, but is 1 for the last combination, since $1 + 1 = 10$, or 0 with 1 to carry.

Using the previous procedures, we can easily write the logical equations for the 'sum' and 'carry' columns of the table. The sum is 1 when either A is 0 AND B is 1, OR when A is 1 AND B is 0. Hence, the corresponding logical equation for the arithmetic sum,

$$S = \bar{A} \cdot B + A \cdot \bar{B}$$

which you will recognize as the exclusive-OR function or the circle-sum, $A \oplus B$, described in the previous chapter. Similarly, the 'carry' column obviously corresponds to the logical product (AND) function, so that we can write

$$\text{Carry, } C = A \cdot B$$

This result, which illustrates the equivalence of logical and arithmetic operations, is 'mechanized' by a circuit known as the half-adder, which we shall later study in detail.

ELECTRONIC DEVICES FOR PERFORMING ARITHMETIC AND LOGIC OPERATIONS

As we have seen in the last chapter, most logic functions can be performed by combinations of AND and OR gates and by inverters, the latter being used for the NOT function. We must also become acquainted with the 'flip-flop' circuit, which is one type of an ubiquitous device known as the multivibrator, that performs a host of useful functions, such as counting, temporary storage, delay, etc.

AND Gates

You will recall that in accordance with the truth table definition an AND gate produces a true (= one) output signal only when all its input signals are simultaneously true (one). In other words, an output is obtained only when all inputs are activated at the same time. There are several ways of 'mechanizing' the AND functions, among them relays, diode valves and semiconductor diodes, triode valves and transistors. Although some of these devices have been made obsolete by present-day computer requirements, we shall study them briefly to become familiar with basic electronic apparatus and techniques.

Relays. In Fig. 67 we have shown the AND function represented by two normally open, series-connected switches (A and B). When both switches are placed in the closed (= one) position a 'true' output current flows through the

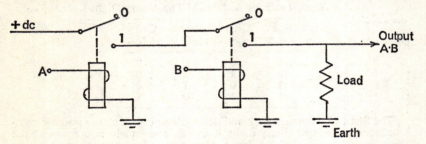

Fig. 88. Typical relay AND circuit.

circuit, representing the logical AND function. All AND gates must accomplish this basic 'switching' operation, although manual switches are, obviously, of no use in automatic computers. Any remote-controlled switch will do, however. One type of remote-controlled switch is the electromagnetic relay, which consists of switch contacts actuated by a magnetic coil. Present-day relays are extremely rugged devices, which can operate in a few thousandths of a second (milliseconds) and have a life up to a thousand million operations, or about forty years. Relays are still used to some extent in relatively slow computers because of their ideal switching characteristics. In the OFF position relays require no power and their open contacts effectively isolate the output from the inputs. In the ON position the contacts can carry heavy currents and the resistance across the contacts is almost zero. Electronic devices are speedier, but none has these advantages.

Fig. 88 illustrates a relay AND circuit. Note that the circuit is essentially the same as that shown for the series-connected switches (Fig. 67), except that the inputs (A and B) are applied to the relay actuating coils, which then close the current-carrying circuit to provide an output, $A \cdot B$.

Diode AND Circuits. We have already studied diodes in the section on function generators in the analogue computer portion of this book. You will recall that a diode is essentially a voltage-sensitive on–off switch that has two operating elements. One, called the cathode (emitter), emits a stream of electrons that constitutes a current, while the other, called the anode or plate (collector), collects the electron current and makes it available to an external circuit. A diode conducts and represents a closed switch whenever its anode is positive with respect to the cathode; the diode is in a non-conducting (open-circuit) condition whenever the anode is negative with respect to the cathode. Thus the diode acts essentially as a switch that closes a circuit whenever its anode becomes positive with respect to the cathode (or the cathode becomes

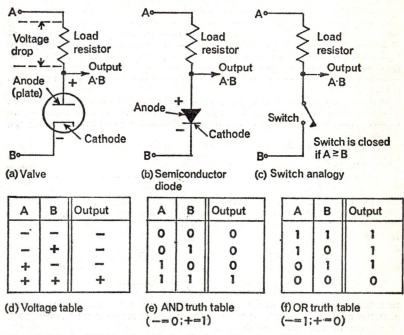

Fig. 89. Diode AND gates with two inputs: (*a*) valve; (*b*) semiconductor diode; (*c*) switch analogy; (*d*) voltage table; (*e*) AND truth table; (*f*) OR truth table.

negative with respect to the anode) and opens the circuit whenever its anode becomes negative with respect to the cathode, or equivalently, when the cathode becomes positive with respect to the anode (see Fig. 57).

Consider a series combination of a diode and a load resistor, as illustrated in Fig. 89. Parts (*a*) and (*b*) of the illustration are equivalent; (*a*) utilizes a diode valve that must have a heated cathode for a stream of electrons to be emitted, while (*b*) illustrates a semiconductor diode (silicon or germanium crystal) that does not require a separate heater. Part (*c*) shows a switch analogy of the circuit. The switch is considered closed whenever input voltage A is positive with respect to input voltage B. When the diode conducts (switch closed), current flows through it and the series load resistor, thus developing a voltage drop across the resistor. This voltage drop separates the input voltage A from the output, which is taken from the junction of the load resistor and the diode anode. In contrast, the voltage drop across the conducting diode may be taken as practically zero. Thus, when conduction takes place the output is connected by the diode directly to input voltage B, and hence equals B. When the diode is non-conducting (switch open), however, no current flows and no voltage drop is developed across the load resistor. Input voltage B is then effectively disconnected from the output by the non-conducting diode, while input voltage A is connected to the output through the load resistor. Therefore, in the absence of a voltage drop across the resistor, the output voltage for the non-conducting condition equals input voltage A.

Let us now consider four possible combinations of input voltage levels. We shall use the plus ($+$) sign to designate a 'high' voltage level and the minus ($-$) sign to designate a 'lower' level. This is a relative matter and does not refer to absolute level a negative voltage, or the high ($+$) level may actually be zero volts and the 'low' ($-$) level a negative voltage, or the high ($+$) level could represent the presence of a positive voltage pulse while the low ($-$) signifies its absence, or zero volts. For the first combination shown in the voltage table [Fig. 89 (d)], input voltages A and B are both low ($-$). Conduction does not take place; if it did, the voltage drop across the resistor would make the anode of the diode lower in voltage, or negative, with respect to the cathode, which in turn would cut off the current. The output thus is equal to input voltage A, which is low ($-$).

The second combination in the table shows input voltage A low ($-$) and input voltage B high ($+$). With the cathode positive with respect to the anode of the diode, conduction cannot take place, and the output voltage again equals input voltage A, which is low ($-$). In the third case A is high ($+$) and B is low ($-$). The diode, therefore, conducts and connects input voltage B directly to the output. Thus, the output voltage equals B, which is low ($-$). The final combination occurs when both A and B are high ($+$). Again the voltage drop across the resistor prevents conduction, and the output voltage is equal to either input, or high ($+$). Note that the output is high ($+$) only for this combination, where A and B are both high ($+$). The entire voltage table may be summarized by the statement that the output is always equal to the 'lower' ($-$) of the two input voltages, or to both, if they are the same.

If we assign the binary digit '0' to a low or negative ($-$) voltage, and the digit '1' to a high or positive ($+$) voltage, the voltage table becomes transformed into the truth table for the logical AND function, as Fig. 89 (e) shows. The circuit then conforms to the definition of the AND gate: the output is 1 only if inputs A and B are both 1.

However, the logical designer has the choice of which voltage level is to be identified as '1' and which is to be identified as '0'. Let us assume he had assigned the binary '0' to a high ($+$) voltage level and '1' to a low ($-$) voltage level. The voltage table [Fig. 89 (d)] would then be transformed into the truth table shown in Fig. 89 (f), which you will immediately recognize as the truth table for the OR function (logical sum), written upside down. In this case the identical diode circuits (of Fig. 89) conform to the definition of the OR gate in that the output is 1 whenever one or more inputs are 1 (negative) and the output is 0 only when both inputs are 0 (positive). The example illustrates that the AND and OR functions can be mechanized in at least two ways by a choice of the truth value and the polarity of the input signals.

Multiple Inputs. Fig. 90 illustrates the extension of the diode AND gate to three or more inputs. Although crystal diodes are shown, the set-up would be the same for vacuum-tube diodes. In its simplest version the magnitude of the positive anode voltage is made equal to that of the positive input voltages or

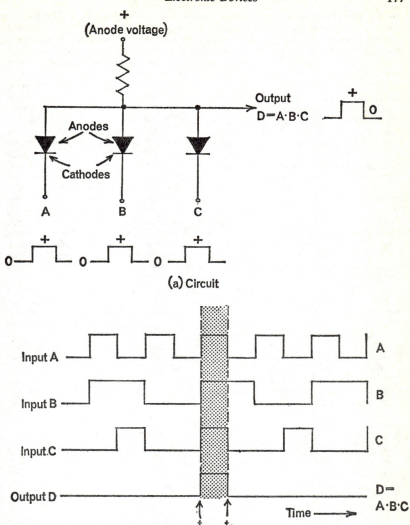

(a) Circuit

(b) Possible sequence of input and output signals

Fig. 90. Three-input diode AND gate.

pulses. Thus, whenever the input voltage to any diode is low (−), signifying the absence of a pulse, that diode conducts and shorts the output to the low level (−) input. The gate is then said to be inhibited. When all three inputs (A, B, and C) are high (+), or positive pulses are simultaneously present, none of the diodes conduct and the output equals the positive anode voltage. The gate is then said to be enabled. A possible sequence of positive input pulses is shown in Fig. 90 (b). A positive output pulse (D) occurs only during the time interval from t_1 to t_2, when all input pulses are simultaneously positive. If a binary '0' is assigned a low (negative) input and a '1' is assigned to a high (positive) input the circuit is seen to perform the logical AND operation for three input variables.

Although the circuit operates satisfactorily at low speeds for equal anode and input voltages, unavoidable stray capacitances prevent a rapid response. This effect is overcome by making the positive anode voltage several times greater than the positive peak of the gate input signals. This changes the operation as follows: if all input signals are simultaneously high (positive pulses present) all diodes conduct (because of the more positive anode voltage) and the output is connected to the high level of the input voltages. The gate is then enabled, since the output is high (binary 1) when inputs A and B and C are simultaneously high (1). Assume now that one of the inputs—A, for example—is low (zero volts or negative). With the cathode negative and the anode positive, the corresponding diode (A) conducts and switches the output to the low level of input A. Since the anodes of the remaining diodes (B and C) are tied to the common output lead, theses anodes are also at a low level. Hence, positive pulses present at inputs B and C are blocked by the corresponding non-conducting diodes B and C, and the gate is inhibited. The gate is similarly inhibited when any two inputs are low and a third is high. Only the simultaneous presence of high voltage levels at all three inputs produces a high output.

Positive AND Gate Equals Negative OR Gate. You can easily see what happens if a negative, or low, input is considered true (1) and a positive, or high, input is considered false (0). Since any combination of low (1) inputs A or B or C (or any two or three taken together) produces a low (1) output, and only a combination of three high (0) inputs produces a high output, the conditions for the logical OR function are fulfilled. Stated differently, the output is true (low) whenever one or more of the inputs are true (low), and the output is false (high) only when all three inputs are simultaneously false (high). Thus, the same circuit that functions as an AND gate when positive (high) inputs are considered true, operates as an OR gate when negative (low) input signals are considered true.

Transistors

Transistors have largely replaced valves in computers because they need only a few volts for operation and do not require any heater power at all. Transistors operate as efficiently as valves and are capable of performing up to several million switching (logical) operations per second. As the semiconductor (crystal) diode has largely replaced the equivalent diode valve, the semiconductor transistor is rapidly replacing its valve equivalent.

Fig. 91 illustrates in schematic form the two commonly available types of transistors, the PNP and the NPN types. While we cannot go into transistor details in this book, the following are some essentials of their operation:

1. Any transistor has three operating elements: emitter, base, and collector. The emitter, analogous to the cathode of a valve, injects the current carrier at one end of the transistor; the collector, analogous to the anode (or plate), collects the current at the other end; the base, analogous to the grid, controls the amount of current flow.

2. A transistor is formed as a three-section sandwich of P-type germanium and N-type germanium. (Sometimes silicon is used.) A PNP transistor has a

P-type emitter, an N-type base, and a P-type collector. An NPN transistor has an N-type emitter, a P-type base, and an N-type collector.

3. P-type germanium has a deficiency of electrons in its atoms; the missing electrons are called 'holes', and they behave like real, positively charged particles, capable of carrying a positive current. N-type germanium has an excess of electrons (negative particles) in its atoms; these free electrons are available as negative charge (current) carriers. Thus, P-type germanium has positive charge carriers, and N-type germanium has negative charge carriers available.

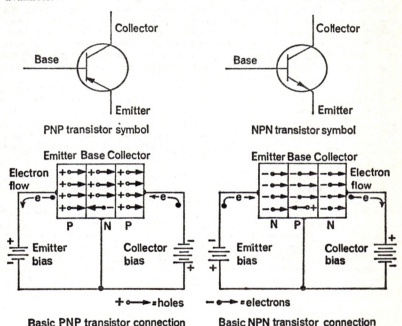

Fig. 91. Transistor schematic symbols and basic connexions.

4. Main current flow in a transistor is from emitter to collector and passes through the base. The emitter is always forward-biased, in order to propel the charged current carriers towards the collector. This means that the emitter of a *P*NP transistor has a positive voltage applied to it (with respect to the base) so as to repel the positively charged holes in the forward direction. Similarly, the emitter of an *N*PN transistor has a negative voltage applied to it so as to repel the negative electrons in the forward direction. The first letter of the transistor type indicates the polarity of the emitter voltage with respect to the base. (*P*NP = *p*ositive emitter; *N*PN = *n*egative emitter.)

5. The collector is reverse-biased in order to attract the current carriers. In a P*N*P transistor the collector is *n*egative with respect to the emitter (to attract the positive holes), while in an N*P*N transistor the collector is *p*ositive with respect to the emitter (to attract the negative electrons). The second letter of the transistor type, thus, indicates the polarity of the collector voltage with respect to the emitter. (P*N*P = *n*egative collector; N*P*N = *p*ositive collector.)

6. Current flow in the circuit external to the transistor is carried on by means of electrons. The direction of electron current flow is always against the direction of the arrow on the emitter; that is, out of the emitter of the PNP transistor in Fig. 91 and into the emitter of the NPN transistor. (The arrow indicates 'conventional' current flow, which is opposite in direction to electron current flow.) If electrons flow into the emitter (in the NPN transistor), they must flow out from the collector, and if electrons flow out of the emitter (in the PNP transistor), they must flow into the collector.

7. Any input voltage that assists (increases) the forward bias of the emitter (with respect to the base) increases the emitter-to-collector current flow. An input voltage that opposes (decreases) the forward bias decreases the current flow. Thus, when a positive input signal is applied between base and emitter of a PNP transistor, it opposes the normally negative base bias (with respect to emitter), and reduces the collector current. In contrast, a negative signal applied to the base of a PNP transistor assists the base bias and, hence, increases the collector current.

Inverters

The logical NOT function (not-A or $\bar{A}$) is performed by inverters. An inverter is any device whose output is opposite in polarity with respect to that of an input signal. Fortunately, triode valves and transistors automatically invert the polarity of an input signal, when connected as ordinary amplifiers. Fig. 92

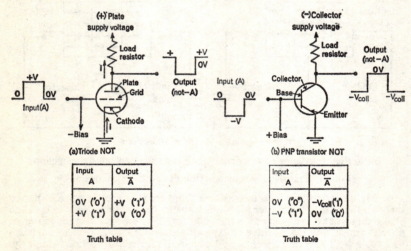

Fig. 92. Comparison of (*a*) triode valve inverter and (*b*) transistor inverter.

illustrates a comparison of a triode valve and a transistor inverter, showing their basic similarity. The grid in the triode valve performs the same function as the base in the transistor: it controls the amount of current flowing from cathode to anode.

Triode inverter. Consider first the action of the triode valve. When the grid is made negative in potential it repels electrons and impedes the electron current flow from cathode to plate. Assume that a sufficiently negative bias voltage is applied to the grid to prevent the flow of plate current, a condition known as plate-current cut-off. In the absence of current flow, the valve is an open circuit, no voltage drop occurs across the load resistor, and the voltage at the output terminal is equal to the highly positive plate-supply voltage. This condition is undisturbed by a low-level (negative or zero) input signal to the grid, since the tube remains at cut-off. Assume now that a highly positive input signal (such as a positive pulse) is applied to the grid, which overcomes the negative grid bias and allows a large plate current to flow. This current flows through the plate load resistor and develops a voltage drop across it that subtracts from the available plate-supply voltage. Moreover, since electrons flow from minus to plus, the bottom end of the resistor becomes negative with respect to the top end.

With the output connected to the bottom end of the load resistor, the output voltage, therefore, becomes negative (or less positive) with respect to its previous value. If the positive input signal is sufficiently great a large 'saturation' current flows, using up the entire plate-supply voltage and the output voltage drops to zero. (It cannot become negative, of course.) Thus, a high (positive) input signal at the grid causes a low (or zero) output voltage at the plate, while a low-level input signal, or the absence of a pulse, causes a high (positive) output voltage. This is equivalent to the logical NOT operation, since an input A is changed to not-A ($\bar{A}$), or vice versa, as the truth table shows.

Transistor inverter. Fig. 92 (*b*) shows the equivalent inverter (NOT) circuit for a PNP transistor. Like the triode, the transistor is usually operated at collector current cut off in the absence of an input signal to the base. This is accomplished by applying reverse (positive) bias to the base. In the absence of current flow, the output voltage, taken off the collector, is equal to the negative collector supply voltage (typically -3 to -5 V). If a positive or zero-voltage (low) input signal is applied to the base the transistor remains cut off, and the output voltage stays at its high (negative) level.

Assume now that a highly negative input signal is applied to the base of the transistor. This overcomes the positive bias at the base and makes it negative with respect to the emitter (or the emitter positive with respect to the base), the condition required for conduction in a PNP transistor. The resulting collector current flow produces a large voltage drop across the load resistor, which makes its bottom (collector) end more positive, or less negative than before (since electrons flow into the collector from minus to plus). If the negative input signal is sufficiently great the transistor is driven to 'saturation', and the output voltage swings from its high negative value to almost zero volts. Thus, a low (zero or positive) input signal results in a high (negative) output signal, while a high (negative) input signal results in a low (zero) output signal. Depending upon the truth values assigned, therefore, a true (1) input signal is changed to a false (0) output signal, or vice versa, as the truth table shows. An NPN transistor inverter operates in exactly the same manner, except that the polarity of all voltages is reversed.

OR Gates

In accordance with the the truth table definition of the inclusive OR (logical sum), an OR gate produces a true (1) output signal whenever one or more of its input signals are true (1), and produces a false (0) output only when all of its input signals are simultaneously false (0). We have already seen that positive AND gates function as OR gates for negative input signals, and similarly, negative AND gates operate as positive OR gates. Thus, a particular circuit configuration may represent either gate, depending only upon which input condition (low or high, minus or plus) is considered the 'true' (1) input.

Relay OR. For the sake of comparison with the relay AND circuit,

Fig. 93 (*a*) illustrates a typical relay OR circuit. The arrangement is basically the same as that of Fig. 68, showing two switches connected in parallel, except that the inputs are applied to the relay actuating coils. With both relays de-energized as shown, the contacts are open ($A = 0, B = 0$) and the output

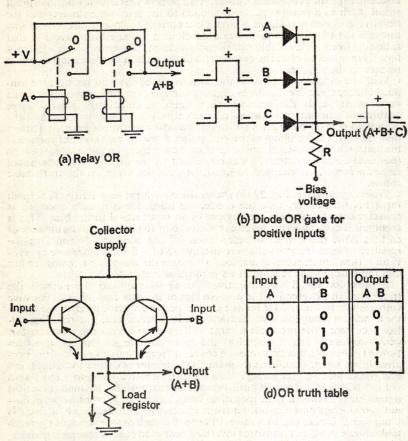

Fig. 93. OR gates and truth table: (*a*) relay OR; (*b*) diode OR gate for positive inputs; (*c*) transistor OR gate; (*d*) OR truth table.

current is zero (0) or false. When either coil *A* OR coil *B*, OR both, are actuated, a current flows through the contacts, thus making the output true, or 1. The circuit, therefore, fulfils the conditions for the logical OR function, as shown in the OR truth table [Fig. 93 (*d*)].

Diode OR Gate. The diode OR gate for three positive inputs [Fig. 93 (*b*)] should be compared to the three-input AND gate illustrated in Fig. 90. Note that the two circuits are essentially the same, except that the diodes have been turned around to reverse their polarity and a negative bias voltage has been applied to the cathodes. The common output is taken from the diode cathodes (across R), rather than from the anodes.

The operation of the OR gate is very similar to that of the AND gate. The common cathode bias voltage (applied through resistor R) is made more negative than the low (negative) level of any input voltage, so that the diodes conduct when all three inputs are simultaneously low. With the diodes conducting, the output is connected to the inputs and hence is also low. If the low input level (absence of pulse) is considered false (0), the output is also false (0), which is the condition portrayed by the first row in the OR truth table [Fig. 93 (d)]. Similarly, if all inputs are simultaneously high (i.e. positive voltages or pulses) the anodes of the diodes are highly positive with respect to the cathodes and, again, all three diodes conduct. The output is now connected directly to the high-level inputs, and hence is also high. If the high level is considered true (1) the output is 1 when all inputs are 1, the condition portrayed in the last row of the OR truth table.

Consider now what happens if one or two of the inputs are high (positive) and the others are at a low (negative) level. Let us assume that A is high (pulse present) and that both B and C are low (no pulse). With input A high, its diode conducts and connects the output to the high input voltage. Thus, for A being true (1), the output is also true (1), the condition shown in the third row of the OR truth table. However, with the output high, or positive, the cathodes of diodes B and C—connected to the output—are also positive with respect to their anodes and, hence, these diodes cannot conduct. Inputs B and C, therefore, are disconnected when input A is high. The same is true, of course, if input B is high and A and C are low. The output will again be high and inputs A and C will be disconnected. Thus, any high (1) input or combination of high input results in a high (1) output; the output is low (0) only if all inputs are simultaneously low (0). This corresponds, of course, to the definition of the OR function, as portrayed in the truth table (d).

Positive OR Equals Negative AND. In the discussion of the positive OR gate we have identified the low (negative) level with 0 (false) and the high (positive) level with 1 (true). Let us do the opposite for a moment and consider the low (negative) level as true (1) and the high (positive) level as false (0). This means that we require negative (low-level) input signals for the gate circuit shown in Fig 93 (b). With this new assignment of truth values the positive OR gate automatically becomes a negative AND gate. We have already seen in the previous discussion that the output from the gate is low (or 1) only when all inputs A AND B AND C, are simultaneously low (or 1). On the other hand, any high (0) input or combination of high inputs produces a high (0) output. Thus, the output is false (low) for any combination of false and true inputs, and it is true (high) only when all inputs are simultaneously true (high). This operation corresponds to the definition of the logical AND and complements our earlier observation that a positive AND gate is equivalent to a negative OR gate.

Transistor OR Gate. In the circuit shown in Fig. 93 (c) two transistors have been connected in parallel to provide alternate current paths, similar to the relay OR circuit. The inputs (A and B) are applied to the bases of the transistors, as before, but the output is taken across a common load resistor inserted into the emitters, rather than from the collectors, the more usual connexion [see Fig. 92 (b)]. The emitter output connexion is known as emitter follower and is analogous to the cathode follower circuit used with valves. Like the cathode follower, the emitter follower does not provide amplification of an input signal, but it does not invert the phase (polarity) of a signal. This latter characteristic is more important than amplification in a computer, since it avoids inverting true input signals into false output signals. Inversion can be useful, however, as in the NOR circuit, which we shall consider presently.

With the PNP transistors shown connected as in Fig. 93 (c) and the collectors tied to a common negative supply voltage, the emitters must be positive with respect to the base for the transistors to conduct. Equivalently, with the emitters at a fixed voltage (earth), the bases must be made negative with respect to the emitters for conduction to occur. This particular arrangement, therefore, operates with negative input signals applied to the respective bases. When a low-level input signal is applied to the base of either transistor the transistor remains cut off and no emitter current flows. If both inputs are low (0) both transistors are cut off, and no current flows through the common load resistor. The output from the load resistor, therefore, remains at ground or zero (0) potential. However, if a high negative input signal (negative pulse) is applied to the base of either transistor the base becomes negative with respect to the emitter, and the transistor is turned on. Current then flows out of the emitter of the conducting transistor (against the arrow) and into the load resistor, making its top (output) end negative, since electrons flow from minus to plus.

The output voltage, taken from the top of the load resistor, becomes negative. That is, it has the same polarity as the input signal. A high (negative) input signal, therefore, produces a high (negative) output signal and no inversion takes place. If both input signals are high (negative) both transistors will be turned on, of course, and the output voltage across the resistor is increased, but again is high. Consequently, if either input *A* OR input *B* OR both are high (negative), the output will be high; the output is low only if both inputs are simultaneously low (zero voltage). Assigning the truth value '1' to the high (negative) level, and '0' to the low (ground) level, it is apparent that the circuit fulfils the conditions of the logical OR function summarized in the truth table [Fig. 93 (d)]. The transistor OR gate can be extended to additional inputs by connecting an extra transistor in parallel for each input.

NOR (NOT OR) Gates. NOR is the contraction of NOT OR, or the inverted OR function. Consequently, the logical OR function can be performed by any OR gate whose output is inverted in some way. A NOR gate produces a true (1) output whenever all of its input signals are simultaneously false (0), and produces a false (0) output whenever one or more of its input signals are true (1). Fig. 94 illustrates several possible circuits for mechanizing the NOR function.

Transistor NOR. The PNP transistor NOR gate illustrated in Fig. 94 (a) is essentially the OR gate of Fig. 93 (c), except that the emitters are tied to earth and the output is taken from a common load resistor (*R*) to which both collectors are tied. With the output taken from the collectors, polarity inversion is obtained and the circuit functions as two transistor inverters (NOT circuits) connected in parallel. [See also Fig. 92 (b)]. Both transistors are initially cut off by the application of a positive bias voltage to the transistor bases. The gate requires negative input signals to be enabled. When both input signals are low (zero volts or positive) the transistors remain cut off (open circuits) and the output from *R* is connected to the high negative collector supply voltage ($-V_{cc}$). If a low voltage level is considered to have a truth value of 0, and high equals 1, the output is 1 (true) when both inputs are 0 (false). This fulfils the condition stated in the first row of the NOR truth table [Fig. 94 (d)]. When either or both of the inputs are high (negative), however, the corresponding transistors are turned on, a large current flows through load resistor *R*, and the output drops essentially to zero (earth) potential. Stated differently, whenever one or more of the inputs are high (true, or 1), the output drops to the low (false, or 0) level. This fulfils the conditions listed in the remaining rows of the NOR truth table [Fig. 94 (d)].

Resistor–Transistor NOR. The circuit illustrated in Fig. 94 (*b*) is essentially a transistor amplifier inverter (NOT circuit) with a multiplicity of base resistor inputs. As such it is basically similar to the summing amplifier we have studied in the analogue portion of this volume, except that the summation in this case is logical rather than arithmetical. The circuit is the most versatile of those illustrated and can be extended to any number of inputs simply by the addition of resistors. Note that any one input taken by itself together with the transistor inverter operates as the simple NOT circuit illustrated in Fig. 92 (*b*). Thus, a negative (high) signal applied to input *A*, for example, will overcome the positive base bias and hence turn the transistor on. The output, taken from

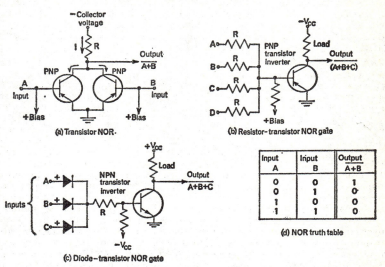

Fig. 94. NOR (NOT OR) gates and truth table. (*a*) Transistor NOR; (*b*) resistor-transistor NOR gate; (*c*) diode-transistor NOR gate; (*d*) NOR truth table.

the collector junction of the load resistor, then drops from its normally negative (high) collector voltage value to the low (ground) potential, in the manner previously explained. Hence, signal *A* applied to the input becomes not-*A* (or $\bar{A}$) in the output. The parallel input resistors permit this action to happen for any combination of true (high or negative) inputs. Thus, the inputs *A* OR *B* OR *C* OR *D* ($A + B + C + D$) become the outputs not-*A* NOR *B* NOR *C* NOR *D* (i.e. $\bar{A} + \bar{B} + \bar{C} + \bar{D}$). Only when all inputs are simultaneously low (or false) will the output be at its high negative collector potential (i.e. true). All conditions for the logical NOR function are thereby fulfilled. It can be shown that by changing the polarities of the bias and the input signals the circuit can also be made to perform the logical AND and OR functions.

Diode–Transistor NOR Gate. Finally, the circuit illustrated in Fig. 94 (*c*) is a combination of the positive diode OR gate, shown in Fig. 93 (*b*), and the transistor inverter (NOT) circuit [Fig. 92 (b)]. To make the inverter operate with positive input signals, an NPN transistor has been selected, although the circuit works equally well with a negative diode OR gate and a PNP transistor inverter.

You will recall from the earlier discussion that an NPN transistor requires a positive collector voltage and negative emitter bias (or positive base bias) in order to conduct. Applying a negative bias to the base of the NPN transistor in Fig. 94 (c) makes the emitter positive with respect to base, and therefore the transistor is normally cut off. If a high (positive pulse) input signal is applied to either or all of the diode inputs (A,B,C), the corresponding diodes conduct and pass the positive pulse through R to the base of the transistor. This in turn overcomes the negative base bias and turns the transistor on. The output, taken from the collector, then drops from its normally high positive (collector voltage) value to the low (zero volts or earth) level in the manner previously explained. Any high (positive) input is, therefore, inverted to a low (zero or earth) output and, since the inputs are in parallel, any combination of inputs is equally inverted to a NOT output. Only when all inputs are simultaneously low (no pulses) is the output at its high (positive collector) value. If low is identified with the truth value '0' and high with '1' all conditions for the logical NOR function [Fig. 94 (d)] are fulfilled.

NAND (NOT AND) Gates

NAND is the contraction of NOT AND, or the inverted AND function. The NAND function can, therefore, be performed by inverting the output of any AND gate. Note also that by De Morgan's rules the logical NAND function $\overline{A \cdot B} = \bar{A} + \bar{B}$ (not-A or not-B), and hence can also be mechanized in this fashion. A NAND gate produces a true (1) output whenever one or more of its inputs are false (0), and it produces a false (0) output only when all its inputs are simultaneously true (1). Two possible ways for mechanizing the NAND function are shown in Fig. 95.

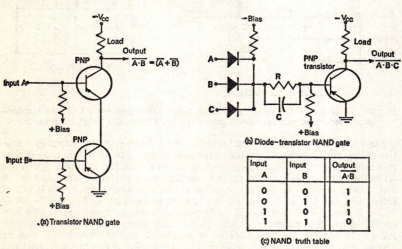

Fig. 95. NAND (NOT AND) gates and truth table. (*a*) Transistor NAND gate; (*b*) diode–transistor NAND gate; (*c*) NAND truth table.

Transistor NAND Gate. By connecting two transistors in series and taking the output from one of the collectors, as shown in Fig. 95 (*a*), an inverted output will be obtained only when current flows through both transistors—that is, when both inputs A and B have turned on their corresponding transistors. As shown in the last row of the NAND truth table [Fig. 95 (c)], this is the condition required for obtaining a false (inverted) output when both inputs *A* AND *B* are high or true (1). When either input is low (0) the corresponding transistor is at cut off, no current flows through the load resistor, and the output is at the high (negative) collector potential. That is, for any combination of low (false) inputs the output will be high (true), fulfilling the conditions of the remaining rows in the NAND truth table. With the PNP inverting transistors shown, the collector voltage is negative, the cut-off base bias must be positive and, therefore, negative input signals are required to turn the transistors on. Hence, a negative voltage represents 'high' or 1, while zero volts, or earth potential, represents 'low' or 0. By adding transistors in series, additional inputs may be accommodated.

Diode–Transistor NAND Gate. The circuit illustrated in Fig. 95 (*b*) is a combination of a negative diode AND gate (or positive OR gate) together with a PNP transistor inverter. Of course, a positive diode AND gate with an NPN transistor inverter could be used equally well. Negative (high) input signals at *A, B,* AND *C* turn on the transistor by overcoming its positive base bias and the output drops to a low (earth) potential, as previously explained. When any input is low (zero voltage) the transistor is not turned on, and the output remains at the high (negative) collector potential. Thus, the conditions listed in the NAND truth table [Fig. 95 (*c*)] are fulfilled. The capacitor (C) is required to block the negative diode bias voltage from the positive bias at the transistor base.

Multivibrators (Flip-Flops)

The multivibrator is one of the most useful electronic tools employed in digital computers. Depending upon the circuit design, multivibrators can be used to count binary numbers or store them in 'registers'; they can perform various logical operations and make up 'shift registers' used in multiplication; they can generate rectangular pulses of any frequency, and they can delay these pulses by a short time interval. Multivibrators may be made up of either valves or transistors (Fig. 96), with the latter prevailing in more recent designs.

Basic Multivibrator: Two Amplifier–Inverters Coupled 'Back to Back'

As shown in Fig. 96, the multivibrator consists essentially of two amplifier–inverters (valves or transistors) coupled back to back so that the output of the first drives the input of the second and the output of the second drives the input of the first. With the output taken from the plates of the valves [Fig. 96 (*a*)], or from the collectors of the transistors [Fig. 96 (*b*)], each stage acts as a phase (polarity) inverter; that is, it produces a 180° phase reversal between its input and output signals. This means that the output of the second stage is in phase with the input to the first stage, and the output of the first stage is in phase with the input to the second. You will realize immediately that this connexion results in 100 per cent positive (or regenerative) feedback, an arrangement that usually results in oscillations, as you may recall from the discussion in Chapter 4 on feedback. Whether the circuit actually will oscillate or not depends upon the nature of the coupling networks (Z_1 and Z_2) between the stages.

In the illustration of the basic multivibrator circuits (Fig. 96) we have not indicated the components within the two coupling networks, and we have omitted other circuit details as well, for clarity. We now must distinguish between three basic types of coupling that determine the nature and operation of the resulting multivibrator.

Type 1: Both coupling networks (Z_1 and Z_2) are resistors. This results in the bistable multivibrator, or flip-flop, so-called because it has two stable states:

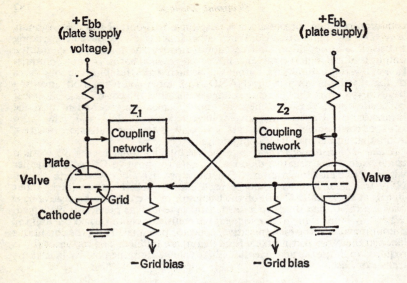

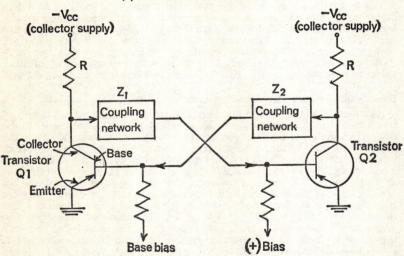

Fig. 96. Basic multivibrator circuits. (a) Valve multivibrator; (b) PNP transistor multivibrator.

either stage 1 is conducting (ON) and stage 2 is non-conducting (OFF), or stage 1 is OFF and stage 2 is ON.

Type 2: Both coupling networks are capacitors. This is the astable, or free-running, multivibrator. Having no stable states, the free-running multivibrator is an oscillator, generating a continuous train of rectangular pulses.

Type 3: One coupling network is a capacitor, the other a resistor. This mixed capacitive–resistive coupling results in the monostable, or 'one-shot', multivibrator. The circuit has one stable state and always reverts to it when temporarily disturbed. Since it has the ability to delay an input pulse by the period of its operation, the circuit is also known as a delay multivibrator.

Of these three types, the bistable multivibrator, or flip-flop, is the most important for digital computers. However, the other types are also useful, so we shall briefly analyse all of them.

Type 1: Resistive Coupling—The Bistable Multivibrator (Flip-flop)

Fig. 97 illustrates a PNP transistor multivibrator using d.c. coupling, with the output of each stage fed back through a resistor to the input of the other stage. The following analysis, which applies equally well to a triode valve circuit (except for the changed polarities), will show that if one stage is conducting the other is cut off, and vice versa, so that operation is bistable.

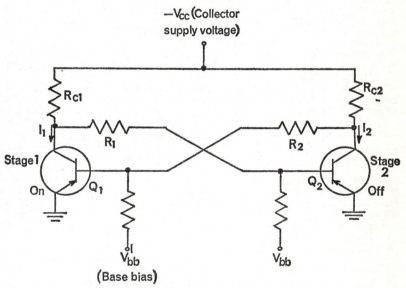

Fig. 97. Bistable multivibrator (flip-flop) circuit.

Suppose that both transistors are conducting and are drawing equal collector currents (I_1 and I_2), a plausible assumption because of the symmetry of the circuit. Although theoretically possible, it is a state of unstable equilibrium, and to preserve it would be a feat equal to that of trying to balance a large ball on the tip of a pencil; it cannot be done. In practice, various factors, such as unequal transistor characteristics and thermal fluctuations, tend to unbalance the equilibrium and cause one current to be greater than the other. Hence, assume that a slight fluctuation causes the Q_1 collector current (I_1) to increase momentarily. With I_1 increasing, the voltage drop across the collector resistor R_{c1} increases and the negative collector voltage of transistor Q_1 is reduced, or becomes more positive. This positive rise is coupled through resistor R_1 to the base of transistor Q_2, making it more positive. This, in turn, opposes the forward bias (emitter positive, base negative) of transistor Q_2 and causes its collector current (I_2) to decrease. The decrease in I_2 reduces the voltage drop across collector resistor R_{c2} and causes the negative collector voltage of transistor Q_2 to rise. The negative rise of the Q_2 collector voltage, in turn, is fed back through coupling resistor R_2 to the base of transistor Q_1, thus assisting the forward bias there and further increasing the Q_1 collector current I_2. You can see that this regenerative feedback action will continue until transistor Q_1 is driven to maximum, or saturation collector current (fully conducting), while the collector current of transistor Q_2 is cut off (non-conducting). Nothing further can then happen. Although it takes long to describe, the cycle takes place very rapidly, usually in a fraction of a millionth of a second (microsecond).

Two Stable States. A moment's reflection shows that the condition described above is a stable state. If transistor Q_1 is fully conducting (stage 1 ON), its collector must be essentially at earth (zero volts) potential, since the transistor represents practically a short circuit to earth. With the base of Q_2 coupled (through R_1) to the Q_1 collector, the Q_2 base is also at ground potential, keeping the transistor cut off. (The Q_2 base must be negative with respect to its emitter for conduction.) With stage 2 in OFF condition, transistor Q_2 is essentially an open circuit and its collector is connected to the highly negative collector supply voltage, $-V_{cc}$. This high negative voltage is coupled back through R_2 to the base of Q_1, keeping transistor Q_1 conducting (stage 1 ON). Evidently this is a stable condition.

It is also apparent that a small increase in I_2, rather than in I_1, would have swung the regenerative action in the opposite direction, with transistor Q_2 becoming fully conducting (stage 2 ON) and transistor Q_1 being cut off (stage 1 OFF). Otherwise the action would have been identical. There are, therefore, two stable states for resistive (d.c.) coupling: either stage 1 (Q_1) is ON, thus keeping stage 2 (Q_2) cut OFF, or stage 2 is ON and stage 1 is OFF. We shall see presently how the circuit can be 'switched' from one stable state to the other.

Setting and Resetting (Triggering) the Flip-Flop. There are obviously two things that can be done to change the state of a flip-flop circuit. Either the transistor that is conducting can be turned OFF and the regenerative action (described above) will turn the non-conducting transister ON; or the non-conducting transistor can be turned ON, whereupon the regenerative action will cause the conducting transistor to be turned OFF. The conducting transistor can be turned OFF by applying a positive pulse to its base input, while the non-conducting transistor can be turned ON by applying a negative pulse to its base. This is called 'triggering' or 'setting' the flip-flop. Another pulse of the same polarity applied to the stage that has just been turned ON or OFF cannot change the state of the flip-flop, of course, but a reset pulse delivered to the other stage will 'flop' the circuit back to its original state.

A somewhat more practical transistor flip-flop circuit is illustrated in Fig. 98 (*a*). The circuit is essentially identical to that shown in Fig. 97, except that base input and collector output connexions have been provided for each stage, and capacitors C_1 and C_2 have been added in parallel with coupling resistors R_1 and R_2, respectively. While not affecting the d.c. coupling paths, the purpose of the two capacitors (known as 'commutating' or switching capacitors) is to speed up the switching action from one state to the other by sharpening the voltage pulse coupled from the collector of each stage to the base of the other. The capacitor in conjunction with the resistor acts like the differentiating circuit we have described in our discussion of analogue computers; it puts a 'spike' in the voltage pulse that rapidly turns on the stage.

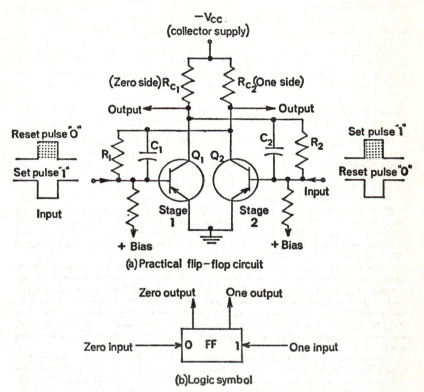

Fig. 98. Setting and resetting the flip-flop: (*a*) practical circuit; (*b*) logic symbol.

To define the state of a flip-flop, certain arbitrary conventions are adhered to, which are shown in the logic symbol of the circuit [Fig. 98 (*b*)]. Let us assign the binary 0 (false) to a collector that is low, or at zero (earth) potential, which means, of course, that the stage is ON, or conducting. We shall assign the binary 1 (true) to a collector that is at a high (negative) potential to indicate that the stage is OFF, or non-conducting. Furthermore, let us call one stage of the circuit (the left in Fig. 98) the 'zero side' and the other (right) stage the 'one side'. Thus, the input going into the left, or zero side, is the zero-side input and the output coming out of the top of the same side is the zero-side output. Similarly, the input going into the right, or one side, is the one-side input and the output coming out of the top of that side is the one-side output, as shown in the logic symbol (*b*). With these conventions in mind, we can define the zero state and the one state of the flip-flop as follows:

ZERO STATE:

The flip-flop is considered in the zero state whenever its zero side (stage 1) is non-conducting and generates a true (negative collector voltage) output, while its one side (stage 2) is conducting and generates a false (zero volts collector) output.

ONE STATE:

The flip-flop is in the one state whenever its zero side is conducting and generates false (zero volts) output, while its one side is non-conducting and generates a true (negative) output. These two definitions can be summarized briefly as follows:

ZERO STATE:

Zero side is OFF, negative collector output = 1 (true)
One side is ON, zero-volts (earth) collector output = 0 (false)

ONE STATE:

Zero side is ON, zero-volts collector output = 0 (false)
One side is OFF, negative collector output = 1 (true)

Fig. 98 illustrates the two ways a change in state can be accomplished. Assume the flip-flop is originally in the zero state, with zero side (stage 1) OFF and the one side (stage 2) ON. The circuit may be flipped to the one state by applying either a negative set pulse to the zero-side input, thus turning this stage ON, or a positive set pulse (shown shaded) to the one-side input, thus turning that stage (2) OFF. In either case, the one-side output terminal generates a true (1) or enable signal, while the zero output terminal generates a false (0) or inhibit signal. Both these outputs are available for driving AND or OR gates with either positive or negative signals. The circuit will then remain indefinitely in the one state until a reset or 'clear' pulse 'flops' it back to its original zero state. This can be accomplished by applying either a negative reset (clear) pulse to the one side (stage 2), thus turning it back ON, or a positive reset (clear) pulse (shown shaded) to the zero side (stage 1), thus turning it OFF again. In either case the one-side output terminal now generates a false (0) or inhibit signal, while the zero-side output terminal generates a true (1) or enable signal. Of course, an opposite convention with respect to true (enable) and false (inhibit) can be equally well adopted.

Type 2: Capacitive Coupling—The Astable (Free-running) Multivibrator

If we use two coupling capacitors, C_1 and C_2, to connect the collector of each transistor stage to the base of the other, we obtain the circuit of the astable, or free-running, multivibrator shown in Fig. 99 (*a*). This again consists of two PNP transistors coupled back to back. With the exception of the polarities involved, the following analysis applies also to NPN transistor or valve circuits.

Since positive feedback is operative, we would expect the circuit of Fig. 99 (*a*) to oscillate, but let us see in a little more detail just how this comes about. To assist the analysis, recall the following fundamental properties of (PNP) transistors, capacitors, and electrical circuits:

1. A PNP transistor must have a negative base voltage (i.e emitter positive with respect to base) in order to conduct. A rise in the negative base

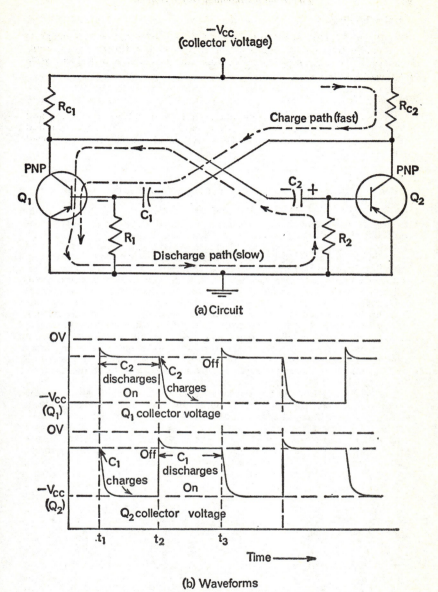

Fig. 99. Circuit and waveforms of astable (free-running) multivibrator.

voltage causes an increase in the collector current, while a fall in the negative base voltage (i.e. a positive rise) causes a decrease in the collector current.

2. An increase in current through the collector load resistor causes the negative voltage on the collector to decrease, or become more positive (being always less than zero, however). A decrease in the collector current causes the collector voltage to increase, or become more negative.

3. Since a capacitor requires a definite time to charge or discharge through a resistor, the voltage across a capacitor in an R–C circuit cannot change instantaneously. A measure of the charge (or discharge) time, called the time constant (T), is the product of resistance and capacitance ($T = RC$).

4. The polarity of the voltage developed across a resistor is such that electrons traverse it from the negative end to the positive end.

Detailed Analysis. Assume that the multivibrator has been operating for a while and that at a certain time [t_1 in Fig. 99 (*b*)] the collector current of transistor Q_1 is increasing. The Q_1 collector voltage, therefore, is becoming less negative, or equivalently, is rising in the positive direction. This positive rise is coupled through capacitor C_2 to the base of transistor Q_2 and causes its collector current to decrease. The decrease in the Q_2 collector current causes a rise in the negative Q_2 collector voltage, which is coupled back through capacitor C_1 to the base of transistor Q_1. The resulting rise in the negative base voltage of Q_1 further increases the Q_1 collector current and makes the Q_1 collector voltage still less negative, or more positive. By the cumulative regenerative feedback action we have described earlier, the collector current of Q_1 is abruptly driven to saturation (fully conducting), with the Q_1 collector voltage being close to zero volts (ground potential), while the collector current of Q_2 is cut off (non-conducting) and its collector voltage rises to its maximum negative value ($-V_{cc}$). This is the situation portrayed at time t_1, where transistor Q_1 has just been turned on, and Q_2 has been turned off. Essentially it corresponds to the rapid switching action of the bistable multivibrator (flip-flop) when 'flipped' from one state to the other.

Charge Path. Although rapid, the switching action is not instantaneous, since the voltage across a capacitor requires time to change. When the collector voltage of Q_2 rises, capacitor C_1 must charge to the higher negative collector voltage. As shown in Fig. 99 (*a*), the charge path is through conducting transistor Q_1 (practically zero resistance) and through collector load resistor R_{c2}. Since both C_1 and R_{c2} are relatively small in value, the time constant $R_{c2}C_1$ is very brief (a fraction of a microsecond) and the Q_2 collector voltage rise is very fast, with only a slight 'exponential' rounding off, as shown in Fig. 99 (*b*).

Inactive (Quiescent) Period. Although Q_1 is now ON and Q_2 is OFF, the circuit cannot remain in this inactive (quiescent) state indefinitely because of the action of capacitor C_2. This capacitor has previously been charged to the Q_1 collector voltage, but now must discharge, since the collector voltage has dropped abruptly from its negative peak ($-V_{cc}$) to almost zero (earth potential). As shown in Fig. 99 (*a*), the C_2 discharge path is through conducting transistor Q_1 and through resistor R_2. Neglecting the resistance of the transistor (practically a short circuit), the time constant of the discharge curve is given by R_2C_2. In contrast to the charging time, this time constant is considerable, since base resistors R_1 and R_2 are usually high, perhaps several megohms. The charge on capacitor C_2, therefore, leaks off slowly through R_2, and as it does, the capacitor voltage decreases and the base of Q_2 becomes less positive (more negative).

At a certain time, t_2, the conducting level of Q_2 is reached and the transistor starts to conduct. With increasing Q_2 collector current the Q_2 collector voltage starts to become less negative or more positive, and this positive rise is coupled through capacitor C_1 to the base of transistor Q_1. The regenerative action is now reversed with respect to the previous condition. The Q_2 collector voltage drops abruptly from its negative maximum to almost zero (earth) level and drives the base of Q_1 rapidly to collector current cut off. With the Q_1 collector current cut off, the Q_1 collector voltage rises to its maximum negative value ($-V_{cc}$). At the same time the Q_2 collector current reaches its saturation (maximum) value. The result of this second switching action, at time t_2, is to turn transistor Q_2 fully ON, while turning transistor Q_1 OFF.

During the time interval from t_2 to t_3 [in Fig. 99 (b)], while capacitor C_1 discharges, the circuit is in its second quiescent state. As soon as the Q_1 base voltage has reached a value that permits conduction (at time t_3), the entire cycle is repeated again, with transistor Q_1 going ON and transistor Q_2 going OFF. It is evident from this analysis that each stage is alternately ON and OFF for relatively long periods (determined by the time constants R_1C_1 and R_2C_2), followed by a very rapid switchover to the opposite state. The circuit, therefore, oscillates from one unstable state to the other, and the collector voltage waveforms are a series of rectangular oscillations, or pulses, which will be symmetrical (square) if the respective time constants are equal. For this reason the circuit is also known as a square-wave generator.

Application. If the collector of either stage of the free-running multivibrator is tapped as output a reactangular pulse train is made available. This is useful in the timing (control) section of a digital computer. The pulses can be synchronized with a submultiple of the frequency of a master (clock) oscillator that determines the timing sequence of the entire computer. The synchronized pulses can also serve either as binary-number inputs to the arithmetic circuits or can initiate binary-coded instructions to be carried out by the logic circuits.

Type: 3 Mixed Capacitive–Resistive Coupling—The Monostable (One-shot) Multivibrator

Look back for a moment at the two cases we have discussed so far. In the first case we employed two d.c. (resistive) coupling paths. This resulted in two stable operating states; that is, a bistable multivibrator, or flip-flop. In the second case we used two a.c. (capacitive) paths, and this arrangement resulted in two unstable states, with the multivibrator oscillating (free-running) between the two. You would expect that a combination of one a.c. (capacitive) and one d.c. (resistive) coupling path would result in one stable and one unstable state, or a monostable (one-shot) multivibrator. This is indeed the case, as shown by the circuit and waveform of the monostable multivibrator (Fig. 100).

The basic circuit of a PNP transistor one-shot multivibrator is shown in Fig. 100 (a). The collector of transistor Q_1 is coupled through capacitor C to the base of transistor Q_2, and the collector of Q_2 is coupled through resistor R to the base of Q_1. The Q_1 base resistor, R_1, is connected to a high positive bias voltage, which normally keeps the Q_1 collector current cut off (non-conducting). Under these conditions transistor Q_1 is OFF, while transistor Q_2 is ON (conducting), since its base is returned to zero (earth) potential. In the absence of an input 'trigger' the circuit remains indefinitely in this stable state.

Starting Trigger. Consider what happens if a negative input trigger pulse is applied to the base of transistor Q_1. If the negative amplitude of this pulse is greater than the positive (cut-off) bias, the base voltage is momentarily raised above cut-off and the stage starts to conduct. This decreases the negative Q_1

collector voltage, or makes it more positive, and this positive rise is coupled through C to the base of Q_2, causing the Q_2 collector current to decrease. Again the cumulative regenerative feedback action described earlier causes Q_1 to be turned ON abruptly, while stage Q_2 is simultaneously cut OFF. This is the situation depicted at time t_1 [in Fig. 100 (b)], when the negative starting trigger has just been applied to Q_1 and consequently the Q_2 collector voltage has abruptly risen to its maximum negative value ($-V_{cc}$).

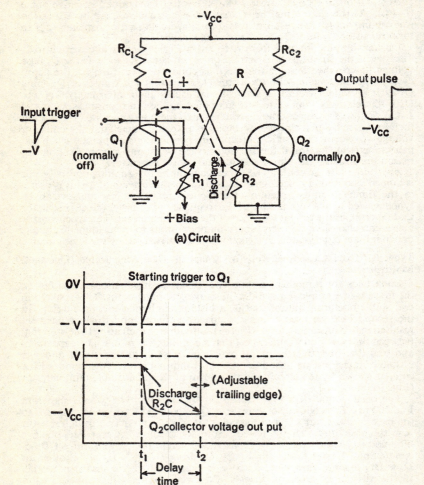

Fig. 100. Circuit and waveforms of monostable (one-shot) multivibrator.

In the time interval from t_1 to t_2 the circuit is in its inactive (quiescent) condition, similar to the free-running multivibrator. Again, this is not a stable state because the capacitor (C), which had previously been charged to the highly negative collector supply voltage $-V_{cc}$, is now connected to the abruptly reduced (zero) voltage at the collector of conducting transistor Q_1. The capacitor, therefore, slowly discharges through conducting transistor Q_1 and base resistor R_2, as indicated in Fig. 100 (a). Since Q_1 is practically a short circuit, the time constant of the capacitor discharge is essentially equal to R_2C. This time is represented approximately by the interval from t_1 to t_2. While the capacitor discharges, its voltage drops, and hence the cut-off voltage applied to the base of Q_2 decreases. When the conducting bias is reached, at time t_2, transistor Q_2 starts to conduct again and rapidly drives Q_1 towards cut off. Thus, the circuit flops over again to its initial stable condition and will remain there unless another trigger pulse repeats the entire cycle.

Application as Delay Circuit. It is apparent that the one-shot multivibrator provides a single (square or rectangular) output pulse upon receipt of each brief trigger pulse. Moreover, the repetition rate (frequency) of these output pulses is determined entirely by the trigger source, while the pulse duration (interval t_2-t_1) is determined by the circuit constants, such as the cut-off bias and the time constant R_2C. By making the latter constants adjustable, a variable time delay can be attained between the trigger and the 'trailing edge' of the output pulse (t_2), which is frequently used to delay pulses in a digital computer.

LOGIC CHAINS FOR ARITHMETIC

We have considered some of the basic electronic devices for carrying out logical operations, storing a binary digit in a flip-flop, or delaying a pulse (digit) by means of a one-shot multivibrator. Let us now combine these elements in logical chains to perform some of the fundamental operations of computer arithmetic. Since the reader already knows the functioning of the electronic devices, we need only explain the logical interconnexions of the 'black boxes' to perform various required jobs.

Binary Counters

As we saw in our discussion of analogue computers, addition is essentially a matter of counting. If you can count forward in any numbering system you know how to add one digit at a time. The early mechanical counting wheels added numbers by counting the gear teeth that represented each number. Electronic counters add by counting pulses instead. Of course, any counter must have a method of storing the discrete digits that represent the number in the particular system used. Thus, a decimal counter must be able to store ten counts, from 0 to 9, before the next count resets the counter to zero again (with '1' to carry). Such a counter needs at least ten elements to store each of the ten digits, like the ten teeth of a mechanical counting wheel. Early digital computers, which operated on the decimal system, actually had electronic counters consisting of ten elements each (known as ring counters). The use of the binary number system, however, has simplified matters considerably. A binary counter need store only two digits, 0 or 1, in any column. A single flip-flop, with its set (one) state representing binary '1' and its reset (zero) state representing binary '0', can act as a binary counter if the appropriate connexions are made.

Fig. 101 (a) illustrates a binary counter made up from a flip-flop circuit and two AND gates. Its operation is very simple. You will recall that the ordinary flip-flop remains indefinitely in one of its two possible states, the 'one' (set) state or the 'zero' (reset or clear) state. It can be thrown from one state to the other by application of a pulse to the proper input, provided you know the state in which it is at the time. The binary counter fulfils this function: specifically, each incoming count pulse reverses the previous condition of the flip-flop, so that the circuit is alternately thrown from one state to the other. If the flip-flop is in the zero state an incoming count pulse sets the circuit to the one state; if the flip-flop is already in the one state an incoming pulse resets it to the zero state. In the latter case (reset to zero) the counter also develops a 'carry' output pulse to the next column counter, since 0 and 1 exhaust the digits of one binary column. Thus, the binary counter is essentially a flip-flop with a conditional switching action provided by AND gates.

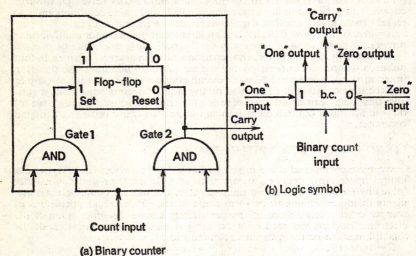

Fig. 101. Binary counter and logic symbol made up from a flip-flop and two AND gates.

Pulse Steering. The illustration [Fig. 101 (*a*)] shows how the AND gates are employed to steer the incoming count pulses to the proper input terminal (0 or 1) of the flip-flop circuit. One of the inputs of AND gate 1 is connected to the zero output side of the flip-flop, while one input of AND gate 2 is connected to the one output side. Similarly, the output of gate 1 feeds to the one input side of the flip-flop, while the output of gate 2 feeds to the zero input side. Incoming count pulses are applied simultaneously to the second inputs of gates 1 and 2. This arrangement automatically steers the incoming pulse to the flip-flop input that will change its state.

You can see how this comes about. Assume the flip-flop is initially in the zero state, so that its zero side provides a one output pulse and the one side provides zero (earth) output. AND gate 1 (connected to the zero output of the flip-flop) will therefore be enabled by the one output of the flip-flop, while AND gate 2 (connected to the one output side) will be inhibited by the zero output. The next count pulse applied to the input of the counter, therefore, cannot pass through gate 2 but is routed through gate 1 to the one side of the flip-flop, thus setting it to the one state. The situation is now reversed: the zero side of the flip-flop provides zero output, inhibiting AND gate 1, while the one side of the flip-flop emits a one output pulse that enables AND gate 2. The next incoming count pulse, therefore, is blocked by gate 1 and routed through gate 2 to the zero input side of the flip-flop, resetting it to the zero state. Simultaneously, AND gate 2 provides a 'carry' output pulse for use in the next counter. It is evident that the AND gates always steer the count pulse to the flip-flop input that will reverse its state, as required for binary counting. A logic symbol of a binary counter is shown in Fig. 101 (b).

Practical Counters. Though binary counters are frequently made up by combining a flip-flop with two AND gates, as Fig. 101 shows, separate AND elements are not always used. A pair of diodes may be built right into the flip-flop to perform the pulse-steering in the simplest possible way.

Fig. 102 illustrates the schematic diagram of a four-column binary counter made up of four separate binary counters, with the 'carry' output of each connected to the 'count' input of the next counter. This particular counter counts the 16 binary digits from (decimal) 0 to 15 which, as you recall, requires four binary columns.

The chart of the normal binary progression from 0 to 15, shown in Fig. 102 (b), illustrates the operation of the four-stage counter. The count pulses are applied to the input of the first, or one's column, binary counter. Incoming pulses alternately set this counter to 1, then reset it to 0, back to 1, and so on. On every other pulse, when being reset to 0, the unit counter emits a 'carry' pulse to the two's column counter. The two's column counter also alternates between 1 and 0 each time it receives a count, but this happens only for every other input count, when the unit counter develops a carry pulse. On every fourth count the two's column counter is reset to 0 and simultaneously transmits a carry pulse to the four's column counter. This counter, therefore, alternates between 1 and 0 every four counts (that is, whenever it receives a 'carry' input). After every eight counts, the four's column counter is reset to 0 and emits a carry pulse to the eight's column counter. After every 16 counts, all counters are simultaneously reset to 0 (cleared) and the eight's column counter emits a carry pulse to the next higher column counter, if available. With the four counter stages set up as shown by the encircled outputs in Fig. 102 (a), the binary number 1011 (decimal 11) is being stored.

The four-column binary counter shown in Fig. 102 is somewhat slow in practice. Since each counter stage requires a definite, though brief, time to change state, it may take quite a while before the last stage on the left receives its carry pulse, especially if more than four stages are involved. In a 15-column binary counter, for example, this propagation delay may add up to several microseconds, which by computer standards is considerably more than can be permitted. To reduce the propagation delay, binary counters are usually interconnected with AND gates in a manner that permits simultaneous sensing of a carry pulse from the lowest-column to the highest-column stage. A count pulse at the input of the first (lowest-column) stage will set all required stages at the same time, so that there is no waiting for the individual transitions to occur. There are also counters that do not count in a binary sequence, but shift a '1' carry pulse to the left during each count pulse; these are known as 'shift counters'.

Binary Adders

Instead of counting off binary digits until the required number is obtained, we can add two binary digits by performing the logical operations inherent in the binary addition table. If two binary numbers have a number of digits (or columns), the digits in each column must be added together two at a time. In general, for each column three binary digits (bits) must be added. These are the two original bits in each column (i.e. the addend and the augend) plus the

200 *Electronic Computers Made Simple*

'carry' digit from the next lower-order column. A device that can perform this addition and produce the correct sum and carry term for each column is known as a full adder. There is a simpler unit, called a half-adder, that adds only two binary input bits (addend and augend) to produce a sum and a possible carry output. Two of these half-adders make up a 'full adder'.

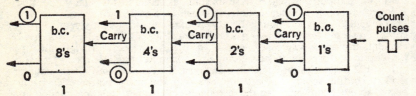

(a) Schematic diagram of 4-column binary counter

	Four-column binary progression						
	Eight's column	Carry	Four's column	Carry	Two's column	Carry	One's column	Pulse count
	0		0		0		0	
	0		0		0		1	1
	0		0		1		0	2
	0		0		1		1	3
	0		1		0		0	4
	0		1		0		1	5
	0		1		1		0	6
	0		1		1		1	7
	1		0		0		0	8
	1		0		0		1	9
	1		0		1		0	10
(Counter setting)	1		0		1		1	11
	1		1		0		0	12
	1		1		0		1	13
	1		1		1		0	14
	1		1		1		1	15
Carry to 16's column	0		0		0		0	(Reset)

(b) Binary progression in 4-column counter

Fig. 102. Four-column binary counter.

The Half-adder

In the beginning of the chapter we developed two very simple logical relations for the sum and the carry of the binary addition table. For convenience, the addition table for two binary digits, an addend A and an augend B, is reproduced in Fig. 103 (b). You can see that the sum of the two digits is 1 when either A is 0 and B is 1, or when A is 1 and B is 0. The logical expression of this statement is the exclusive OR (circle-sum):

$$\text{SUM} = \bar{A}B + A\bar{B}$$

Checking back in the binary addition table [Fig. 103 (b)], observe that a 'carry' is developed only when A is 1 and B is 1. This corresponds, of course, to the AND function, or logical product:

$$\text{CARRY} = A \cdot B$$

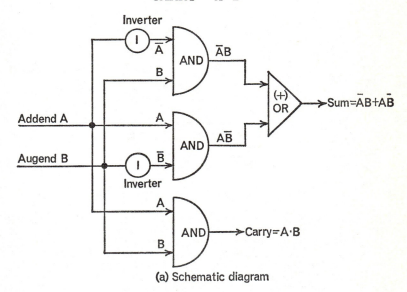

(a) Schematic diagram

Addend A	Augend B	Sum	Carry
0	0	0	0
0	1	1	0
1	0	1	0
1	1	0	1

(b) Binary addition table

Fig. 103. One form of the half-adder.

Fig. 103 (a) illustrates in schematic form one way of mechanizing these logical expressions by means of a half-adder. The carry term is obtained by applying the addend (A) and the augend (B) to the inputs of an AND gate, so that the logical product at the output of the gate is $A \cdot B$, as required. The

$\bar{A}B$ term of the sum is developed by inverting the addend A (resulting in $\bar{A}$) and applying it together with the augend B to the inputs of another AND gate, whose output, thus, is $\bar{A}B$. Similarly, the second term of the sum is obtained by inverting B (resulting in $\bar{B}$) and applying it, together with A, to the inputs of a third AND gate; the output of this gate, thus, is $A\bar{B}$. These two sum terms are added together by an OR gate, whose output is the required logical sum, $\bar{A} \cdot B + A \cdot \bar{B}$. The gates themselves may be designed in the manner previously shown.

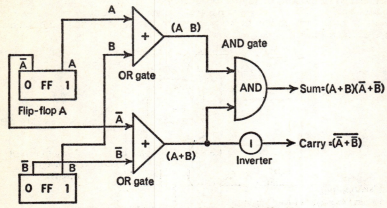

Fig. 104. An alternate form of the half-adder.

Alternative Half-adder. Fig. 104 illustrates an alternative way of mechanizing a half-adder with two OR gates, one AND gate, and a inverter, thus saving a gate and an inverter over the previous circuit (Fig. 103). The flip-flops that produce the addend A and the augend B are not part of the half-adder, but illustrate the derivation of the NOT functions ($\bar{A}$ and $\bar{B}$) from the regular zero-side outputs. The circuit is of interest because it is one of several possible methods for determining when a pair of flip-flops are in opposite (complementary) states. Two flip-flops may have four possible combinations of states: 00, 01, 10, or 11. The sum portion of the logical array shown in Fig. 104 senses the state of flip-flops A and B by determining when A OR B AND (not-A OR not-B) are true (1). In equation form:

$$(A + B) \cdot (\bar{A} + \bar{B}) = 1$$

This is one of several possible ways of deciding the complementary states of the flip-flops.

You can easily convince yourself that in determining the complementary states of two flip-flops the circuit of Fig. 104 performs the function of a half-adder. By one of De Morgan's rules, the 'carry' term $\overline{(\bar{A} + \bar{B})}$ equals $A \cdot B$, which is the same as previously established. Multiplying out the sum term in the circuit of Fig. 104:

SUM term
$$(A + B)(\bar{A} + \bar{B}) = A(\bar{A} + \bar{B}) + B(\bar{A} + \bar{B})$$
$$= A\bar{A} + A\bar{B} + B\bar{A} + B\bar{B}$$

We have seen earlier that the product of a variable and its complement (negation) is always false, so that $A\bar{A} = 0$ and $B\bar{B} = 0$. Hence, the SUM

$(A + B)(\bar{A} + \bar{B}) = A\bar{B} + B\bar{A} = \bar{A}B + A\bar{B}$ (by the commutative law), which is identical with the result previously obtained (Fig. 103).

Binary Addition with Half-adders

As its name implies, the half-adder does only half the job, since even though it adds two input bits, it does not take into account a possible 'carry' digit from the next lower-order (less significant) column of the numbers to be added. The complete job consists of combining the sum resulting from adding the addend and the augend with the 'carry' from the next lower-order column. The result yields a final sum and a possible carry to the next higher-order (more significant) column.

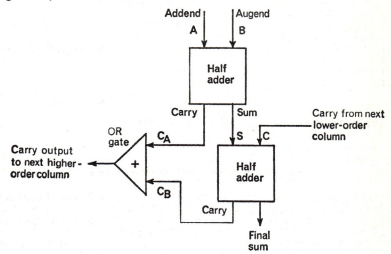

Fig. 105. Parallel binary adder using two half-adders per column.

One way of doing this is to use two half-adders for every column to be added, as illustrated in Fig. 105. The first half-adder adds the two bits (addend and augend) of the column, as described above. The second half-adder combines the sum of the first with the possible carry from the next less significant (lower-order) column. The final 'carry' output from the second half-adder may be the result of either the first or the second addition. Consequently, an additional OR gate is required to 'sense' the carry at either the first or second half-adder output, as the illustration shows. The full addition with two half-adders, illustrated in Fig. 105, is known as parallel addition, since all three bits (addend, augend, and carry) are added at the same time. In the next chapter we shall consider a circuit with two half-adders that adds the bits of each column in sequence, one at a time—a method known as serial addition.

Full Binary Adder

Since the complete addition of any column of two numbers involves three bits—the addend, the augend, and the previous (lower-order) carry—a full, or three-input binary adder can be used in place of the two half-adders. To visualize the logical rules for the three-input full adder, we must consider the complete binary addition table for all possible combinations of the addend (A), augend (B), and the lower-order carry (C). A chart giving all the possible combinations of the three inputs is shown below:

COMPLETE (THREE-INPUT) BINARY ADDITION TABLE

Addend—A	0	1	0	0	1	1	0	1
Augend—B	0	0	1	0	1	0	1	1
Lower-Order 'Carry'—C	0	0	0	1	0	1	1	1
Sum—S	0	1	1	1	0	0	0	1
Forward Carry	0	0	0	0	1	1	1	1

Using the chart above, which may be considered a truth table turned on its side, we can easily form the conditions of the inputs which make either the sum or the forward (higher-order) carry true (1). The following logical equations for the sum and the forward (higher-order) carry may be written. Each represents alternate conditions (logical sums) that make the sum or carry expression equal to '1' (true):

$$\text{Sum} = A\bar{B}\bar{C} + \bar{A}B\bar{C} + \bar{A}\bar{B}C + ABC$$
$$\text{Carry} = AB\bar{C} + A\bar{B}C + \bar{A}BC + ABC$$

The logical equations indicate that the sum is 1 whenever either one of the inputs (A, B, or C) is 1, or when all three are 1; the carry is 1 whenever any two or three of the inputs are 1.

Fig. 106 illustrates one way of mechanizing these logical expressions for the full adder. Four AND gates and a four-input OR gate must be used for each of the sum and carry expressions. The complementary addend (A) and augend (B) inputs are obtained from two flip-flops, or registers, while the lower-order carries are derived from the next lower-order column full adder. The operation of the circuit is evident from the diagram. The number of gates required can be cut down considerably by algebraic manipulation of the 'sum' and 'carry' expressions.

Generation of Time Delay

To carry out the operations in a digital computer in proper sequence, binary instructions and numbers must frequently be delayed by a certain interval until they can be used. Since all digital information is in pulse form, this means in practice that we must be able to generate an output pulse that is delayed by a known time interval after an input pulse has been applied. There are many ways in which such a time delay can be generated; for example, electrical and acoustic delay lines, in which information travels back and forth indefinitely until called for use. You will also recall the one-shot multivibrator, whose variable pulse duration permits generating a variable time delay, as we have briefly described earlier. Let us look at this process a little more closely.

Delay Multivibrator. A practical arrangement of a delay multivibrator is shown in block-diagram form in Fig. 107. A negative input pulse (a) is used to 'trigger' the one-shot multivibrator at time t_1 [see Fig. 107 (b)]. The coupling network of the circuit can be adjusted to provide a pulse of variable width (duration), as indicated by the variable capacitor. The rectangular output pulse from the multivibrator extends in duration from time t_1 to t_2. This rectangular pulse is fed through a differentiator circuit, whose time constant (RC) is very short compared with the pulse duration t_2–t_1. The effect of differentiating the rectangular pulse is the appearance of two brief 'spikes' (trigger pulses); one is a positive spike at the beginning of the rectangular pulse, at time t_1, the

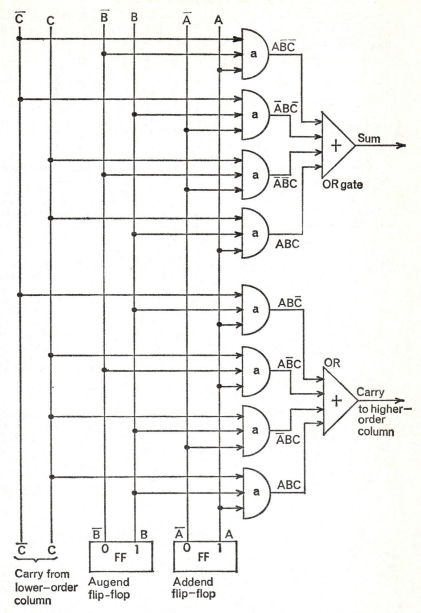

Fig. 106. Three-input, or full adder, for complete binary addition.

other a negative spike at the end of the pulse, at time t_2. Thus a time delay, the interval t_2-t_1 between the two differentiator output spikes, has been created. Subsequent circuitry normally responds only to the negative (delayed) pulse emerging at time t_2, but if necessary the positive spike can be removed by means of an electronic 'clipper', such as a diode.

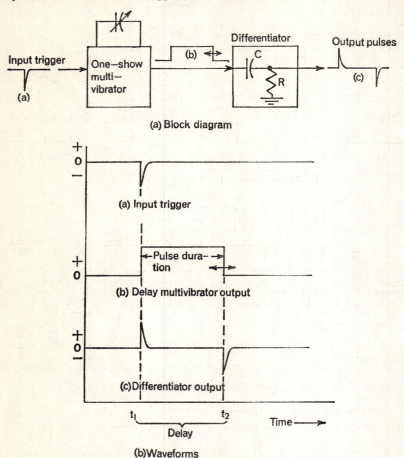

Fig. 107. Generation of time delay by one-shot multivibrator.

Shift Register

In Chapter 8 we explained that binary (or any other) multiplication consists of forming partial products, then shifting each of these products by one place before adding them. Whether the partial products are shifted to the left or to the right depends upon the order of multiplication, as you recall. The device that performs this shifting operation is known as a shift register. Shifting is also required when converting from serial to parallel computer operation, or vice versa. In parallel operation all digits of an arithmetic problem are processed simultaneously, while in the serial mode the arithmetic operation is carried out in sequence, using the bits of one column at a time. The input information to a computer is generally in serial form, consisting of a time sequence of ones and zeros coming from an electric typewriter, radio, magnetic tape, or other source. If the computer itself operates in the parallel (simultaneous) mode, some device is needed which can accept each input digit in turn, store it temporarily, and then place it in its proper position for processing. Again, a shift register can perform this serial-to-parallel conversion. It is, therefore, a very useful device.

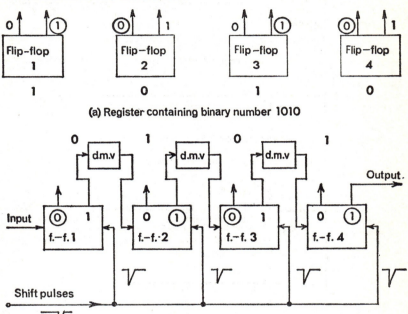

Fig. 108. From register to shift register. (*a*) register containing binary number 1010; (*b*) shift register using flip-flops and delay multivibrators.

From Register to Shift Register. You will remember that we can represent a binary digit (bit) by the state of a flip-flop. In the zero state the right side of the flip-flop is ON and its collector voltage is low, representing a binary 0, while in the one (1) state the right side is OFF and its collector voltage is high, representing a binary 1. Furthermore, applying a negative pulse to the right (1) side will turn it back ON, thus resetting it to 0, as we have explained before. For example, we can represent the binary number 1010 by means of four flip-flops, whose right (one) sides are alternately set to 1 and 0, as shown by the encircled numbers in Fig. 108 (*a*). Such a set-up of flip-flops is called a 'register', since it contains data—the digits 1010 in this case. (Later you will find out how such a register is actually set up.)

Suppose we want to shift each binary digit in the register of Fig. 108 (*a*) one place to the right and at the same time clear (reset to 0) the first flip-flop on the left, so that it is ready to receive another input pulse. After shifting, therefore, the register should read—from left to right—0101. A little analysis shows that we must actually do three things to carry through the shifting process:

1. We must 'clear', or reset, each flip-flop to zero by a pulse of the proper polarity to make room for the next bit (pulse).
2. We must hold, or store, the data from each flip-flop until the clearing process is completed.
3. We must reapply the stored data to the flip-flops, with each bit shifted to the adjacent element at right.

The block diagram of the shift register [Fig. 108 (*b*)] shows how these three things can be done by a single negative shift pulse applied to the set-up. The one (right) output of each flip-flop is connected through a delay multivibrator (d.m.v.) to the zero (left) input of the next flip-flop to the right. A common bus delivers a negative shift pulse to the one-side input of each flip-flop. The negative pulse affects only the flip-flops that are in the one state (that is, those whose right sides are set to one). Consequently, flip-flops 1 and 3, which are in the one state, are reset to the zero state. Flip-flops 2 and 4 are already in the zero state and hence remain unchanged.

As flip-flops 1 and 3 are reset to zero, they each emit an output pulse from their one side to the adjacent delay multivibrator, which provides a predetermined time delay sufficient to complete the operation. At the end of the delay interval, the stored digits (pulses) emerge from the delay multivibrators and set each of the flip-flops immediately to the right to the one state. This means that flip-flops 2 and 4 will now be set to one, while flip-flops 1 and 3 remain at zero. From left to right, the register then reads 0 1 0 1, as required. A new digit may now be placed into flip-flop 1 at the left, which has been cleared. Upon the application of four shift pulses to the register, the original number (1010) will have been moved out of the register by shifts to the right, and if desired, a new number can be moved in from the left.

REVIEW AND SUMMARY

The rules of the algebra of logic (Boolean algebra) control binary arithmetic; hence, the arithmetic portion of a digital computer consists of logic circuits.

An AND gate produces a true (one) output signal only when all its inputs are simultaneously true (one), or activated. AND gates may be mechanized by relays, diode valves or semiconductor diodes, triodes, and by transistors.

In the two-input AND gate the output is always equal to the lower of the two input voltages, or to both if they are the same. If the truth value 0 is assigned to a high (+) voltage level and 1 to a low (−) voltage level, a negative AND gate may be considered as a positive OR gate, and vice versa. Also, a positive AND gate is equivalent to a negative OR gate, and vice versa.

Transistors: (1) A PNP transistor has a P-type germanium emitter, an N-type base, and a P-type collector; an NPN transistor has an N-type germanium emitter, a P-type base, and an N-type collector. *P*-type germanium has *p*ositive majority charge carriers (holes), while *N*-type germanium has *n*egative charge carriers (electrons) available to conduct a current.

(2) Main current flow in transistors is from emitter to collector and passes through the base. The emitter is always forward-biased, to propel the current

carriers towards the collector, while the collector is reverse-biased, to attract the current carriers.

(3) The polarity of the emitter voltage with respect to the base is indicated by the first letter of the type (PNP = + emitter; NPN = − emitter). The polarity of the collector voltage with respect to the emitter is indicated by the second letter of the type; i.e. PNP = $n$egative collector, NPN = $p$ositive collector.

(4) Electron current flow is always against the direction of the arrow on the emitter; that is, out of the emitter (and into the collector) of a PNP transistor, and into the emitter (and out of the collector) of an NPN transistor (see Fig. 91).

(5) An input signal that assists the emitter forward bias increases the emitter-to-collector current flow; a signal that opposes the forward bias decreases current flow. For example, a negative signal applied to the base of a PNP transistor assists the forward bias and thus increases the collector current, while a positive signal reduces the collector current. (Opposite polarities apply to NPN types.)

The logical NOT (negation) function is performed by inverters. An inverter is any device whose output is opposite in polarity with respect to the polarity of an input signal. Triodes valves and transistors automatically invert the polarity of an input signal, provided the output is taken from the plate or collector, respectively. A low (0) input signal to the inverter results in a high (1) output signal, and vice versa.

An OR gate produces a true (one) output signal whenever one or more of its input signals are true (one), and produces a false (zero) output only when all its input signals are simultaneously false (zero). OR gates can be mechanized by relays, diodes, triodes, or transistors.

A positive input OR gate is equivalent to a negative input AND gate, and a negative OR gate is equivalent to a positive AND gate.

A NOR (NOT OR) gate produces a true (one) output whenever all its input signals are simultaneously false (zero), and it produces a false (zero) output whenever one or more of its input signals are true (one). NOR gates may be mechanized by transistors in parallel, by a combination of parallel input resistors with a transistor, or by a combination of parallel input diodes with a transistor.

A NAND (NOT AND) gate produces a true (one) output whenever one or more of its inputs are false (zero), and produces a false (zero) output only when all its inputs are simultaneously true (one) (see Fig. 95).

A multivibrator consists essentially of two amplifier-inverters (valves or transistors) coupled back to back, so that the output of one stage is in phase with, and drives, the input of the other stage. This results in 100 per cent positive (regenerative) feedback. The operation of the multivibrator is determined by the type of coupling paths between the stages. There are three types:

(1) *Two d.c., or resistive, coupling paths*: results in two stable states (bi-stable multivibrator or flip-flop), with either the left stage conducting (ON) and the right stage non-conducting (OFF), or vice versa.

(2) *Two a.c., or capacitive, coupling paths*: results in no stable states, or an oscillator that generates a train of rectangular pulses (known as astable or free-running multivibrator).

(3) *One d.c. (resistive) and one a.c. (capacitive) coupling path*: results in one stable state to which the circuit always reverts when temporarily 'triggered' by an input pulse (known as monostable, or one-shot, multivibrator). It can be used to delay an input pulse by a known interval (delay multivibrator).

A flip-flop (bistable multivibrator) is in the zero state if its one (right) side is ON with the collector low (zero) and is in the one state if its one side is OFF with the collector high (one). The flip-flop may be set to the one state by applying either a negative pulse to the zero-side input or a positive pulse to the one-side input. It can be reset to the zero state by applying either a negative pulse to the one-side input or a positive pulse to the zero-side input.

A binary counter is essentially a flip-flop with two pulse-steering AND gates (diodes), so that incoming count pulses alternately throw the circuit from one state to the other. Thus, the counter stores the two binary digits 0 or 1. A count to any desired binary number is made possible by connecting binary counters in cascade ('carry' output to 'count' input) to represent all significant columns of the number.

A binary adder performs the logical operations of the binary addition table. A half-adder adds two binary input bits (addend and augend) to produce a sum and possible 'carry' output. The sum is the exclusive OR function ($\bar{A}B + A\bar{B}$), while the carry is the logical product (AB). These expressions can be mechanized by three AND gates and an OR gate.

A three-input, or full binary adder, performs complete addition of three bits, the addend (A), the augend (B), and the lower-order column carry (C). A full adder can be made up by connecting together two half-adders (Fig. 105) for parallel (simultaneous) or serial (sequential) addition, or by the mechanization (Fig. 106) of the logical expressions for the sum and higher-order (forward) carry outputs, as follows:

$$\text{Sum} = A\bar{B}\bar{C} + \bar{A}B\bar{C} + \bar{A}\bar{B}C + ABC$$
$$\text{Carry} = AB\bar{C} + A\bar{B}C + \bar{A}BC + ABC$$

A shift register (Fig. 108), used for serial-to-parallel conversion or multiplication, shifts the binary digits contained in a register (flip-flops) by one place to the right or left, whenever it receives a shift pulse.

CHAPTER 11

BUILDING BLOCKS OF DIGITAL COMPUTERS—III: MAGNETIC AND OTHER DEVICES

Because of their inherent high-speed response, electronic devices are the most numerous in present-day digital computers and will probably remain so for some time to come. Next in popularity and almost equally indispensable are a variety of magnetic devices. All these have one important property in common: 'remembering' the magnetic state they were last in, a property that is extensively exploited in the memory (storage) functions of digital computers. (Without a 'memory', computers—like people—would be incapable of carrying out lengthy logical and mathematical operations.) We shall see that magnetic devices have essentially two states—they can be magnetized in either one or the opposite direction (or they may not be magnetized at all, comprising a third state). This bistable characteristic is similar to that of flip-flops and electronic gates, and it is natural, therefore, that magnetic devices are frequently used for binary logic operations, just like the electronic devices we have studied.

Before the advent of magnetic memory devices, memory functions were performed by electrostatic storage tubes which were capable of preserving logical patterns by corresponding charge patterns on a screen similar to that of a television set. In addition to these, there are various electrical and acoustic delay lines for short-term data storage, as was mentioned in the last chapter. Finally, we shall take a look at a more recently introduced device that holds great promise for the future. This is the cryotron, which makes use of the superconducting (zero-resistance) properties of some metals near the absolute zero of temperature to perform high-speed switching functions.

REVIEW OF MAGNETISM

We all have observed that magnets have the ability to attract iron and similar ferromagnetic substances. Although some already magnetized materials are found in nature, most magnets are made artificially by magnetization with another magnet or by placing them in the 'field' of a strong electromagnet. Magnets made of hard steel, ferrites, and other ferromagnetic materials (such as cobalt, nickel, and their alloys) retain their magnetism permanently, even after the magnetizing field is removed, while other materials, such as soft iron, are magnetized only temporarily, while under the influence of the magnetizing field. The magnets used in computers are all of the permanent variety.

Magnetic Theory. Magnetism is not a fundamental phenomenon, but only one aspect of electrical behaviour. The Danish physicist Oersted discovered in 1820 that an electric current in a conductor is always surrounded by a magnetic field. This is the basis of electromagnetism. More recently it has become evident that magnetism itself must be attributed to electrical charges in motion, specifically the spinning of electrons while orbiting the nuclei of atoms. In the atoms of ferromagnetic materials more electrons spin in one direction about their axes than in the opposite direction. These uncompensated electron spins create small magnetic 'twists', or moments, which make each atom of a magnetic material a tiny magnet. The atomic magnets are oriented in random directions, so that their magnetic moments cancel out and no observable overall magnetic effects are produced. However, throughout a tiny region of some 10^{15} (a thousand billion) atoms, known as a domain, all the uncompensated electron spins have the same direction; i.e. they are parallel. The parallel electron spins within a single domain produce an intense magnetic field in its vicinity. Again the domains within a magnetic material are oriented in all conceivable directions, and hence the internal fields of the domains ordinarily cancel out.

Magnetization. Fig. 109 illustrates what happens when a material is magnetized. Fig. 109 (*a*) shows the random orientation of the domains in an

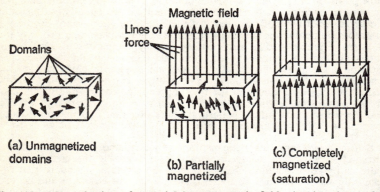

Fig. 109. Magnetization of material by a magnetic field, showing the gradual alignment of the domains.

unmagnetized rectangular-shaped material. The arrows indicate the directions of the internal fields created by the domains. You can see that these cancel out. When the material is placed in the external field of a strong permanent or electromagnet the domains begin to rotate—a few at a time—to align themselves with the external field, as is shown in Fig. 109 (*b*). (The external magnetic field is indicated by the parallel flux lines all going in the same direction.) The material is now partly magnetized. If the field is sufficiently strong and applied long enough all the domains will eventually jump into alignment with the external field, as is illustrated in Fig. 109 (*c*). The material is then completely magnetized, and magnetic saturation is said to occur. No further magnetization can then be produced by an increase in the intensity of the external magnetic field.

Field Intensity and Flux Density. As shown in Fig. 109, imaginary 'lines of force' are used to represent the invisible magnetic field and calculate its effects. A single line of force, called the maxwell, represents the unit of magnetic flux in a magnetic field. The total number of lines of force (or maxwells) issuing from the north pole of a magnet and terminating on its south pole is a measure of the total magnetic flux. The strength of the field in any particular region is determined by the number of lines of force traversing a unit area in that region. The flux per unit area is known as the flux density (symbol B) and is expressed in gauss (representing lines of force per square centimetre). The flux density, hence, is a measure of the flux per unit area induced in a material by a magnetic field of a certain intensity (symbol *H*). Flux density also depends on the permeability (symbol µ), which is a measure of the ease of magnetization of a material. The permeability of a material is defined as the ratio of the flux density achieved to the intensity of the field applied, numerically:

$$\text{Permeability } \mu = \frac{\text{flux density}}{\text{field intensity}} = \frac{B}{H}$$
and flux density $B = \mu H$

The permeability of vacuum, or air, is taken as unity (1), so that the flux density numerically equals the field intensity for these media. The permeability of magnetic substances, however, such as iron, cobalt, ferrites, etc., is many thousand times that of air, which permits large values of the flux density to be attained.

The Hysteresis Loop. We have seen (in Fig. 109) that magnetization is a gradual process with the induced flux density (B) slowly increasing as the domains align themselves with the external magnetizing field (H). The energy

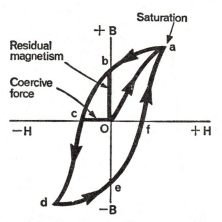

Fig. 110. Hysteresis loop of a typical ferromagnetic material.

wasted in producing magnetization of a material is known as hysteresis. It can be represented by a diagram that plots the induced flux density against the applied magnetizing force, or field intensity. Such a plot is called a hysteresis loop or B–H curve, since it shows the dependence of the flux density B on the magnetizing force, or field intensity, H. For a given electromagnet with a specific number of turns in the winding, the magnetizing force (H) is proportional to the current flowing through the winding and can be expressed as the product of the current and the number of turns (ampere-turns). The hysteresis loop, therefore, also shows the amount of flux density (magnetization) as a function of the current flowing through the magnet. The direction of current flow through the winding determines the relative polarity (north and south) of the magnet; reversing the current also reverses the magnetic polarity.

The hysteresis loop of a typical ferromagnetic material is shown in Fig. 110. The material, which is originally completely unmagnetized, is made the core of an electromagnet. An increasing current is applied to the magnet winding, which establishes an increasing magnetic field intensity, H. As the current and field increase in one direction, the flux density induced in the material increases slowly along the S-shaped curve, *o–a* in Fig. 110. The curve levels off when magnetic saturation of the core material is attained; no appreciable increase in flux density results from further increases in the magnetizing current.

If the magnetizing current and resulting field H is now reduced gradually to zero, the flux (B) does not collapse along the same curve (*o–a*), but rather follows curve *a–b*. A certain amount of magnetism appears to have been stored in the material and remains even after the magnetizing current and field have been removed. The amount of magnetism retained (segment *b–o*) when the field has been reduced to zero is known as residual magnetism, or more technically, remanent flux density.

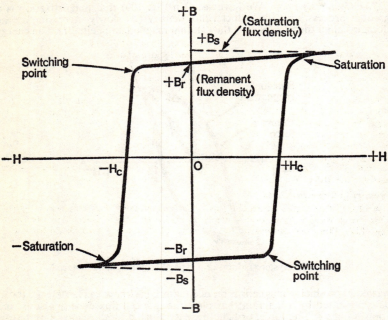

Fig. 111. Square hysteresis loop of ferrite core.

If the direction of the magnetizing current and the resulting field is reversed (i.e. becomes negative) the flux density continues to decrease along curve *b–c* until it reaches zero and the material is completely demagnetized. The value of the negative magnetizing force required to completely demagnetize the material is called the coercive force, and it is represented by segment *o–c*. Further magnetization in the negative direction along curve *c–d* eventually results in magnetic saturation in the negative direction, at point *d*. If the magnetizing current and field is once again reduced to zero the flux falls off along curve *d–e* and a part of the magnetism (segment *e–o*) is again retained. To reduce the flux to zero, along *e–f*, the magnetizing field must be increased in the positive direction. Further increases of the current result in saturating the core in the positive direction, along curve *f–a*. This completes one cycle of the hysteresis loop. If the magnetizing current (and field H) is carried through another cycle the flux density will follow along the outer loop *a–b–c–d–e–f*, and the original magnetization curve *o–a* is never repeated.

The Square Hysteresis Loop. Let us now consider one of the special types of powdered ferrite core materials used in computers. These materials retain most of the magnetism once imparted, as is shown by the almost square hysteresis loop in Fig. 111. Note that the remanent flux density, B_r (the magnetism retained), is almost equal to the maximum flux density at saturation, B_s. This is true for either the positive or the negative direction of magnetization. As a result, variations in the field intensity less than a certain critical switching value, H_c, have practically no effect on the flux density, B, while a change in H greater than the critical value abruptly reverses the magnetization of the core material. If the core has been magnetized in the positive direction, for example, the removal of the magnetizing field reduces the flux density by only a small amount, from $+B_s$ to $+B_r$. Even a negative magnetizing current has little effect as long as the variation in H is less than the critical value, H_c. If the negative magnetizing current is made sufficiently large, however, so that the value of the field exceeds $-H_c$ (the switching point), the magnetization of the core is abruptly reversed from positive to full negative saturation. The flux density declines only slightly, to $-B_r$, when the negative magnetizing current and field are then removed.

Because of this square-loop effect, ferrite and similar magnetic cores are always in one of two stable states, either near positive saturation or near negative saturation. Such a magnetic core can store a binary digit similar to the flip-flop circuit. We simply identify one direction of magnetization—say positive—with the binary digit 1, and the other (negative) direction with the binary digit 0. To change from one state to the other we apply a reverse current. To do this we must know, of course, the state of magnetization of the core at any time. We shall see presently how this can be determined.

Practical Magnetic Cores

A closed flux path makes the most efficient magnet. For this reason, magnetic cores used in computers are either ring- or spool-shaped, as shown in Fig. 112. The core shown at (*a*) consists of a small plastic bobbin that is

(a) Tape—wound bobbin   (b) Moulded ferrite core

Fig. 112. Types of magnetic cores.

wrapped with several turns of very thin magnetic tape. The core at (*b*) is moulded directly into a tiny ring from magnetic material consisting of powdered ferrite and a chemical binder. The smaller the cores, the faster they can be switched from one state to the other. Cores used in magnetic memories, which must make their stored information readily accessible, are especially tiny, in the order of a few hundredths of an inch in diameter. Those used for logic are a little larger. To obtain even greater switching speeds, very thin magnetic films have been used (instead of cores) in the memory and logic circuits of more recent computers.

The Two Binary States. Fig. 113 illustrates the two possible states of magnetization for a moulded ferrite core. A few turns of wire (sometimes only a single loop) are wound around the ferrite ring to permit magnetizing the core. If a current pulse is applied to the winding at A, so that electron current flows from A to B, the core becomes magnetized in the positive direction, with the magnetic flux flowing anticlockwise, as shown. Since we have previously identified positive saturation with the binary digit 1, it is evident that the core has been set to 1 by the input pulse.

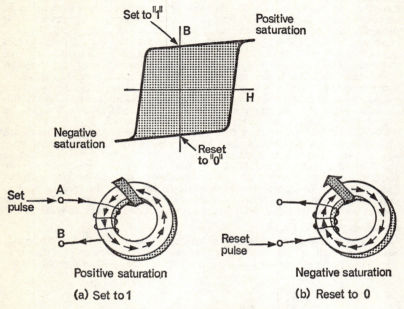

Fig. 113. The two binary states of a magnetic core.

To switch, or reset, the core to 0, either a pulse of opposite polarity must be applied to A or a pulse of the same polarity must be applied to the opposite end of the winding, at B. The latter is shown at the right in Fig. 113. With electron current now flowing from B to A, the core becomes saturated in the negative direction and the direction of flux is clockwise. As previously defined, this condition represents the binary digit 0, or zero state.

A simple rule determines the direction of magnetization: with the fingers of the left hand wrapped around the coil and core in the direction of electron flow, the thumb points in the direction of magnetic flux. (Anticlockwise flux = positive saturation; clockwise flux = negative saturation.)

Magnetic Core versus Flip-Flop

One advantage of the magnetic core over the flip-flop circuit, which performs the same basic function, is immediately apparent. Once the set or reset pulse has been applied, the magnetic core remains in an almost saturated condition for an indefinite period. It can thus store a binary digit without consuming any power. This is in contrast to the flip-flop circuit, which must be powered continuously to provide its memory; if the power is interrupted, even for a moment, the flip-flop immediately forgets the condition to which it was set. Magnetic core memories, besides being considerably cheaper than flip-flops, remain in the condition to which they were last set even over long power shutdowns. The magnetic core, however, has a serious disadvantage compared with the flip-flop. The flip-flop provides two complementary outputs, which can be used to provide an indication (such as a lamp) of its state. No such indication of the direction of magnetization in a magnetic core is possible. As we shall see, only a pulse obtained from a separate winding on the core, can 'sense' its state. Pulses must, therefore, be applied both for switching (setting and resetting) magnetic cores and for sensing their binary condition.

Determining the State of a Magnetic Core. A practical magnetic core usually has three windings, as is illustrated in Fig. 114. One or more 'write'

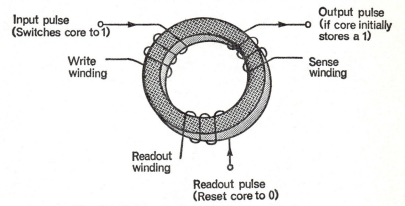

Fig. 114. Sensing the binary state of a magnetic core.

windings receive the input pulses that magnetize the core. A 'read-out' winding permits the application of a pulse to sample the condition of the core; a 'sense' winding produces an output pulse if the core initially stored a binary one. The method used to determine the state of the core is very simple:

1. A pulse is applied to the read-out winding in a direction that will reset (clear) the core to the zero state.
2. If the core is already in the zero state practically no change in flux takes place and there is no appreciable output from the sense winding. (A small change in flux density from the saturation value, B_s, to the remanent value, B_r, does take place, however, producing a tiny 'noise' voltage from the sense winding.)
3. If the core is initially in the one (1) state, the read-out pulse switches it to zero and the resulting large change in flux produces an output pulse from the sense winding. Thus, the presence of an output pulse indicates a stored 1, while its absence represents a 0.

Destructive Read-out. The only way we are able to sense the state of the core is by changing it. The read-out pulse always clears the core to zero, regardless of its previous state, and no 'memory' of the previous condition remains. Such a method is known as destructive read-out, and it is highly undesirable when the cores are used for data storage in the main computer memory. To circumvent this difficulty, the output data 'read' from the cores is temporarily stored and 'written' right back into the cores to restore them to their former condition.

There is also a method of reading out the data in a core that eliminates the sense winding. This is based upon the fact that a read-out pulse that switches the core makes the impedance of the read-out winding look very high, while a pulse that does not switch the core makes it look like a low impedance. The read-out winding thus acts somewhat like a diode or switch. The action may be used to develop an output pulse from the winding when the impedance is low (core in zero state), giving a true (one) output for a false (zero) input. This type of operation, hence, results in inverted logic.

MAGNETIC CORE LOGIC

The simplicity of the magnetic core makes it suitable for a variety of gating and logic circuits. The functions of these are the same as those described for diodes and transistors. A few examples will illustrate the ready implementation of magnetic core logic.

OR Gate. A three-input magnetic core OR gate is shown in Fig. 115. Three input (write) windings, A, B, and C, are placed on the core. An input current flowing into the dotted input terminal A or B or C (or any combination) will write a ONE into the core. A read-out pulse applied in the opposite direction

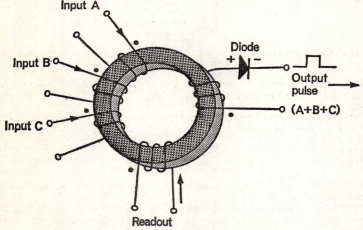

Fig. 115. Three-input magnetic core OR gate.

(into the undotted read-out terminal) restores the core to ZERO and results in an output pulse if the core was previously set to ONE. This output pulse makes the undotted output terminal positive. However, any input signal also results in an output pulse, which will make the dotted terminal positive; this pulse may be blocked by inserting a diode in series with the output terminals, as shown in the figure.

NOT Function (Inverter). We have mentioned that an inverted (negated) logic results if the signal output is taken from the read-out rather than from the sense winding. This, obviously, is one way of implementing the logical NOT function. Another way consists of applying the negating signal to a winding with a polarity opposite to that of the 'true' (one) input. If this inhibiting winding has a sufficient number of turns it is possible to block, or negate, any combination of input signals.

AND Gate. Since the windings on a magnetic core are essentially parallel, it is slightly more difficult to mechanize a series (AND) function. One way of implementing the AND function is to convert it to an OR function by means of De Morgan's law. You will recall that

$$A \cdot B = \overline{(\overline{A} + \overline{B})}$$

The right side of this equation can readily be implemented by inverting both inputs A and B, and then applying the inverted inputs ($\overline{A}$ and $\overline{B}$) to an inverting OR gate, as is illustrated in Fig. 116 in schematic form. Note that the diagonal

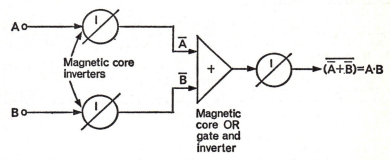

Fig. 116. Two-input magnetic core AND gate assembled from two inverters and an inverting OR gate.

lines through the cores in Fig. 116 indicate inversion of the input, or the logical NOT. The two-input AND gate requires three inverting magnetic cores.

Shift Register. A shift register can easily be made up from magnetic cores, as Fig. 117 shows. Note the similarity to the flip-flop register of Fig. 108. The

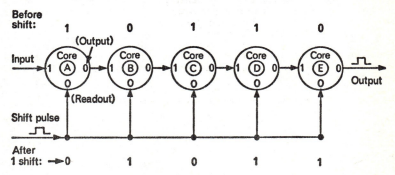

Fig. 117. Magnetic core shift register containing binary number 10110.

cores are shown schematically with the input (write) side identified as 1, the output side as 0, and the reset-to-zero shift pulse applied to the read-out winding (labelled 0). For simplicity, we have omitted the necessary delay elements.

Assume that the binary number 1 0 1 1 0 (decimal 22) is originally stored by the five magnetic cores in the shift register. Each time a shift pulse is applied simultaneously to the core read-out windings the digits are shifted one position to the right and the adjacent element at the left is cleared to 0 for the insertion of additional data. You can readily see why this is so. The application of the first shift pulse initially resets all cores to 0. Core A, which contained a 1, is reset to 0 and applies an output pulse to core B, which sets it to 1. Core B is now at 1, but since it was originally at 0, it does not provide an output pulse. Core C, which stored a 1, is reset to 0 by the shift pulse and applies an output pulse to core D. Core D, originally at 1, is at first reset to 0 by the shift pulse, but the output pulse from C immediately sets it back to 1. While being reset, however, core D applies an output pulse to core E. Core E, which was at 0, is not affected by the shift pulse, but is set to 1 by the output pulse from core D. Recapitulating, the line-up from left to right is now 0 1 0 1 1; that is, the digits have been moved one place to the right. Simultaneously, the last digit at the right (0) of the original number has been dropped, while core A at the left has been cleared to 0 for the insertion of a new digit.

A second shift pulse again shifts all the digits one place to the right and clears core B, resulting in the binary number 0 0 1 0 1. Two new input bits may be inserted into cores A and B at this time. The application of three additional shift pulses results in shifting out the remaining digits to the right and, at the same time, clearing all cores to 0 0 0 0 0 for the insertion of a new five-bit number. In practice, the insertion of a new bit through an input pulse applied at the left takes place each time a shift pulse clears out one of the digits at the right, so that new data is continuously being inserted as old data is read out at the right.

Magnetic Storage (Memory)

To perform its function, a digital computer must be able to store large masses of data consisting of input information, partial results of computations, and final results at the computer output. A large computer uses a variety of internal and external 'memories' for storing this information until it can be used. Magnetic storage has proven very popular because magnetic elements are relatively inexpensive, can be switched rapidly, and require little power. (Incidentally, the term 'store' is preferred to the subjective-sounding term 'memory') Three types of magnetic store are in extensive use, each with its own advantages and limitations. These are magnetic cores, magnetic drums, and magnetic tapes. Cores are ideal for high-speed internal (main) storage, but their capacity—in terms of the number of bits stored—is somewhat limited. Magnetic drums, though not nearly as fast as cores, are preferred in small and intermediate-scale computers because of their large storage capacity. Finally, magnetic tape, which is based on the familiar principle of the tape recorder, is used for the permanent storage of large amounts of input and output data. There are also non-magnetic types of data store, such as electrostatic storage tubes and delay lines, which we shall consider later in the chapter.

Magnetic Core Storage. Static (non-moving) storage, rapid access time, and the fact that information is preserved when power is turned off, are three powerful points in favour of magnetic core storage. Since each core can store only one bit, however, many thousands of cores are needed to store a significant amount of information. Cores are arranged in a rectangular matrix of rows and columns, as shown for 16 elements in a '4 × 4' matrix in Fig. 118. To accommodate the large number of elements, the two dimensional planes (matrices) are further stacked side by side (vertically or horizontally), so that a three-dimensional cubic array results. It must be possible, of course, to 'write in' a bit of data into any selected core and 'read out' this bit from the core at any later time. You can imagine the staggering wiring and switching problem that would result if each core were equipped with complete write, read-out, and sense (output) windings and associated wiring, such as shown in Fig. 114. Fortunately, separate core writing is rarely necessary; considerable simplification has been achieved by the coincident-current method described below.

Writing In with Coincident Currents. The wiring of individual cores was initially simplified by the observation that it is not necessary to provide windings of many turns, but that a single wire passing through the core (entering at the bottom and leaving at the top) has a sufficient magnetizing effect, if the current pulse is strong enough. Thus only three separate wires (write, read, and sense) would be required for each core. However, as you can see in

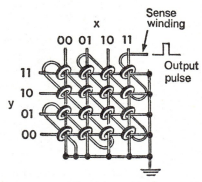

Fig. 118. Two-dimensional magnetic core '4 × 4' matrix that can store sixteen bits of information.

Fig. 118, all the cores are strung on a few continuous wires. Only one wire is used for each row of the matrix, one for each column, and one wire—threaded diagonally through all the cores—acts as the 'sense' (output) winding. This vast simplification, over the 48 wires that 16 cores would normally require, is achieved by the principle of coincident currents.

You will recall that a fully magnetized core can be switched to its opposite state only by a change in the field intensity of more than the critical switching value (H_c) (see Fig. 111). Anything less than that will cause an insignificant change in the flux density from its saturation value (B_s) to the slightly lower remanent value (B_r). This is the key to core selection. Each core has two 'write' (input) wires associated with it, one for the column (x) and one for the row (y) of the matrix in which it is located. The currents passing through these wires are only half of the value required to switch a core and are, accordingly, known as half-select currents. As long as only one of the wires passing through a core carries a current pulse, the core will not be affected by it, except for an inconsequential change in flux density. Thus, one half-select current pulse applied to any row or column does not affect any core in that row or column. However, whenever both (row and column) wires passing through a core carry half-select current pulses simultaneously the coincident pulses double the change in field intensity and the core is switched to its opposite state. Whether the core is switched to one or zero depends, of course, on the direction of the coincident current pulses.

You can see from Fig. 118 that only the core at the intersection of a selected row and column can be switched by coincident pulses. Suppose you wanted to store a one in the core that is at the intersection of the column, $x = 01$, and the row, $y = 10$. You would simply apply a half-select current pulse of the proper polarity to the column $x = 01$ and one of the same polarity to the row $y = 10$. The flux density of all the cores in this row and column would be momentarily slightly changed, but would return to its original state as soon as the pulse had passed. The core at the intersection of the selected row and column ($x = 01$, $y = 10$), however, would receive coincident half-select current pulses and, hence, have its state changed to one (if at zero).

Reading Out. The binary state of any core is sensed (read out) in the manner we have previously described. The selected core is first cleared to zero by application of half-select read-out pulses of the proper polarity to the appropriate row and column. If the core had stored a one it will be switched to zero, and the resulting large change of flux generates an output pulse, which is picked up by the diagonal sense (output) wire passing through the core. The output pulse is strengthened by a sense amplifier and then applied to a register for later use. If the core had been at zero, however, the clearing pulse will cause little change in flux and the tiny resultant 'noise' output voltage will not be registered. The diagonal output winding itself cancels most of these small unwanted noise outputs. Note that the sense line passes through all the cores, so that an output pulse from any core is fed to the common output terminal. The read-out process is destructive, since the affected core is cleared to zero and no longer stores the original information. To preserve the stored information the output data is temporarily stored by some device (a delay line, for example) and then immediately rewritten into the core matrix. A read-out operation thus is automatically followed by a writing operation that restores the original core information.

Writing in a Zero. A writing cycle starts with all cores cleared to zero by the previous read-out. Writing a one into any core is easily accomplished by coincidence selection of the proper row and column, as we have seen. You would think that writing a zero could be accomplished in an easier manner, simply by not writing in a one. For practical reasons concerning wiring and switching, this is not done, however, a positive action being required to write in a zero. The core in which zero is to be stored is chosen by half-select coincident current pulses from the appropriate row and column in the same manner as if a one were to be stored. Then an additional half-select pulse is sent in an opposing direction through an 'inhibit' wire (also known as F-winding), which cancels out one of the half-select write pulses and thus prevents writing a one. This 'inhibit' wire is threaded through every row of the matrix in opposition to the selecting current. No current is sent through the inhibit wire when a one is to be written into a core.

Magnetic Drums

Magnetic drums and magnetic tapes utilize the principle of magnetic recording, which has become familiar through the tape recorder. Instead of recording speech or music, however, magnetic drums record and read out binary digits, or pulses. Like a core, small magnetic particles or groups of particles can be magnetized to saturation in one direction or the other by a magnetic field. The field can be produced by a nearby coil in a magnetic head (Fig. 119). This is simply an electromagnet with a tiny (thousandth of an inch) air gap between its pole pieces. The head is brought very near to the magnetic particles contained on a revolving drum. Whenever a current is applied to the coil of the magnetic head, the leakage flux fringing out between the pole pieces magnetizes the particles on the drum immediately beneath the head (see

Fig. 119). Thus, a series of magnetically polarized spots are produced on the drum, in accordance with pulses applied to the coil of the magnetic head. A spot polarized in one direction is assigned the binary value 1; one polarized in the opposite direction, a 0.

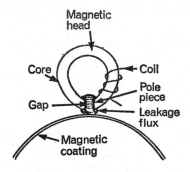

Fig. 119. Magnetic recording head shown magnetizing coating on drum through leakage flux.

Fig. 120 illustrates a typical magnetic drum with four heads for simultaneous recording of four separate information channels. Depending on its storage capacity, the drum may be up to 2 ft. long and, perhaps, a foot in diameter. Each channel can store a certain number of bits around the periphery of the

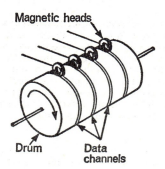

Fig. 120. A four-channel magnetic drum.

drum, and these repeat endlessly as the drum rotates. The faster the drum rotates, the more rapidly information is stored or read out. Practical drums may rotate at speeds from 3600 rev/min up to 10,000 rev/min. For a drum rotating at 3600 rev/min, for example, each binary bit is positioned under the magnetic head 3600 times every minute, or 60 times each second. The bit can, therefore, be stored or read out in $\frac{1}{60}$ second.

Writing and Reading. The same magnetic head is usually used for writing and reading as well as erasing binary information. For reading out (sensing), the magnetic head is used as an output device. As the magnetized spot with its stored digit moves past the head, the tiny magnetic field of the spot induces a small voltage in the coil of the head, whose direction indicates whether a 1 or a 0 has been stored. Note that this is non-destructive read-out, since the information remains on the drum. Occasionally two heads are positioned on the same channel, displaced by a short distance on the drum's periphery. By using one head for writing and the other for reading, the access time for part of the information can be considerably shortened, since the magnetized spot need only move through the short distance between the two heads. This scheme is also utilized to produce short time delays.

The same magnetic head can also be used to erase the information on the drum by the application of a strong demagnetizing field, which wipes the drum clean.

Magnetic Tapes

Paper or plastic tape coated with a magnetic material may be magnetized with binary information by the same type of magnetic head and in the same manner as that described for the magnetic drum. In tape recorders tape is transferred from one reel to another by means of a tape transport mechanism, while moving past a stationary magnetic head. A single head is generally employed for writing, reading, and erasing the information, and up to 10 channels may be arranged side by side across an inch-wide tape. More than 200 bits per inch of length may be stored in the form of polarized spots, so that a tape moving at about 15 ft. per second can store tens of thousands of bits each second. The big disadvantage of tapes is, of course, their slow access time. Anyone who has handled a tape machine knows the difficult and time-consuming process of finding a desired spot on the tape. For this reason, tapes are used only for the storage of input and output data, but not inside the computer itself.

Electrostatic Storage

In Part I on analogue computers we discussed the cathode-ray tube (see Fig. 60), in which an electron gun shoots out a beam of electrons which makes a bright spot on the screen of the tube. Recall that the electron spot on the screen may be deflected in any direction to the left or right and up or down by appropriate voltages placed on the horizontal and vertical deflection plates, respectively. A slightly modified tube can be used for electrostatic storage of binary data in a digital computer. The bright spot on the screen of a cathode-ray tube is actually an accumulation of electrons, which are negative electric charges. By making the phosphor coating on the face of the tube persistent, the charges accumulated on a spot may be held for a few minutes, thus providing a short-term memory. The presence of a negatively charged spot on the screen can represent a binary 1; its absence, a 0. Depending on the size of the tube, a thousand binary spots or more can be stored on the screen.

To write in the information to be stored on the screen of the tube, the electron beam is swept in a regular pattern from left to right and from top to bottom so that it covers all parts of the screen. At the same time the beam is pulsed (turned on and off) in accordance with the binary data, so that a charge impinges on a spot on the screen where a 1 is to be written and no charge (beam turned off) is placed on those spots which are to store a 0. Since the charges have a tendency to wander away and leak off, they must be frequently regenerated, or rewritten, by a holding beam that sprays the pattern of electrons on the screen before they are lost.

To read the charges stored on the screen of a cathode-ray tube, the same technique is used as reading out from magnetic cores. To read a bit, the beam attempts to write a specific charge into that bit position. If the position is already charged to a 1 no further charge is accepted and the excess charge is diverted to the output, thus registering a 1. If the position is uncharged (0), however, the additional charge of the beam is accepted, and no change is registered in the output, signifying a 0.

Williams Tubes. The Williams tube memory system uses a cathode-ray tube to store binary data in the form of dots and dashes. The electron beam in the Williams tube is given sufficient energy so that it knocks off more electrons from the screen than it supplies. As a result, the spot on the screen has a deficiency of electrons, or becomes positively charged, at the instant of writing. Each memory area consists of a dot and a dash. When the dot is written it becomes positive by the emission of electrons. As the beam moves sideways to form the dash, some of the knocked-off electrons fall upon a portion of the dash and make it less positive (or negative). Thus, a positive dot is followed by a dash whose first half is negative and whose second half is positive. The location of the higher positive charge determines whether a 1 or a 0 is stored.

To read the data, the beam is focused on a memory area to clear it to 0; if there is a change in state it is sensed and appears as a 1 in the output. If there is no change a 0 had been stored.

DELAY-LINE STORAGE

We have already become acquainted with devices that can temporarily store and delay information. The one-shot multivibrator, for example, can store and delay a single pulse by the period of its operation, as you will recall. Delay lines are an extension of this principle; they can store the binary information contained in an entire train of pulses and make it available a short time later. The ordinary echo is an example of the delay principle, since it delays a spoken word by a definite interval.

There are two main types of delay lines, electrical and acoustic. An electrical delay line is an assembly of coils and capacitors which simulates a long transmission line. As pulses travel along the line they are slightly delayed—by a few millionths of a second. Because of the extremely rapid propagation of electrical pulses, an electrical delay line must be quite long (in simulated length) to store just a few pulses, which makes it somewhat uneconomical. Acoustic delay lines, in contrast, make use of the relatively slow speed of sound waves to store a great many pulses (a thousand or more) in a comparatively short length of line. Since the computer pulses are electrical, however, acoustic delay lines must convert the pulses first into mechanical (sound) vibrations at the input end and then reconvert them into electrical impulses at the output end of the line. There are three major types of acoustic delay lines that can do this job. These are mercury delay lines, quartz delay lines, and magnetostrictive delay lines.

Mercury Delay Lines. As shown in Fig. 121, a mercury delay line is essentially a tube filled with mercury, with a piezoelectric crystal at each end. Piezoelectric crystals, such as quartz and tourmaline, have the ability to expand and contract mechanically when a changing electrical voltage is placed across the crystal faces, and conversely, generate a pulsating electrical voltage when subjected to mechanical vibrations. Thus, when a pulse train representing the binary data is applied to the transmitting crystal at one end of the mercury tube it is transformed into corresponding mechanical pressure waves, similar to the action of a loudspeaker. These mechanical waves travel through the mercury until they impinge upon the receiving crystal at the other end of

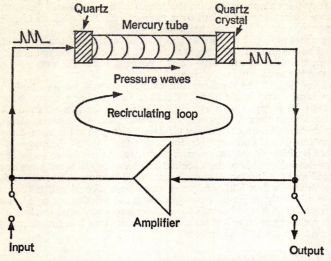

Fig. 121. Mercury delay line that recirculates a train of waves.

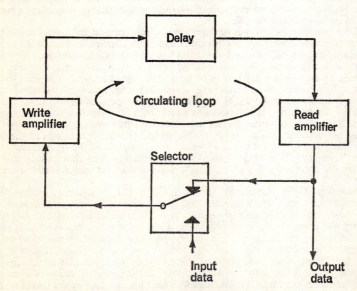

Fig. 122. Block diagram of regenerative delay.

the tube. The action of this piezoelectric crystal is similar to that of a microphone and results in translating the mechanical vibrations back into the original electrical pulse train. The considerably weakened output pulses from the receiving crystal are then strengthened by an amplifier and, if desired, may be reinserted at the input of the line so that the pulse train recirculates indefinitely around the loop. To avoid 'blurring' of the pulses, the temperature of the mercury delay lines (which affects the sound velocity) must be maintained precisely. Sometimes a number of delay lines storing information are contained in a single temperature-compensated mercury tank.

Regenerative Delay. The reinsertion of data into the mercury delay lines illustrates the important principle of regenerative delay (Fig. 122). A delay line can delay binary information by a few hundred microseconds at most. If the pulses cannot yet be utilized at the end of this interval, they must be strengthened, restored to their original shapes, and reinserted at the beginning of the line. The use of such a regenerative delay loop permits recirculating an entire pulse train indefinitely, until either the data is read at the output and a new pulse pattern is inserted, or the power ceases. (A power stoppage results, of course, in the complete loss of information.) With the selector switch in the position shown in the block diagram (Fig. 122), the pulse information is recirculated continuously, but can be read into a register (or tape) any time it appears at the output. If new information is to be stored, the recycling loop is broken and the new input data is sent to the write amplifier for storage in the delay line. The regenerative delay principle can be applied to any delay device in which stored information travels physically from input to output.

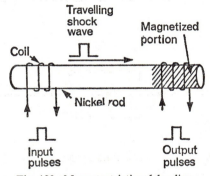

Fig. 123. Magnetostrictive delay line.

Quartz Delay Lines. Instead of mercury, an acoustic delay line can be made of a solid, such as quartz or magnesium. In a quartz tube the electrical impulses fed in at one end are converted by the piezoelectric material into mechanical vibrations, as was described for the mercury delay line. The vibrations are transmitted along the length of the line and reconverted into the original electrical pulses at the far end of the line. Commercial quartz lines with delays of 100 microseconds and more are available. For longer delays a quartz polygon is used instead of a tube. The wave pattern is reflected from face to face of the polygon, following a long folded path from input to output, thus making the assembly more compact than a long tube.

Magnetostrictive Delay Line. Some materials, such as nickel and nickel-iron alloys, deform lengthwise when subjected to a magnetic field, a phenomenon known as *magnetostriction*. This effect has been exploited for obtaining acoustic delays. As shown in Fig. 123, coils are wrapped around the ends of a bar of nickel or other magnetostrictive material. A pulse through the input coil creates a magnetic field, which develops magnetostrictive stresses in the bar. The stress pattern is propagated through the bar as a shock wave. When the wave reaches the magnetized output end of the bar it sets up corresponding variations in the magnetic field, which are picked up by the output coil. The current flowing through this coil consists of the original, though weakened, electrical input pulses. Short delays can be secured in this manner.

CRYOTRONS

An unusual bistable device has been developed for high-speed logical gating and memory functions. This is the cryotron, whose operation is based upon the principle of superconduction. Elementary electricity teaches that the resistance of all conductors decreases gradually with decreasing temperatures. In materials known as superconductors the resistance decreases with lower temperatures until a critical temperature, T_c, within a few degrees of absolute zero is reached, at which the resistance drops to zero. This is shown by the solid curve in Fig. 124. If the superconductor is subjected to a magnetic field

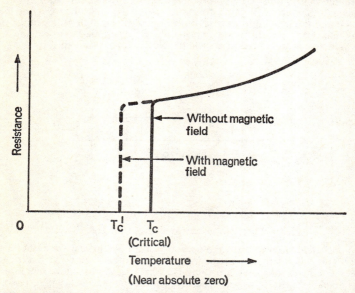

Fig. 124. Resistance of a superconductor as a function of temperature in the presence and absence of a magnetic field.

the transition to zero resistance occurs at a lower than critical temperature, as shown by the dotted curve in Fig. 124. This is the effect which is exploited in the cryotron. By applying a relatively small magnetic field to the superconductor at a temperature barely below the critical value (T_c), the superconductor is forced back into its normal resistive state. When the field is removed, zero resistance recurs. In practice, then, it is only necessary to apply a small control current through a coil wrapped around the superconductor to establish the

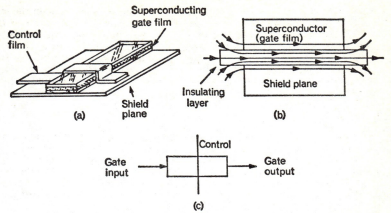

Fig. 125. (a) Thin-film cryotron; (b) its magnetic field and (c) logic symbol.

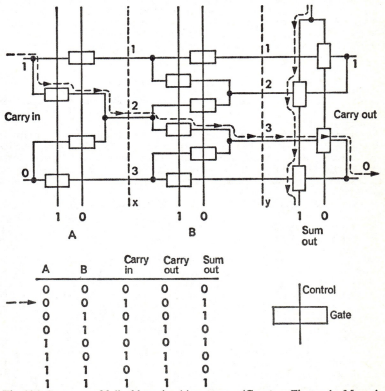

Fig. 126. One stage of full adder using 14 cryotrons. (Courtesy *Electronics Magazine*.)

magnetic field for normal resistance. Turning the current on and off gives complete bistable control over the resistance of the material.

Thin-film Cryotrons. The cryotron acts as an amplifying switching device, or gate. A large current can be sent through the cryotron in its superconducting state. Applying a small current to the control winding of the device causes it to switch to its normal resistive state, thus substantially decreasing the current passing through the cryotron gate. The ratio of the gate current in the superconducting state to the current required to control it is known as the current gain of the cryotron. Ratios of 2 : 1 or 3 : 1 are easily obtained.

To attain very high switching speeds—in the order of a tenth of a microsecond—practical cryotrons are constructed of thin films. The superconducting gate film that carries the gate current is positioned at right angles to a narrow control film, which replaces the wire-wound coil. As Fig. 125 (*a*) shows, the device is mounted on top of a superconducting shield plane but is separated from it by a thin insulating layer. The shield plane drastically reduces the inductance compared to that of a wire-wound (coil) cryotron, making possible the rapid switching speeds. The magnetic field configuration of the thin-film cryotron is shown in Fig. 125 (*b*).

Applications. The fact that cryotron devices must be refrigerated close to absolute zero has so far prevented their extensive commercial use. The bistable nature of the cryotron allows its application in logical circuits, counters, shift registers, etc. As an example, Fig. 126 shows one stage of a full binary adder that uses 14 cryotrons. The intermittent lines show the current path for a forward carry (carry in) of 1, $A = 0$, and $B = 0$, the sum being a 1.

REVIEW AND SUMMARY

Magnets made of hard steel, ferrites, and other ferromagnetic materials are permanent; soft iron can be magnetized only temporarily under the influence of a magnetic field.

Magnetism is caused by uncompensated electron spins in the atoms of ferromagnetic materials. The spins in a domain of about 10^{15} atoms have the same direction and produce intense, but random-oriented, magnetic fields. Magnetization is produced by alignment of the domains under the influence of an external magnetic field. Magnetic saturation occurs when all the domains are aligned.

Lines of force are used to represent the invisible magnetic field and to calculate its effects. A single line of force (1 maxwell) is the unit of magnetic flux. The total number of lines of force, issuing from the north pole of a magnet and terminating on its south pole, is a measure of the total magnetic flux.

Flux density (B) is the flux traversing a unit area, measured in gauss (maxwells/cm^2). Total flux is the product of flux density and area ($B \times A$). Permeability is the ratio of flux density to field intensity ($\mu = B/H$); it is a measure of the relative ease of magnetization.

The lagging of the flux density (B) induced in a magnetic material behind the magnetizing force (field intensity H) signifies an energy loss, called hysteresis. The magnetism (flux density) remaining after the field intensity has been reduced to zero is known as remanent flux density (B_s) or residual magnetism; the coercive force is the negative field intensity required to demagnetize a material completely. The hysteresis loop (B–H curve) shows the dependence of the flux density on the field intensity (magnetizing force) for a complete cycle of magnetization.

A square hysteresis loop, produced by some core materials such as powdered ferrite, indicates that the magnetism retained (remanent flux density B_r) by the material almost equals the saturation flux density (B_s) for either positive or negative magnetization. A square-loop characteristic can be used for switching the core material between two stable states, either positive or negative saturation. A change in field intensity greater than a critical switching value (H_c) abruptly reverses the magnetization of the core from one state to the other. A square-loop magnetic core, thus, is a bistable device that can store a binary digit (0 or 1). Positive saturation usually is identified with binary 1, negative saturation with binary 0.

Practical magnetic cores are either tape-wound bobbins or moulded ferrite toroids (Fig. 112) provided with write (input), read-out, and sense (output) windings. The core is magnetized to saturation (set to one) by applying an input pulse to the write winding. The direction of magnetization is indicated by the thumb of the left hand when the fingers are wrapped around the coil in the direction of electron flow. (Positive saturation = anticlockwise flux; negative saturation = clockwise flux.) Applying a pulse to the read-out winding resets the core to zero. If the core was initially set to one the resulting large change in flux produces an output pulse from the sense winding; if the core was initially at zero no output pulse results. This is known as destructive read-out; to preserve the information during read-out, it must be temporarily stored and then rewritten.

Magnetic core logic circuits can be assembled by providing cores with several input (OR) windings and applying inputs of opposing polarity to inhibit (negate) other inputs. AND gates can be made up from inverters and OR gates. (See Figs. 115 and 116).

Magnetic memory storage can be effected by magnetic cores, magnetic drums, and magnetic tapes. Magnetic cores have high speed, rapid access, and limited storage capacity; they are used in internal (main) memories. Magnetic drums have intermediate speed and access, but have high storage capacity; hence, they are frequently used in small and medium-sized computers. Magnetic tape has low speed and access time, but unlimited storage capacity; it is used for input/output memories.

Magnetic-core memories are assembled in three-dimensional arrays consisting of planar frames in which the magnetic cores are positioned in a regular matrix of rows and columns. Each core stores one bit (0 or 1). A core is selected for write-in or read-out by sending half-select (half-strength) current pulses through the row and column at whose intersection the core is positioned; the coincident current pulses cause the core to switch to its opposite state. During read-out the selected core is cleared to zero, causing an output pulse to appear at the sense winding if the core originally stored a binary one. Read-out is automatically followed by a writing cycle, to preserve the data.

Magnetic drums and tapes make use of magnetic heads (electromagnets) to produce magnetically polarized spots on a moving magnetic coating beneath the head (see Figs. 119 and 120).

In electrostatic storage a cathode-ray tube with a persistent screen is used for temporary storage of electronic charges sprayed on the screen by an electron beam; this is known as a Williams tube.

Electrical and acoustic delay lines can temporarily store and delay a train of pulses containing binary data. Acoustic delay lines can be liquid (mercury) or solid (quartz) pipes or polygons, or magnetostrictive rods; the electrical input pulses are converted to mechanical vibrations inside the line and then reconverted to electrical impulses at the output.

Regenerative delay consists of reshaping and amplifying the delayed output pulses and then reinserting them into the input of the delay device, so that the stored information recirculates until new data is stored or the power ceases.

A cryotron is a bistable device that makes use of the zero-resistance characteristics of superconductors at extremely low temperatures (near absolute zero). In the presence of a magnetic field superconduction takes place at slightly lower temperatures; this effect is used to gate the current through the superconductor by means of a small control current that establishes a magnetic field, thus forcing the superconductor back to its normal resistance. Turning the control current off re-establishes superconduction and zero resistance.

CHAPTER 12

THE COMPLETE DIGITAL COMPUTER—I: OPERATION OF COMPUTER MEMORY AND ARITHMETIC UNIT

A computer is more than the sum of its parts. The building blocks we have studied in earlier chapters can be and are used in many devices and systems not related to computers. In this chapter we shall discover how the electronic, magnetic, and logical building blocks are integrated into a functioning computer system. This involves additional concepts and a strange-sounding computer jargon that applies only to the system as a whole and has little meaning apart from it. Fortunately, the concepts and definitions we must learn can be anchored to a basic computer framework, whose structure and logical design is duplicated essentially by nearly every automatic digital computer. After studying this framework and its interrelations briefly, we shall consider its parts in greater detail and learn how the various building blocks perform their functions within the system.

SYSTEM OPERATION

Present-day digital computers, or electronic data-processing systems (EDP, the term frequently used in the world of business), have essentially five functional elements, illustrated in the system block diagram of an automatic digital computer (see Fig. 127). These elements are: (1) the input devices; (2) the storage or memory unit; (3) the arithmetic unit; (4) the control unit; and (5) the output equipment. Those input and output devices that are not part of the computer proper but are used for auxiliary functions are occasionally lumped together under the term 'peripheral equipment'.

Data. As shown in the block diagram, the actual processing of the data is carried out by the arithmetic and memory units under the supervision of the control unit, while communication with the outside world takes place through the input and output devices. The information to be processed is usually presented to the computer in terms of binary digits (bits). Remember that a binary one may be represented by a pulse of a certain magnitude, a high voltage level, or by a specific direction of magnetization of a particle (dipole) on a magnetic coating; the binary zero can be represented by the absence of that pulse, a low voltage level, or the opposite direction of magnetization. The units of information in computers are groups of binary digits called words. Each word contains a fixed number of bits (more than 50 in some computers), which may represent numerical information, algebraic signs (+ or −), accuracy checks, and alphabetic codes.

There are two basic types of computer words: data words, containing numerical information to be processed, and order or instruction words, which prescribe the manner in which the data is to be processed. The binary data and instructions are carried throughout the computer by the information bus, which may have either one or several channels along which information flows either sequentially (serially) or simultaneously (parallel).

Input. Information from the outside world originates from a number of media and can be conveyed to the input section of the computer in various forms. Data may originate from a notebook, business file, telephone network, radio transmitter, teletypewriter, or perhaps a radar set. This information must first be converted into binary machine language acceptable to the input unit. One common method is to type the data on a keyboard that prepares coded punched paper tape or punched cards. For high-speed input processing the information is encoded and then transferred to magnetic tape handlers, which can process more than 50,000 characters per second. In addition to the data to be processed, a program of operating instructions is prepared that

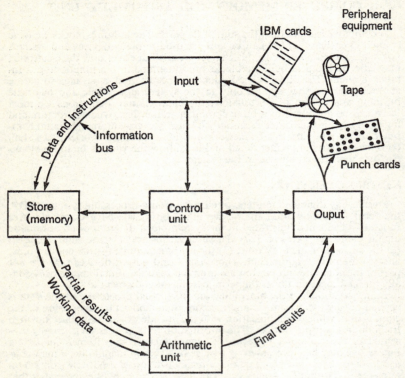

Fig. 127. Functional block diagram of an automatic digital computer.

prescribes the manner in which operations are to be carried out. Both data and instructions are transferred to the computer input section, which selects the information in the order needed and feeds the resulting selection into the internal store (memory).

Store or Memory. The internal store, or memory, unit consists of a large number of locations (magnetic core arrays) where data and instructions are stored. Each location has a unique address where the information can be found. Instructions and data are written into or read out from internal storage without discriminating between them. The only way instructions can be distinguished from working data is by the routing of pulses from internal storage. Pulses routed to the control unit are interpreted as instructions, while those routed to the arithmetic unit are considered data. If data are sent, by mistake, to the control unit instead of the arithmetic unit they are treated as instructions, and since this will undoubtedly result in confusion, an error alarm is activated and the computer is halted. The order in which data are placed into or taken out of the store is governed by the instructions contained in the program.

Arithmetic Unit. The arithmetic unit is the electronic calculator of the computer. All calculations are broken down into the simplest arithmetical processes, preferably addition or looking up a table. You will recall that subtraction can be carried out by adding complements, multiplication by repeated addition and shifts, and division by repeated subtractions and shifts. The arithmetic units also make various logical decisions, such as choosing between two alternatives, the familiar AND, OR, NOT, and other operations we have studied. If the decisions are in terms of numbers this is simply an extension of addition and subtraction. The arithmetic unit must distinguish between positive and negative numbers and modify its behaviour accordingly. For example, in subtracting one number from another, the remainder may come out zero, positive, or negative. Depending upon the remainder, different alternatives exist for the next instruction to be carried out, and the computer must choose the right one. If the data is in the form of letters (alphanumeric), a logical comparison may be required to determine which word is earlier or later in a collating sequence. Partial results are sent back to the store to be called for again at a later stage of the calculations. The final results of processing may be passed directly to the output; more usually they are sent to the store (memory), from which they are fed to the output section in accordance with the program requirements.

Control Unit. The control unit is the master dispatching station and clock of the computer. It directs the rhythmic flow of data through the system and controls the sequence of operations. To do this, it must interpret the coded instructions contained in the program and initiate the appropriate commands to the various computer sections. These successive commands, in the form of control signals, flow along control lines, opening or closing switches (gates) that connect specific registers to the information bus, and thus control the flow of data in accordance with the program. In general, this flow consists of a continuous, rapid transfer of information from sending registers to receiving registers and to the output via the information bus. The control unit automatically times and collates all these activities and ascertains that the computer behaves as a fully integrated system.

Output. The output section is similar to the input, except that the processed information must now be retrieved from memory and reconverted into a form suitable for the output devices. Upon receiving the appropriate commands from the control, the output section transfers the processed data bits from storage in a logical sequence and arranged into characters, for insertion into the selected output device. As with the input, the data may first be transcribed on to magnetic tape for high-speed operation. The tape may, in turn, feed a high-speed printer or a paper or card punch. Modern high-speed printers can print out approximately 1000 lines per minute, each 120 characters long. Thus, more than 100,000 characters can be transferred out of the computer each minute. The printed-out final data, again, can be transmitted to a remote location by radio, telephone, or teletype.

Timing. The flow of information is sped through a computer in a systematic manner by timing all operations in terms of a basic cycle. This cycle consists of the time elapsed between the issuing of a command and the complete execution of this command by the computer. During one cycle the command will be carried out through a series of pulses, or sharp changes of the voltage level, occurring in fractions of a microsecond. The fastest rate at which pulses are transmitted and used throughout the computer is known as the repetition rate. In fixed-cycle or synchronous computers a series of equally spaced pulses from a master clock oscillator starts and controls all operations. In variable-cycle, or asynchronous, computers, the completion of any operation releases a signal to start the next operation. The flow of information through the machine may be either sequentially through one information channel at a time—called serial operation—or it may occur simultaneously through several channels; this is known as a parallel operation. Combinations of these two modes are frequently employed in the same computer. Computers of more recent design tend to utilize almost 'pure' parallel operation to speed the flow of data.

We are now ready to consider the major units of a digital computer in greater detail. We shall proceed from the more familiar elements—the store (memory) and arithmetic units—to the less familiar ones, the control unit and the input–output equipment, which will be covered in the next chapter.

THE STORAGE OF INFORMATION IN THE COMPUTER MEMORY

Before the computer can solve any problem, all the information concerned with the problem must be loaded into the memory. The units of information going into the memory are data words and instruction words, containing respectively, numbers and instructions about operations to be performed. The function of the main, or internal, memory is to receive thousands of these complete 'words' and store them until they are required elsewhere. When the information is needed it is taken out of the memory, word by word, and sent to various special registers for instructions and use in computations. The results of these computations are transferred back to the memory and used again later for additional computations, if needed. The final results are passed from the memory to the output of the computer.

Types of Store

In earlier chapters the functioning of the various types of storage devices were described in detail. It may be helpful at this point to review the main categories and their chief application. In the order of increasing storage capacity, we can distinguish the following types:

1. *Storage of a Single Digit.* All bistable devices can store at least one binary digit, or bit. Among devices we have considered earlier are flip-flop circuits (multivibrators), relays, magnetic cores, and superconducting cryotrons. Also included in this category are devices for temporary storage or delay, such as the one-shot (monostable) multivibrator and delay lines.

2. *Storage of Several Digits* (*One Word*). Devices that can hold a character or a complete word, consisting of data or an instruction, are called registers. Registers can be made up of a number of flip-flops, magnetic cores, or other bistable elements. A dynamic register, in which information moves continually from element to element, can be made up either from the same elements or may make use of a delay line.

3. *Main Internal Store.* The internal 'memory' of the computer, with which we are primarily concerned in this section, stores many thousands of complete words, consisting possibly of a million bits or more. The preferred devices for internal stores are magnetic-core arrays and magnetic drums, which we studied in the last chapter. Electrostatic storage, in the form of Williams tubes and similar devices, is also used in some computers.

4. *Buffer Store.* Intermediate in size between the main internal memory and registers, the buffer or back-up memories are designed to compensate for differences in speed in various parts of the computer; data transferred to buffer

Operation of Computer Memory and Arithmetic Unit 237

storage from a high-speed portion of the computer are stored temporarily until they can be used by a lower-speed portion, such as a print-out device. Any of the bistable elements used for registers and internal storage can also be employed for buffer, or back-up, stores.

5. *External Storage.* The computer program, 'subroutines', and other input and output information are held in external stores as in a filing cabinet. This type of store usually has a capacity of millions of words and is represented by magnetic tape reels, punched paper tape, or punched cards. We shall describe external storage devices in the input–output section in the next chapter.

Attributes of Storage Systems

Regardless of type, all storage systems have certain common attributes, which are reviewed briefly below:

1. *Memory Locations.* Any store must have elementary locations, each of which must be capable of storing at least one digit. These locations are frequently referred to as memory cells; they may be flip-flops, magnetic cores, or any of the bistable devices we have studied.

2. *Registers.* Memory cells are combined into small groups known as registers, each capable of storing one computer word. The word usually consists of a fixed number of digits for any one computer, hence, the number of memory cells, or elements, in any register is also fixed.

3. *Address.* Each register in the memory must have one or more addresses by which the word can be identified. Therefore each word not only contains binary data or instructions but also at least one address indicating where the word may be found in the memory. However, words frequently have more than one and possibly up to four addresses. The first address may specify the location of an operation to be carried out, the second where to find the next instruction, and so on.

4. *Writing In (Storage).* Any store must provide a method for writing in, or recording, information at a desired memory location. The method may be electronic (flip-flop), magnetic (cores, drums, and tape), electrostatic (Williams tubes), or other (cryotrons, etc.).

5. *Reading Out.* A method must be provided for reading out the information from any location in the store. The method may be non-destructive (preserving the information), such as magnetic and punched tapes, or it may destroy the information (destructive read-out), such as occurs with coincident–current magnetic core memories; during destructive read-out the information must immediately be rewritten into the identical memory location.

6. *Permanence.* In a permanent memory, such as magnetic drums and tapes, the information is not lost when the power is turned off. Permanent memories may be erasable, as in magnetic storage devices, or they may be non-erasable, as punched cards and tape, or photographic materials. Storage media, such as flip-flops, electrostatic (Williams) tubes, and delay lines, are non-permanent, or volatile. In volatile storage systems the information disappears if the power is turned off or accidentally interrupted. Definite steps must be taken in such systems to preserve the stored information in the event of power failure. Volatile media are used, for this reason, primarily for short-term, temporary storage.

7. *Storage Capacity.* By adding identical memory cells, any type of store can be designed for almost unlimited capacity. However, the cost of the individual elements determines the capacity that can be built economically into a particular type of storage device. Fortunately, devices that have a relatively low economic store capacity, such as flip-flops, magnetic cores, and electrostatic devices, usually compensate for this disadvantage by extremely high-speed operation and access to the information. In contrast, very high-capacity

storage devices, such as magnetic and punched tapes, are relatively slow in recording and providing access to the information. Intermediate-capacity systems, like magnetic drums, also provide intermediate speed of operation and access. To give some idea what is meant by 'low' and 'high' capacity, the low-capacity, high-speed magnetic core or electrostatic memories used for internal storage may easily have a capacity of 5000 to more than 50,000 words, depending upon the size of the computer. With each word consisting of, perhaps, 40 bits, this would equal from 200,000 to more than 1,200,000 bits for a 'low-capacity' store. The capacity of external storage media, such as magnetic and punched tapes, is, of course, essentially unlimited and may run to hundreds of thousands of bits each second of operation. Magnetic drums typically have some 20,000–30,000 words of storage, amounting to approximately a million bits of information.

8. *Mode of Access.* As previously mentioned, access to information may either be in sequence, bit by bit or word by word, called serial operation, or it may be simultaneous through several parallel channels. The two basic modes of access, therefore, are serial or parallel. Both modes may be used in the same store, depending on the units of information stored. For example, magnetic drums usually employ serial storage of entire words, but make the individual bits of a word accessible in parallel; they are therefore classified as serial by word and parallel by bit. In serial-word storage there is always a waiting time, or latency period, for words to be stored or to be read out. In parallel store all words are equally available without a latency period. There is also a type of access known as random access, which corresponds roughly to the process of looking up names in a telephone directory or cards in a file. You can look up any name or card at random without ever referring to a previous name or card. Similarly, in a random-access memory a bit of information is chosen from the memory location or register at random, and does not depend upon the location of a previous bit of information. This is, of course, the most rapid way to gain access to a specific piece of stored information, and hence much effort is expended by computer designers on the development of a truly random-access memory of high storage capacity.

9. *Access Time.* The time required to store or obtain information from the memory is called access time, and is a most important attribute of a storage system. High-capacity memories, such as magnetic tape reels, are necessarily slow and are therefore always used for external storage of large amounts of information to which rapid access is not required. Intermediate-capacity storage devices, such as magnetic drums, have intermediate access times of a few milliseconds (thousandths of a second), which makes them suitable for buffer and back-up memories. Rapid memories, used for internal high-speed storage, such as magnetic-core arrays and electrostatic devices, have access times measured in millionths of seconds (microseconds), with 4–8 microseconds being typical for magnetic core types and 12 microseconds for electrostatic types (much less in more recent computers). This is of the same order as the time required for actual arithmetic operations, and therefore these high-speed memories do not significantly slow down the operation of the computer.

Some of the more important memory attributes are compared in the chart on page 239.

The Computer Word

We have seen that the registers of the computer memory store units of information called words, which may consist either of numerical (or alphanumeric) data or instructions to be carried out. Let us look at a typical binary data word to be stored in a register of the memory. Numbers used in computers may have from eight to ten decimal digits. Allowing four binary digits

Computer Memory Characteristics

TYPE OF STORE	CAPACITY	PERMANENCE	SPEED OF ACCESS	MODE OF ACCESS
Magnetic core	Relatively low (4,000–12,000 words)	Permanent–erasable	Very high (about 5 microsec)	Parallel
Magnetic drum	High (8,000–50,000 words)	Permanent–erasable	Intermediate (about 10 millisec)	Usually word-serial bit-parallel
Magnetic tape	Unlimited	Permanent–erasable	Low	Serial
Electrostatic	Relatively low	Volatile	Very high (about 12 microsec)	Parallel
Delay line	Very low (a few words)	Volatile	High (a few hundred microsec)	Serial
Punched cards (IBM)	Unlimited	Permanent–non-erasable	Low	Serial or parallel
Punched tape	Unlimited	Permanent–non-erasable	Low	Serial
Flip-flop registers	One word	Volatile	Very high (a few microsec)	Serial or parallel

for each binary-coded decimal digit, from 32 to 40 bits are required to represent the number. In addition, allowance must be made for the sign (+ or −) of the number, which may require one or more bits, and for checking the correct number of pulses, which requires a 'parity bit'. The latter is simply a check pulse, which is added whenever the number of pulses is even, to make the total odd. Such an even-odd, or parity check, avoids errors due to pick-up of a spurious extra pulse or the dropping of a pulse in a combination. Fig. 128 illustrates a data word obtained by translating the decimal digits 1 to

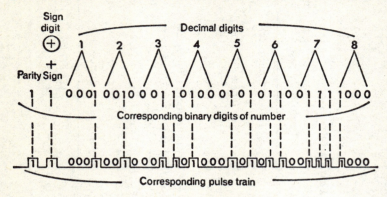

Fig. 128. Typical data word with sign and parity bits.

8 directly into corresponding straight four-digit binaries, and an extra sign bit for +. The resulting 33 digits require 14 pulses (representing ones), to which a check or parity pulse is added to make the total come out odd. The data word, thus, consists of 34 digits, represented by 15 pulses.

Instruction Word. Each instruction word must specify some operation to be performed on the operand or data word. The format of a typical instruction word, again consisting of 34 bits (to be consistent with the data word), is illustrated in Fig. 129. In addition to the usual parity bit, this is seen to be

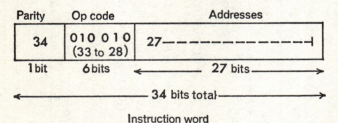

Fig. 129. Typical 34-bit instruction word format.

made up of two major parts, the operation code and the addresses. Six bits have been reserved for the operation (OP) code, which usually consists of binary digits that specify the operation to be performed, such as addition, subtraction, or comparison. All remaining bits are reserved for the addresses. As we shall see in the next chapter, up to four addresses may be required for complete specification.

Word Destination. The words stored in the internal computer memory are sent to different parts of the computer (as determined by the control unit) in accordance with their functions. Thus, data words containing operand (numerical) data are always sent to one of the registers in the arithmetic unit upon the command of the control unit. In contrast, instruction words are sent to the instruction register in the control unit; here they are held in temporary storage until the operation code of the instruction is decoded. The function of the internal memory for either type of word is to store the information until needed and then supply the proper word at the proper time to the correct register, upon command of the control unit.

ADDRESS SELECTION

When an operation is to be performed the control unit requests the operand data stored in the internal memory, and sends it to the appropriate arithmetic register. To do this the control unit must know where to find the data in the memory. This function is performed by the address selector. The address selector is an electromechanical or electronic switching device that locates specific information in the memory in accordance with the program carried out by the control unit. As shown in the general block diagram, Fig. 130,

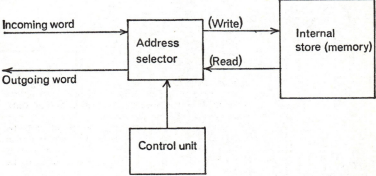

Fig. 130. Memory address selection of incoming and outgoing information.

address selection is required both for reading out and for writing information into the computer memory.

There are essentially three different types of storage systems, and a different method of address selection must be used for each of them. One method is used when information is stored in a fixed location, such as in magnetic-core memories. A second method deals with moving-storage mediums, such as magnetic drums and tapes. Finally, there is a method of address selection for dynamic memories, where information is in continual circulation.

Core Address Selection. Fig. 131 illustrates a possible logic set-up for address selection in a fixed location (core) memory. Individual AND gates control the selection of each row (X-lines) and column (Y-lines). (The logic for the columns is not shown, since it is identical with that used for rows.) Three binary digits, A, B, and C, or a three-element octal digit, have eight possible combinations of truth values, and consequently any one of eight memory rows can be selected. A second octal digit selects any one of eight

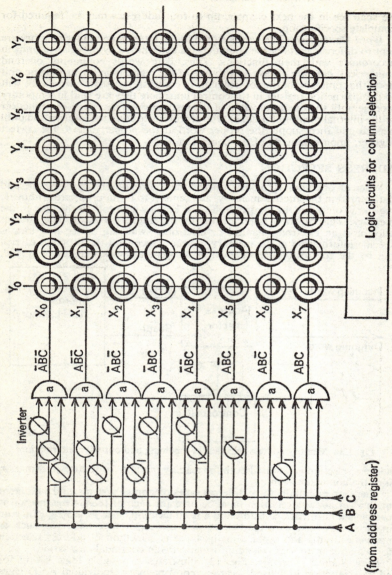

Fig. 131. Address selection in magnetic-core memory.

memory columns, so that the intersections with the selected row determine 64 possible memory locations. Note that each of the three-legged AND gates is provided with one or more inverters at the input so that it can respond to only one combination of truth values. The selection of the proper address combination is determined by a control pulse sent to the address register. Additional address digits are required if the memory contains more than eight rows and columns in a single plane or matrix.

The set-up shown in Fig. 131 permits the selection of only one bit of information from a single core. To select an entire operand (data word), the same address is applied to a number of memory planes in parallel. Thus, a

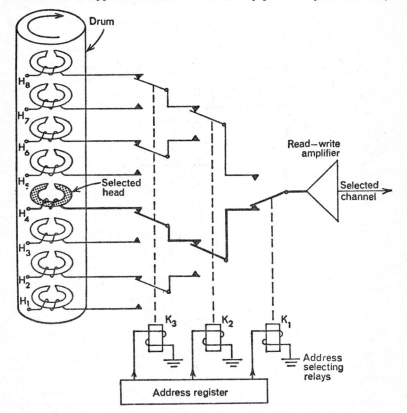

Fig. 132. Address selection in magnetic-drum memory.

binary bit can be read from each of the cores located at the same intersection in each plane. The bits making up a data word are then applied, either all together (parallel mode) or in sequence (serial mode), to the appropriate arithmetic register.

Drum Address Selection. Address selection in a magnetic drum memory or other moving medium consists of two parts. First the channel, or track, containing the desired information must be selected. Second, within the proper channel a search must be made for the desired word. Thus, the address must specify both the information channel on the drum and the memory cell on the channel storing the desired word. The magnetic head must not be turned on until this word actually appears directly beneath it.

Fig. 132 illustrates the first part of address selection, the selection of the required head and memory track. For simplicity, a relay selector circuit is shown, although this is rarely used in practice because of its relatively slow operation. As shown, the common read–write amplifier is coupled to any one of the eight magnetic heads (and corresponding channels) through the contacts of three address selector relays, $K1$, $K2$, and $K3$. Relay $K1$ selects either

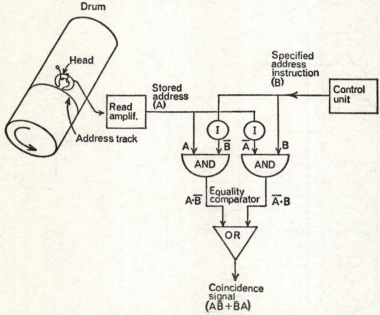

Fig. 133. Address comparison logic for magnetic-drum reading and writing.

the upper or lower four channels, $K2$ chooses between the upper and lower two channels in the selected half of the drum, and $K3$ selects the required channel within each pair. For example, with relay $K1$ energized and relays $K2$ and $K3$ both de-energized, as shown in the illustration, a closed path exists between magnetic head $H4$ and the read–write amplifier. A three-digit binary number, or an octal digit, can select any possible combination of relay modes and hence can control all channels. Relay selection is made at the address register. Instead of relays, diodes are usually employed to apply bias to a special bias winding on the magnetic head to be turned on. All heads are connected in parallel to the read–write amplifier.

The second part of address selection in a drum memory consists of waiting until the desired word appears under the magnetic head of the selected channel and then turning the head on to read out (or write in) this particular word. For identification of the correct word, the drum is usually provided with a special address track, which lists the addresses of all the words in a memory channel. (Alternatively, the words can be counted off in accordance with the drum position.) The search thus consists of comparing the constantly changing address on the address track of the drum with the address specified by the instruction from the control unit. When both are identical, the reading or writing of information can begin.

Fig. 133 illustrates one possible logic set-up for address comparison during read-out from a magnetic drum. The equality comparator consists of two AND gates and an OR gate in a configuration, which you will recognize as the exclusive-OR circuit, or a half-adder without the carry portion. This logic circuit puts out a binary zero whenever its two inputs, A and B, are equal, and a one when its inputs are unequal.

The signals to be compared are the constantly changing address (A) stored on the address track of the drum and the address specified by the instruction (B) from the control unit. When both are in agreement the equality comparator emits a coincidence signal consisting of a binary zero (when A and not-B is true, or B and not-A is true). By inserting an additional inverter into the output of the OR gate, the circuit can be made to put out a one to indicate equality of the two addresses. The reading or writing of data from the selected channel on the drum starts as soon as the coincidence signal indicates the equality of the stored address on the drum with that specified by the instruction.

Dynamic-Memory (Delay-Line) Address Selection. As shown in Fig. 122, the information stored in a regenerative delay line (or other dynamic memory) continuously circulates between the read amplifier at the output and the write amplifier at the input to the line. The information can be read only at the moment it arrives at the output end of the line and hence must be timed for read-out. Since the words are entered into the line at a known time and travel at a known speed, their time of arrival at the output end is also known. The address of a particular word thus consists of its time of arrival at the output of the line. This address time is supplied to a series of logical gates, which form timing signals to interrogate the line. A timing pulse of the proper length, applied at the correct time to the output of the line, permits information to be read out during this interval. As we shall see in the next chapter, all operations are timed by a series of clock pulses from a master (clock) oscillator; hence, the selection of a particular interval is not a difficult matter.

OPERATION OF THE ARITHMETIC UNIT

All arithmetical calculations and most logical operations are performed in the arithmetic (pronounced arithme'tic) unit of the computer. We have already studied the physical elements of which arithmetic units are constructed, among them various logical gates, short-time one-bit storage elements such as flip-flops, and registers made up of flip-flops or magnetic cores. We must now see how these circuit elements can function together as an integrated component to perform arithmetical and logical operations in the simplest possible manner. The four arithmetical operations—addition, subtraction, multiplication, and division—are preferably broken down into simple addition, addition and complementing, repeated additions and shifting, and repeated subtractions (complementing) and shifting, respectively. How this is actually done, however, depends on the particular type of arithmetic used and on the major system features, such as whether serial or parallel operation is used, the type of binary or decimal number representation, etc. Before going into these details, let us take a quick look at the structure of a typical arithmetic unit and its interconnexions with the remainder of the computer.

Overall Operation of Typical Arithmetic Unit

The general organization of an arithmetic unit as part of the computer system is shown in Fig. 134. It consists primarily of three registers (which may be called A, B, and Q) that perform the actual arithmetic and logical operations. The operations are governed by an arithmetic control section, which contains the necessary logic circuits to synchronize the arithmetic unit with the rest of the computer and control the sequence of operations. The computer's memory supplies the operand data to the registers via the information bus and receives, in return, the computed results for storage.

Arithmetic Registers. A minimum of three registers are required to perform arithmetical or logical operations. Two registers are needed to receive and temporarily store the numbers to be operated on (operands), and a third register—called an accumulator—is required to store the results before transfer to the memory. Such an arrangement, however, permits only addition and subtraction of operands, and an additional register (sometimes called Q-register) is required for multiplication and division.

To eliminate one of these registers and permit three to do the work of four, a slightly different arrangement is frequently used, as is shown in Fig. 134.

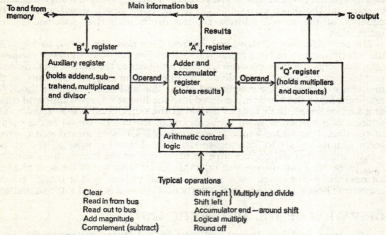

Fig. 134. Functional block diagram of typical arithmetic unit.

An electronic adder is built right into the accumulator register, with the combination generally known as A-register. An auxiliary register (sometimes called B-register) holds one of the arithmetic operands extracted from the internal memory and delivers it, when required, to the A-register. Depending on the operation to be performed, this operand may be an addend, subtrahend, multiplicand, or a divisor. If an addition is to be performed, for example, the auxiliary (B) register delivers the required addend directly to the adder in the accumulator (A) register. The other operand, the augend, is supplied to the adder from the accumulator itself, having previously been stored there. After the addition is completed, the sum is again stored in the accumulator register, thus saving one register. As new input data are supplied and computed by the adder, the register accumulates the partial results into a final total. For addition and subtraction, therefore, only the auxiliary and accumulator registers are used. For multiplication and division, however, one of the operands (multiplicand or divisor) is delivered to the accumulator register by the auxiliary (*B*) register, while the other operand is supplied by the *Q*-register. Three registers, thus, are sufficient for all operations.

Typical Operations. The sequence of arithmetical operations are broken down into small steps by the logic circuits of the arithmetic control. As shown in Fig. 134, coded commands from the control unit to the arithmetic unit may include the following typical operations:

1. Clear all registers for new information to be inserted.
2. Read in new data from information bus.
3. Read out results to information bus.
4. Add magnitude (for arithmetic operations).
5. Complement number (for subtraction).
6. Shift right ⎫ (for division and
7. Shift left ⎭ multiplication).
8. Cycle or end-around shift (shift data out of accumulator register and reinsert into input).
9. Logical multiply (take logical product of terms in *A*- and *B*-registers).
10. Round off numbers to specified number of digits.
11. Double precision—join *Q*- and *A*-registers for computations with double-length words.

Depending on the arrangement and the type of computer, many other operations may be specified in the form of control commands.

Machine Addition

Let us now consider the four basic arithmetic operations in turn. We have already studied the process of addition in various number systems, as well as the means of mechanizing this process with half- and full-adders. (You may want to review these operations at this time.) We need only concern ourselves

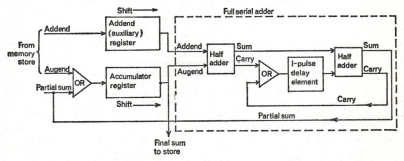

Fig. 135. Serial addition of means of half-adders and delay element.

at this time, therefore, with the two chief methods of machine addition, the serial and the parallel modes. Which method is used in any particular computer depends, of course, on the overall system design of the machine, although the parallel mode is preferred in present-day computers.

Serial Addition. The general scheme of serial addition is illustrated in Fig. 135. Two half-adders (shown in detail in Fig. 103) have been used here to mechanize a full serial adder, although a full adder could have been used equally well.

The original addend and augend data are delivered, upon command of the control unit, from storage to the addend (auxiliary) and accumulator registers, respectively. Both registers are shift registers made up of flip-flops or magnetic cores. As signal (clock) pulses arrive from the control unit, the two registers shift the addend and augend digits to the right and deliver them, bit by bit, to the first half-adder, where they are added immediately. The first pair of digits to enter the adder are from the least-significant (lowest-order) column of the binary number, and hence do not find any previous (lower-order) carry. The output of the first half-adder consists of the sum of the two digits and a possible carry, to be added to the next-higher-order (more-significant) column. The sum is sent to the second half-adder, while the carry bit is passed through an OR gate to a delay element (such as a one-shot multivibrator), where it is stored for one pulse interval.

Upon command of the next control pulse, another pair of operand digits are shifted from the register and added together by the first half-adder. The resulting sum again is sent to the second half-adder, while a possible carry is

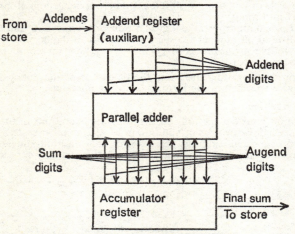

Fig. 136. Scheme for parallel addition.

stored in the delay element. The second half-adder then combines the sum of the first two digits with the delayed carry from the previous, lower-order column addition. The sum output of the second half-adder is fed back through an OR gate to the input of the accumulator register, whose left-hand portion has been emptied by the shift to the right. The possible carry output of the second half-adder is sent through an OR gate to the 1-pulse delay element for combination with the following (higher-order) pair of digits.

You can see that the process of serial addition takes place in a rhythmical pattern. As pulses arrive from the control unit, pairs of addend and augend digits are shifted out of the registers and added by the first half-adder. The sum is combined with the previous (delayed) carry in the second half-adder, whose sum digits are sent back to the accumulator register for temporary storage. A carry formed in either half-adder is stored in the 1-pulse delay element, so that it may be added in with the following pair of digits. When all digits (columns) of the two numbers have been added, the final sum is automatically stored in the accumulator register, while the addend (auxiliary) register is empty. The final sum may then either be placed into internal storage, or may serve as augend if another number (addend) is to be added.

Operation of Computer Memory and Arithmetic Unit

It is evident that serial addition is a rather slow process, and it is used primarily with relatively slow memories, which cannot sustain a faster pace.

Parallel Addition. To save valuable time, all digits (columns) of two numbers can be added together at once, through parallel operation. The price for this speed-up is that a separate adder is required for each column of numbers. Thus, if each number to be added has 16 bits, or columns, 16 binary adders must be supplied in a parallel connexion. The general scheme for parallel addition is shown in Fig. 136. Either half-adders or full binary adders may be used, as described in detail in an earlier chapter. The addend and augend digits of the two numbers are supplied simultaneously to the parallel adder from the respective auxiliary and accumulator registers. (The numbers may have been previously stored in the registers in serial fashion.) Each pair of digits enters a separate adding element in the parallel adder, with the sums and carries formed simultaneously during a single pulse (clock) interval. Since all columns are added at once, the carry digits are not delayed, as in serial addition, but are transferred immediately to the next-higher column adder at right. (See also Fig. 106 for operation of a full parallel adder.) If new carries result from the adding in of the lower-order carries, these are again propagated at once to the next-higher-order column. The entire sum is thus completed within one clock interval.

In one method of operation the accumulator register is initially cleared to zero and the addend is placed in the addend (upper) register. The addend digits are then simultaneously 'dumped' into the parallel adder and combined with the zeros from the accumulator. The sum, which is, of course, the original addend, is then stored in the accumulator register. A second addend is again placed in the upper (addend) register, while the previous sum in the accumulator now becomes the augend. Both addend and augend digits are dumped into the parallel adder, and the sum is returned to the accumulator register. In this manner every addend combines with the previous sum (the new augend) and the grand total is accumulated in the bottom (accumulator) register. The final sum is fed, either serially or in aparallel, to the internal memory for storage until it can be transferred to the output.

Machine Subtraction

Although subtraction by borrowing—the conventional method—can be mechanized in a computer, it is easier to subtract by adding the complement,

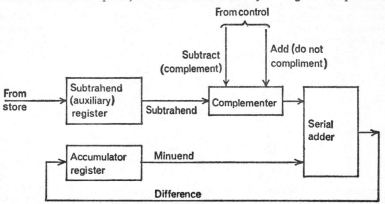

Fig. 137. Serial subtraction with complementer.

as we found out in Chapter 8. By using the method of adding, the complement of the number to be subtracted, we can make use of the already existing electronic adder in the computer and need only construct a device that will automatically find the complement of a number (i.e. a complementer). You will recall that complementing is particularly simple in the binary number system; it consists of putting down a 0 for a binary 1, and a 1 for a 0. The

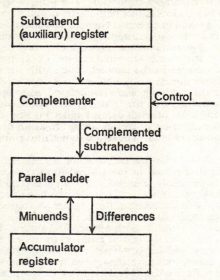

Fig. 138. Parallel subtraction with complementer.

basic element is a binary complementer, thus, is the NOT circuit, or inverter, which we have studied in detail. To obtain the correct result when subtracting by complementing, you will also recall, it is necessary to add a 1 to the extreme right (least significant) digit of the sum and ignore the extreme left (most significant) digit, a process we called 'end-around carry'. For example, subtracting 10001 (decimal 17) from 11001 (decimal 25) by the conventional and complementing methods looks like this:

Conventional Subtraction

```
              1 1 0 0 1  (= 25)
            − 1 0 0 0 1  (= 17)
     (Ans.)   0 1 0 0 0  (= 8)  Diff.
```

Complementing

```
       1 1 0 0 1
     + 0 1 1 1 0   (Complement of 10001)
     ①0 0 1 1 1   (Sum)
     +─────→ 1    (End-around carry)
       0 1 0 0 0   (Ans.)
```

The result, 01000 (decimal 8), is the same for either method.

Incorporating the 'end-around carry' into the complementer is a simple affair. Normally, there is never a previous (lower-order) carry for the extreme right (least significant) column. Hence, the carry circuit always supplies a 0 for the extreme right column during addition. When a complement is to be added, however, the complementer simply changes this 0 carry to a 1. Similarly, the unwanted 1 in the extreme left column can be dropped by applying an inhibit voltage to the flip-flop in the register that stores the result.

Serial Subtraction. Serial subtraction by complementing is illustrated in Fig. 137. Note that we have simply inserted a complementer ahead of the serial adder in the serial addition circuit shown in Fig. 135. As was described, the complementer is essentially a NOT circuit (inverter) with provisions for adding in a 1 carry in the extreme right column and suppressing the 1 in the extreme left column. With proper instructions from the control, the circuit of Fig. 137 can either add or subtract. For addition, the complementer receives the instruction 'do not complement!' while for subtraction it is ordered to complement the subtrahend.

Parallel Subtraction. Parallel subtraction by means of complementing, shown in the block diagram (Fig. 138), is the same as parallel addition (Fig. 136) except that a complementer is inserted between the auxiliary register holding the subtrahend and the parallel adder. Depending upon the control command, this set-up can either subtract (by complementing) or add (by not complementing).

ALGEBRAIC ADDITION AND SUBTRACTION

Arithmetic subtraction of a number is the same as algebraic addition of a negative number. Thus, when the sign of the numbers is taken into account, subtraction is only a special case of algebraic addition. To refresh your memory, these two simple rules govern the addition of algebraic (signed) quantities:

1. If the quantities have like signs (both plus or both minus), simply add them and apply the same sign.
2. If the quantities have unlike signs (one plus and one minus), subtract the smaller from the larger and give the result the sign of the larger quantity.

For maximum flexibility, computers usually deal with signed quantities, which may not always be numbers. As Fig. 128 showed, each data word is provided with a binary sign digit in the extreme left column. The plus sign is generally represented by a binary 1 and the minus sign by a 0, but the converse will do as well.

Sign Comparator. Since either or both numbers may be negative in algebraic addition, complementers must be able to operate upon both input registers when required. Usually separate NOT circuits (complementers) are inserted between the input registers and the adder. If the signs are unlike, the smaller number must be subtracted from the larger (in accordance with Rule 2), and hence the complementer for the proper operand must be turned on. If the signs are alike, however, simple addition is required and the complementer is not turned on. This presumes that the complementer has some way of knowing whether the signs are like or unlike. A sign comparator is used to inspect the sign digits of the operands and give the appropriate order to the complementer. Sign comparison is essentially the same as equality comparison. A flip-flop storing a positive operand will be in the ONE state, while a flip-flop storing a negative operand will be in the ZERO state. Thus, sign comparison amounts to finding out whether the sign flip-flops of the two operand registers are in the same or in complementary (opposite) states. This job can be done by any one of the exclusive-OR logic circuits (i.e. half-adders without carry portion) we have studied. For example, the half-adders shown in Figs. 103, 104, and 133 can act as sign or equality comparators when the carry portion is eliminated. When the outputs of the sign flip-flops are connected to any of these circuits, a coincidence pulse occurs whenever the flip-flops are in unlike (complementary) states. This pulse can be used to turn on the complementer to perform subtraction.

A block diagram set-up for serial addition of algebraic quantities is shown in Fig. 139. Complementers are inserted in series with both operand registers and are controlled by the sign comparator. If both operand signs are alike, the sign comparator does not produce a coincidence pulse, the complementer is not turned on, and the numbers are simply added. (The proper sign is affixed to the sum.) If the operand signs are unlike, a coincidence pulse from the sign

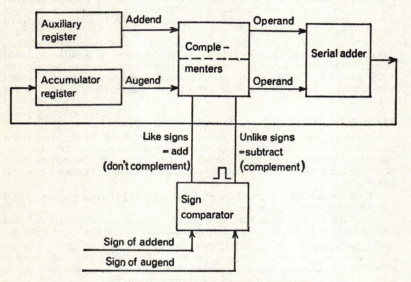

Fig. 139. Serial addition of algebraic quantities.

comparator turns on the complementer for the negative number, and this number is complemented before addition to the other operand. The set-up for parallel addition of algebraic (signed) quantities is the same as shown in Fig. 138, except that an additional complementer must be inserted between the accumulator register and the parallel adder and both complementers must be controlled by a sign comparator.

Magnitude Comparison. You may have noted that we have deliberately ignored an important part of Rule 2 for algebraic addition—the requirement that the smaller number must be subtracted from the larger one if the signs are unlike. To fulfil this requirement would necessitate an additional magnitude comparator that would match up the two numbers, column by column, to detect the larger one. Although some computers have a magnitude comparator, such a device not only is expensive but also wastes valuable time. Fortunately, magnitude comparison is not actually necessary, since, if the smaller number is mistakenly subtracted from the larger, the error is obvious in the result and can easily be corrected. To demonstrate this, suppose we had subtracted 11001 (= 25) from 10001 (= 17) in our previous example. The complement of 11001 is 00110. Adding the complement to 10001, we obtain

```
  1 0 0 0 1   (= 17)
+ 0 0 1 1 0   (Complement of 11001 = 25)
  1 0 1 1 1   (Sum)
```

Note that no additional carry digit is produced in front of the sum (10111), and hence no end-around carry is required, or possible. The sum is, however, obviously false, since the previously obtained result was 01000. We can easily demonstrate that the false sum is the complement of the correct result, for recomplementing:

1 0 1 1 1 yields 0 1 0 0 0 (= 8), which is the correct result.

This suggests the procedure the computer must follow for correct subtraction:

1. If the smaller number is (correctly) complemented and added to the larger one a superfluous carry is produced in front of the number (extreme left column). This carry digit must be eliminated from the left and added to the sum in the extreme right (least significant) column.

2. If the larger number is (incorrectly) complemented and added to the smaller, no superfluous carry occurs. The result is therefore wrong and the complementer must be turned on again to recomplement the number. (The recomplemented result is then passed alone through the adder, thus adding it to zero; this does not change the result.)

Algebraic Subtraction. The simple rule for algebraic subtraction is: change the sign of the subtrahend and add algebraically to the minuend. Thus, subtraction is easily implemented by inverting the sign of the subtrahend before applying it to the sign comparator, as shown in Fig. 140. Any of the inverters

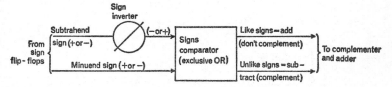

Fig. 140. Algebraic subtraction with sign inverter and comparator.

(NOT circuits) we have studied can be used as a sign inverter. If, after sign inversion, the sign comparator detects like signs, the two operands are added arithmetically; if the signs are unlike, arithmetic subtraction is required. That is, the smaller number must be complemented and added to the larger, as was described before.

Machine Multiplication

Multiplication can always be carried out by repeated addition, that is, by adding the multiplicand as many times to itself as the multiplier specifies. This is a time-consuming process, especially if the multiplier is large. A faster method consists of using the multiplication table to multiply the multiplicand by one multiplier digit at a time, shifting the partial products thus obtained to the left, and then adding all the partial products to obtain the final result. In Chapter 8 we discovered that multiplication by shifting is particularly simple in the binary number system, which is used in most computers. You will recall that binary multiplication (by shifting) obeys these rules:

1. If the multiplier digit is a '1', copy the multiplicand to obtain the partial product; then shift one place to the left.
2. If the multiplier is a '0', do not copy the multiplicand, but shift the previous partial product an additional place to the left.
3. Add the partial products either as soon as they appear or at the conclusion of multiplication.

To illustrate the process, we repeat an example from Chapter 8:

Binary Multiplication by Shifting

```
Multiplicand                1 1 1 1         =   15
Multiplier              ×   1 1 0 1         = × 13
First partial product       1 1 1 1     (Copy multiplicand)
Second partial product  1 1 1 1         (Shift, shift, and copy)
    Sum             1 0 0 1 0 1 1
Third partial product 1 1 1 1           (Shift and copy)
    Sum           1 1 0 0 0 1 1 1       =  195 (Ans.)
```

Note that the answer has twice as many digits as either number. If both numbers have the same number of digits the product will always have twice as many digits, or be of double length.

Parallel Multiplication by Shifting. To take advantage of the high speed of parallel addition, multiplication in most computers is carried out in the parallel mode of operation. A typical scheme for parallel multiplication is

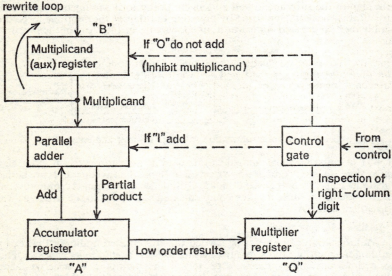

Fig. 141. Parallel multiplication by shifting.

illustrated by the block diagram (Fig. 141) above. Note that an additional multiplier (Q-register) is required over the set-up used for parallel addition.

A possible sequence of operation for parallel multiplication by shifting (Fig. 141) might be as follows:

Operation of Computer Memory and Arithmetic Unit

1. Upon instruction from the control unit, the operands are withdrawn from the store, the multiplicand is initially placed in the auxiliary (B) register, the multiplier is placed into the Q-register, and the accumulator (A) register is cleared to zero.

2. Upon arrival of the next control pulse a gating arrangement (control gate) inspects the extreme right (least significant) multiplier digit, which is stored in the extreme right flip-flop of the Q-register. If this digit is a 0, the multiplicand must not be added and, accordingly, an inhibit pulse is sent to the read-out gates of the multiplicand (B) register to prevent addition. If the digit is a 1, however, the multiplicand must be added, and hence an enable pulse is sent to the parallel adder to accept the multiplicand. The multiplicand is then added to the contents of the accumulator (A) register, which holds zeros for this first addition. (This means that the multiplicand is simply placed into the accumulator register.) Since read-out from the multiplicand register destroys its contents, the multiplicand is immediately 'rewritten' into the register by a 'rewrite loop'.

3. The next control pulse orders both the accumulator and multiplier registers to shift one place to the right. You will recall from the discussion in Chapter 8 that shifting the previous sum to the right is equivalent to shifting the next partial product to the left. Since the sum is contained in the accumulator, which is a shift register, it is easier to shift the sum to the right. Similarly, the multiplier register is of the shifting type; the shift pulse moves the previously examined (extreme right) multiplier digit out of the register and replaces it by the next more significant digit to the left. Simultaneously, the extreme right (least significant) sum digit in the accumulator is shifted out of the register to the right and is placed in the vacant spot at the left of multiplier register. In this way the required sum digit is preserved in storage, while the multiplier digit that is no longer needed is dropped. (If the lower-order sum digits in the accumulator are not required in the final answer they may, alternatively, simply be dropped instead of being stored in the multiplier register. In this arrangement the multiplier register does not need to shift, and the multiplier digits are examined in turn, from right to left. A counter is then needed to determine when all multiplier digits have been examined.)

4. After the multiplier digits have been shifted one place to the right the next control pulse orders the control gate again to examine the extreme right digit stored in the right-hand flip-flop of the multiplier register. This is now the next more significant multiplier digit, of course, because of the right shift. The previous sequence is then repeated. If the multiplier digit is a 0 the multiplicand is rejected; if it is a 1 the multiplicand is added by the parallel adder to the shifted contents (previous sum) of the accumulator register.

5. Again the accumulator and multiplier registers are shifted one place to the right by the next control pulse, the extreme-right multiplier digit is dropped, and the sum digit on the right is shifted from the accumulator to the vacant spot at the left of the multiplier register. The next multiplier digit is then examined. This process is repeated until all multiplier digits have been examined and dropped and all partial products have been added (accumulated) in the accumulator register. The final product is stored in both the accumulator and multiplier registers. The extreme right (lower-order) bits that have been shifted out from the accumulator are stored in the multiplier register, while the left (higher-order) bits are stored in the accumulator. Thus, the accumulator contains the most significant (left) half of the answer, while the multiplier register contains the least significant (right) half of the product. (If the alternative scheme with a non-shifting multiplier register is used only the most significant (left) half of the product is preserved in the accumulator.)

Serial Multiplication. Although not frequently used, the serial mode of multiplication by shifting is essentially similar to the parallel mode. The major difference consists of the method of shifting the partial sums with respect to the multiplicand. The accumulator register is actually shortened to push the partial sums ahead in time.

Machine Division

Division is the process of counting how many times one number (the divisor) can be subtracted from another number (the dividend) while still leaving a positive remainder. The number of times the divisor can be subtracted from the dividend is the result, or quotient. Thus, just like multiplication by repeated addition, division can be carried out by repeated subtraction. However, the process is as clumsy and time-consuming on paper as it is with a computer, and the method of long division by shifting is preferred in both media. You will recall from the discussion in Chapter 8 that binary long division is particularly simple, since we never need to try out the largest divisor that will 'go into' the dividend. The divisor can either be subtracted from the dividend (in a particular position) or it cannot be subtracted. If the divisor can be subtracted, the quotient digit is a 1, and the next divisor is shifted one place to the right. If the divisor cannot be subtracted from the dividend the quotient digit is a 0; the divisor must then be shifted right in position with respect to the dividend until it can be subtracted. The process stops when one more subtraction would make the remainder negative. The answer consists of the quotient digits plus the positive remainder (unless the division is carried into fractions beyond the binary point). You can check the examples in Chapter 8 to convince yourself of the simplicity of the process. In essence, many digital computers perform long binary division in exactly the same way, except that they prefer to shift the dividend to the left (instead of shifting the divisor to the right), and must be equipped with a means for deciding whether or not a divisor can be subtracted (i.e. whether or not the difference is positive).

Parallel Division by Shifting. The parallel mode is the most rapid way of performing binary division by shifting. One possible method is shown in Fig. 142. The same registers that were used for multiplication can be used for division, and only the additional complementer is required for subtracting the divisor (by adding its complement). The auxiliary (B) register now becomes the divisor register, the Q-register becomes the quotient register, and the accumulator (A) register stores the dividend and the final remainder. Instead of shifting right (for multiplication), the accumulator and quotient registers are now required to shift to the left for division. In most computers suitable control pulses permit shifting in either direction.

The sequence of operation for parallel division by shifting (Fig. 142) is, typically, as follows. The divisor is initially inserted into the auxiliary (B) register and the dividend is placed into the accumulator. Upon arrival of a control pulse, the divisor is complemented (for subtraction) and the complemented number is added to the dividend by means of the parallel adder. (This is, of course, the same as subtracting the divisor from the dividend.) Although this step sounds very simple, it really is not, since the machine has no way of knowing whether the divisor is in the correct position with respect to the dividend (i.e. is small enough) to be subtracted from it. This is determined by trial and error. The computer begins the subtraction with the left ends of the divisor and dividend aligned in their respective registers. If the subtraction is successful—leaving a positive difference—everything is fine and the next step can be performed. If the subtraction results in a negative answer, however, the divisor was too large for this part of the dividend; that is, the dividend was not properly lined up. The divisor is then immediately added

Operation of Computer Memory and Arithmetic Unit   257

back to restore the original dividend, and the dividend is shifted one place to the left in the accumulator for another trial. Since the dividend is now larger, the repeated subtraction will probably be successful, and a ONE is stored in the extreme right position of the quotient register.

During the next control pulse both the dividend in the accumulator and the quotient register are shifted one place to the left. This makes room in the quotient register for the next quotient digit and places the dividend in the proper position for the next subtraction. To perform the subtraction, the divisor is again complemented and then added to the dividend by means of the

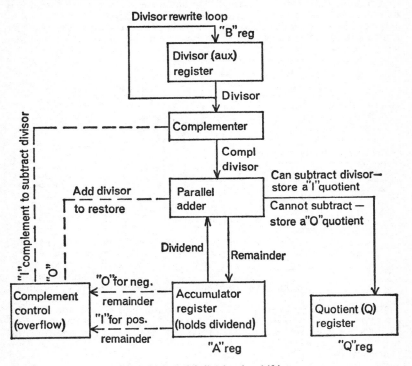

Fig. 142. Parallel division by shifting.

parallel adder. If the subtraction is successful, leaving a positive remainder, a one is placed in the next position in the quotient register, and the entire process of shifting and subtracting by complementing is repeated. If the subtraction results in a negative remainder, however, a 0 is placed in the quotient register, the divisor must be added back to restore the dividend, and the latter is shifted another place to the left for the next subtraction. The process is repeated, with a 1 being recorded in the quotient register for each successful subtraction, and a 0 for each unsuccessful one, until all the quotient digits have been generated and the remainder keeps coming out negative. The activating pulses from the control unit then stop.

You may wonder how the machine can tell when the result of any subtraction is positive or negative. This is the function of the overflow element in the complement control. You will recall from the discussion of machine subtraction by complementing that a correct subtraction produces a superfluous '1' digit (carry) at the extreme left of the result. If, however, the larger number is incorrectly subtracted from the smaller, no extra 1 digit is produced and the remainder comes out negative. This automatic action of the complementer can be used to control the division process. As Fig. 142 shows, an overflow element is provided which accepts any extra digits that have no space at the extreme left of the accumulator register. The overflow element is initially set to 0 and is reset to 0 after each subtraction. For a correct subtraction, resulting in a positive difference, the extra 1 that has no space in the accumulator is automatically placed in the overflow element. An incorrect subtraction, however, results in a negative difference without the extra digit; the overflow element, hence, remains set to 0. The complement control is simply a gate that

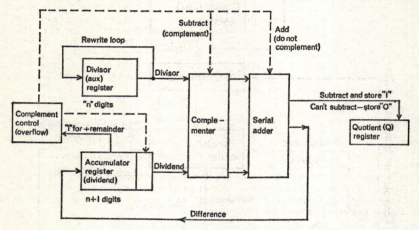

Fig. 143. Serial division by shifting.

inspects the digit in the overflow element and turns the complementer ON (for the next subtraction) if it is a 1 or turns it OFF if a 0 is stored. Moreover, if the overflow digit is 0 because of an incorrect subtraction, the adder is instructed to add the divisor back to the negative remainder (with the complementer turned OFF) to restore the correct dividend. The overflow element is then reset to 0, the registers shift, and the next subtraction takes place. After all the quotient digits have been generated, the dividend is always too small for the divisor, and the remainder stays negative for repeated attempts at subtraction. A control pulse then stops the process.

Serial Division by Shifting. A scheme for serial division by shifting is illustrated in the functional block diagram below (see Fig. 143). Except for the serial mode of operation the general functioning of the serial divider is similar to that of the parallel machine. The computer attempts to subtract the divisor from the dividend (bit by bit). If the difference is positive a 1 is entered in the quotient register and a 1 goes into the overflow element, instructing the complementer to operate during the next subtraction. If the difference is negative a 0 is entered into the left-shifting quotient register, while the 0 in the overflow element turns off the complementer and instructs the adder to add the divisor back to the negative difference to restore the dividend. The accumulator is made longer by one digit ($n + 1$) than the other registers, which permits delaying the repositioning of the dividend after each subtraction and effectively accomplishes a left shift. When the divisor is added back during a false subtraction the extra storage space at the right end of the accumulator is dropped and a shift is avoided.

Fixed and Floating Point

The arithmetical operations do not depend on the location of the binary or decimal point in the number. The rules of addition, subtraction, multiplication, and division are the same no matter where the point is located. However, the range of numbers accommodated and the ease of mechanizing an operation are affected considerably by the location of the point. There are two major methods of locating the point in digital computers, the fixed-point and the floating-point systems. In the fixed-point system the location of the binary or decimal point is constant (with respect to one end of the number) in each of the machine registers. All numbers may be considered either greater than one, with the point always located at the right end of the register, or less than one (i.e. fractions), with the point always located at the left end of each register; in either case the location of the point is fixed. Fixed-point operation is easiest to mechanize, but the location of the point must be recorded and occasional compensating shifts are necessary to preserve the relative size relationships of the numbers.

In floating-point operation the location of the binary or decimal point does not remain constant, but is regularly recalculated. The point is located by expressing each number by a series of significant digits and the power of its base. (The sign must, of course, also be specified.) For example, the number 1,678,593 can be written $1 \cdot 678593 \times 10^6$ and may be expressed in a 13-bit register as +167859300006, with zeros being inserted between the significant digits and the power of the base. Alternatively, the number could be written as +060001678593, with the exponent being recorded ahead of the significant digits. The floating-point method allows greater flexibility of operation and easier handling of numbers differing greatly in magnitude. For this reason it is usually preferred in scientific computers.

CHAPTER 13

THE COMPLETE DIGITAL COMPUTER—II: COMPUTER PROGRAMMING, CONTROL, AND COMMUNICATION

The arithmetic unit in conjunction with the computer's internal 'memory' can carry out a long string of calculations without the need to obtain more information from outside the computer. To do this automatically, however, the computer must continually instruct itself what to do next to solve the problem. All necessary instructions must be prepared in advance in the form of a complete program that is stored (usually on tape) and later inserted, step by step, into the computer's memory. The instructions are then selected in the proper sequence from the stored program under the direction of the control unit, which co-ordinates all operations, times them, and assures a smooth flow of information throughout the computer. To insert the problem data and the program of instructions into the computer and retrieve the answers later on, we must communicate with the computer in its own (binary) language. This communication with the outside world and the coding of all information into computer language is the function of the input and output equipment.

PROGRAM SELECTION AND COMPUTER CO-ORDINATION BY CONTROL UNIT

As we have mentioned earlier, the control unit is the master dispatcher of the system. It must select the instructions from the stored program and interpret them, and it must co-ordinate the flow of information in a properly timed sequence of operations through the memory, arithmetic, and input–output elements of the system. Essentially, the control unit must do everything a human operator would do in carrying out the same computations manually, with pencil and paper or using a desk calculator. The control unit thus provides the all-important automatic feature, which was listed as an essential part of the definition of present-day computers at the beginning of the book. Specifically, the following functions must be performed by the control element of every automatic computer:

1. Start and stop the computer.
2. Generate the computer (clock) pulses that control timing.
3. Transmit pulses to memory (store) to obtain instructions and operands.
4. Decode (interpret) instruction words.
5. Transmit pulses to the arithmetic unit and other parts to execute the operations called for by the instructions.
6. Keep track of memory locations (addresses) to be used.

Starting and Stopping

A computer springs to life when its clock pulse generator is turned on. A steady stream of pulses then flows to every part of the machine; they assist in carrying out the predetermined program. Thus, starting or stopping the computer is equivalent to turning its pulse generator on or off. This is done by pressing the START or STOP button at the computer keyboard, or console, and it is one of the few manual operations required.

After the computer has been started, its first task is to 'read' the information (instructions and data) stored in the program from an input device, such as a card reader or tape machine. This information is transferred into the internal computer storage (memory) for later use. The calculating part of the computer remains idle until at least a portion of the instructions have been stored in the memory.

Generation of Clock Pulses and Timing

Unless the computer is asynchronous (i.e. the completion of one operation starts the next one), all its operations are controlled by a series of equally spaced timing pulses from a master clock oscillator. With huge quantities of logical and arithmetical operations to be carried out, careful timing is obviously necessary to synchronize the sequence of operations and set aside a sufficiently long interval for each operation. Every transfer command, addition, subtraction, comparison, etc., must occur precisely on time and fit into the overall schedule. Moreover, the amount of time allowed for any operation is extremely small, being usually measured in microseconds or fractions of microseconds. The rate at which timing pulses are being generated and used by the computer is known as the repetition rate or clock rate. This may run to several million pulses per second. (A million pulses, or complete cycles, comprise one megacycle; one megacycle per second is written as 1 Mc/s.) The faster the computer operates, the higher its clock rate will be. The timing pulses are generated by the clock generator and are distributed to all parts of the system.

The Clock Generator. Any stable oscillator that can generate signals of a sufficiently high frequency can serve as a clock generator. A crystal-controlled radio-frequency oscillator that generates extremely stable sine waves is frequently used. Special electronic wave-shaping circuits then convert these high-frequency sine waves into the basic rectangular timing pulses of the desired frequency. An alternate approach consists of generating the rectangular timing pulses directly by means of a free-running, or astable, multivibrator, thus eliminating the wave-shaping circuits. Recent design improvements permit the use of direct crystal control to stabilize the frequency of such a square-wave clock generator (see Fig. 144).

The circuit diagram of a typical square-wave clock generator is shown in Fig. 144 (*a*). This is basically a transistorized astable multivibrator of the type described in Chapter 9, except that one of the capacitive coupling networks has been replaced by a 1 Mc/s quartz crystal. The constants of the circuit are so chosen that oscillations occur normally at approximately 1 Mc/s; the crystal in one coupling path forces the oscillations to take place at exactly 1 Mc/s and prevents any drift in frequency due to temperature or other changes. Thus, the necessary frequency stability for an accurate clock is assured.

Basic Timing Pulses. The basic square-wave output of the clock generator, shown at the top of Fig. 144 (*b*), is seen to be a train of equally spaced pulses. The clock waveform is available as a voltage when it is required to activate voltage-sensitive devices, such as diodes and flip-flops; a clock current is provided for magnetic cores and other current-operated devices. In either case the clock output consists of the basic timing pulses, which establish the entire time scale of the computer. You may think of the individual pulses as the smallest units of time in which a single action can be peformed or a bit of information can be represented. The presence of a timing pulse may represent a binary one, while its absence represents a zero. A series of these pulses, perhaps 30 or 40, can represent all the bits in a computer word, consisting of ones and zeros. Note that the time of one pulse, or one complete cycle, includes the pulse itself as well as the interval between adjacent pulses.

Derived Timing Functions. The basic timing pulses are rarely sufficient to carry out all necessary operations and functions in a computer. Additional pulses and waveforms derived from the basic pulses are usually required. The timing waveforms used in the MOBIDIC (Mobile Digital Computer), for example, shown in Fig. 144 (*b*), are illustrative of one type of system. Since pulses have a tendency to deteriorate when transmitted over long paths, the MOBIDIC uses voltage levels, with a high level representing a binary one and a low level representing a zero. Accordingly, the 1 Mc/s clock pulses are divided into high (*p*) levels and low (*t*) levels, each recurring at one-half the

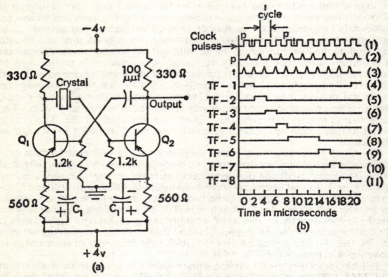

Fig. 144. Transistorized 1 Mc/s square-wave clock generator. (*a*) Circuit diagram; (*b*) clock output and derived timing wave forms. (After *Sylvania Technologist*.)

clock rate, or 500 kilocycles per second. These 500 kc/s timing levels are available as separate square-wave outputs and are transmitted to every part of the computer. Each of these timing levels lasts for 2 microseconds (since $\frac{1}{500} \times 10^{-3} = 2 \times 10^{-6}$ sec = 2 microsec), which is too long a period for starting or stopping a particular operation. Consequently, very sharp timing pulses are generated locally, where needed, by differentiating the *p* and *t* timing levels. You will recall that an R–C circuit whose time constant is short compared to the period of an applied pulse waveform acts as a differentiator; that is, it generates sharp spikes approximately proportional to the rate of change of the waveform at its beginning and end (called leading and trailing edges). The *p* and *t* pulses at the output of the differentiator, shown in lines (2) and (3) of Fig. 144 (*b*), are only 0·1 microsecond in duration. They are used for switching flip-flops and in pulse logic circuits. The 500 kc/s pulses can also be combined in an OR-gate, which acts as an electronic mixer, to re-establish an effective clock rate of 1 Mc/s.

In addition to the *p* and *t* output pulses, a special timer unit generates eight timed gating levels, which are referred to as timing functions TF-1 to TF-8.

[See lines (3) to (11) in Fig. 144 (b).] These timing functions represent the eight periods of a MOBIDIC 'basic cycle', which consists of all the steps required by the computer to execute a single instruction. Note that each timing function, which is normally 2 microseconds in duration, is synchronized with the t pulses at the leading and trailing edge. Thus, each step of a computer operation is properly synchronized with the basic timing rhythm of the system. Simple arithmetic operations, such as addition and subtraction, are performed within the time of a basic computer cycle, or a total of 16 microseconds. More complex operations, such as multiplication and division, require considerably longer basic cycles. The normal timing sequence must therefore be interrupted, and one or more of the eight timing functions must be 'stretched' to the required operating time. This is illustrated by timing function TF-5 (line 8), which is shown expanded to 6 microseconds. The timing sequence must also be interrupted whenever a complete computer word is transferred to or from the internal memory, which requires approximately 8 microseconds in the MOBIDIC computer.

Control Methods. It is not sufficient, of course, to merely specify the time intervals in which the operations are to occur. The various operations must be forced to take place in the specified periods. One way of assuring this is to make the clock time an integral part of all computer operations. The clock voltage and the command pulse for a particular operation are applied together to the input of an AND gate, which acts as a coincidence circuit, as you will remember. The gate is 'open' only when both inputs are present at the same time, and only then is a coincidence pulse produced at the output. By applying a gating level of the required duration and at the proper time, the AND gate is opened during this interval and forces the operation to be carried out at that time. For example, assume that the command 'add magnitude' is the fourth step in a complete eight-step, 16-microsecond addition operation. If the steps are of equal length the 'add magnitude' command (fourth step) must be carried out in the interval from 6 to 8 microseconds after commencing the addition operation. Timing function TF-4 in Fig. 144 (b) would therefore be the required gating level for this step, and it provides one input to an AND (coincidence) gate. The other input is the actual 'add magnitude' command, which will be carried out only if it is present during the 2-microsecond interval of timing function TF-4. Thus, each operation, in turn, can be completed in a strict time sequence.

An alternate method does not require the clock signal to be present during each step of an operation, but it relies on all operations being accurately timed in advance by a central timer, with the end of each step automatically signalling the start of the next one (asynchronous operation). In either case static timing signals of the proper duration are required for the transfer of data words and digits from one computer location to another.

SELECTION OF INSTRUCTIONS

We turn now to an important function of the control unit—the selection and execution of instructions. The essential element that differentiates the modern computer from the 'garden variety' desk calculator is that the computer automatically selects the instructions for its operation from a stored program and then proceeds to carry out these instructions until the problem is solved. A single instruction (sometimes referred to as command, order, or program step) specifies in machine language an action to be carried out by the computer. Such a command or instruction may originally be written on a single line (coding line) of the program tape; during the computer run the line of coded instructions is transferred as an instruction word to the internal memory (store) of the computer. The job of the control unit, thus, is to fetch an instruction from internal storage, interpret (decode) it, execute the instruction, and then go back to fetch the next instruction specified in the program. The basic operating rhythm of any digital computer is an endless repetition of this cycle of fetch–interpret–execute and then fetch again.

Instruction Structure. Just as an address is needed for the selection of an operand from storage, an address is also needed for the selection of an instruction. We have already explained in detail the selection of operand

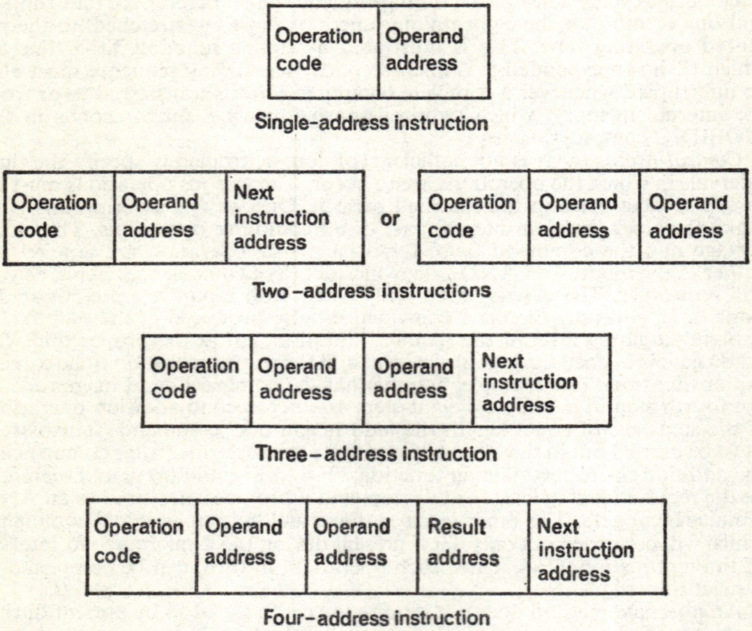

Operation code	Operand address

Single-address instruction

Operation code	Operand address	Next instruction address

or

Operation code	Operand address	Operand address

Two-address instructions

Operation code	Operand address	Operand address	Next instruction address

Three-address instruction

Operation code	Operand address	Operand address	Result address	Next instruction address

Four-address instruction

Fig. 145. Typical instruction word formats in digital computers.

addresses from various types of storage systems; we must now do the same with regard to the selection of instructions. The manner in which the addresses are specified determines the basic format of an instruction word. Essentially, any instruction must specify the memory location of the operands for a particular operation and also a way of obtaining the next instruction to be carried out. If this information were to be completely specified a typical instruction might read: 'Take out the number (operand) from store address 311 in the memory; also take out the operand from storage address 745. Send both operands to the arithmetic unit and perform the operation indicated by the operation code. Store the result in a (specified) register. Then take out the next instruction from the proper register.' Obviously, it would necessitate a very lengthy instruction word to specify such a complete instruction, even when binary digits are used. To prevent this waste of storage space and read-out time, various simplified instruction word formats have been devised, the most common of which are illustrated in Fig. 145.

Four-Address Instruction. The nearest thing to specifying a complete arithmetic operation is the four-address instruction, illustrated at the bottom of Fig. 145. This type of instruction contains the addresses of both operands, the code for the operation to be performed, an address where the result is to be stored, and the address of the next instruction. The operands are sent to the proper registers in the arithmetic unit, the operation code is decoded and the operation is performed, the result is stored in still another register, and finally a new instruction is located. You can see that this format is wasteful for simple types of instructions. If, for instance, a single number is to be taken from storage and is to be added to another number already in the arithmetic unit, only one operand and the operation need be specified; the rest of the instruction is filled in with zeros, and hence is wasted. A few early computers were built with four-address instruction formats.

Three-address Instruction. As shown in Fig. 145, the three-address instruction is the same as the four-address type, except that no address is specified where the result is to be stored. In this case either the result is always stored in the same place or another instruction tells where to store the result. A variation of this type (not shown) provides an address for the result, but omits the address of the next instruction. In the latter type the new instruction is selected from storage automatically, according to a predetermined program sequence, as we shall see in more detail later on. A few computers—the NORC, for example—are three-address machines.

Two-address Instructions. There are two types of two-address instructions, with the one shown at the left in Fig. 145 being used in a great many computers. The instruction consists of an operation code, the address of one operand, and the address of the next instruction. The operand on which the operation is to be performed is selected from storage and sent to the appropriate arithmetic register, the operation is decoded and performed; the address of the next instruction is then looked up. This new instruction most likely contains the address of a second operand, the code for an operation to be performed on it, and the address of the next (third) instruction. The third instruction will then contain the address of a register where the result of the operations is to be stored. Evidently, in this type of instruction format at least three instructions are needed to perform a complete operation: two for looking up the operands and one for storing the result.

A second category of two-address instruction, shown at the right in Fig. 145, lists the addresses of two operands and the code for an operation to be performed on them. Neither the address of the next instruction nor the address where the result is to be stored appear. Although this format specifies a complete operation, at least one additional instruction is needed, specifying where the result is to be stored. The address of the next instruction is implied; it is selected sequentially from storage. The ERA 1103, the IBM 650, and many other computers have two-address instruction formats.

Single-address Instruction. The single-address instruction is the most common, and is used in most of the recent machines. Each instruction is limited to looking up a single operand and contains the code for performing an operation on that operand. This is not as confining as it might appear at first, since many simple operations require only one operand, a second one being already contained in the accumulator register of the arithmetic unit. For example, in a consecutive addition operation the first number (operand) brought to the arithmetic unit is simply added to the zeros in the accumulator register, which has previously been cleared. The second number, or operand, is then added to the first, already in the accumulator, the third to the previous sum, and so on. Thus, only one operand and operation—specified by a single address instruction—is needed for each consecutive addition. Of course, a new instruction is needed to locate each additional operand and perform the required operation. Moreover, when the result is finally compiled in the accumulator another instruction is required which tells where the accumulated result is to be 'put away' in the memory. Since the address of the next instruction does not appear in the single-address system, it will again be necessary to provide some method for selecting new instructions in the proper sequence from internal storage. We shall deal with this topic later.

Partial Address. It is not always necessary to completely specify the addresses of the operands or the next instruction. In machines where the number of instruction word bits is limited, several operand addresses and a new instruction address may have to be squeezed into a small word space. A type of shorthand known as partial address or floating-reference addressing is used in this case. The address of the next instruction is specified with a few digits by reference to the operand address. For example, a two-address instruction may state (in decimal form): 'Subtract, 257 (operand address), 2 (instruction address).' In binary form, the equivalent would be: 'Code for subtract; 100000001 ($= 257$); 10 ($= 2$).' The control unit would then interpret that a subtraction is to be performed on an operand located at storage address 257, and that the address of the next instruction is 259 (i.e. $257 + 2$). To distinguish it from a conventional two-address format, this scheme is frequently referred to as '$1 + (1)$ address'. You can see that this same idea can be applied to single-, triple-, or quadruple-address instruction formats.

Another way of simplifying addressing is the use of a REPEAT order. If the same instruction for a particular operation is to be carried out a number of times, successive operand addresses can be specified by incrementing each by a digit or two, and keeping track of the total by means of an address counter. Moreover, instead of looking up the address of a new instruction each time, the order is simply repeated.

Control Sequence for Executing an Instruction

Let us now see how the computer actually carries out an instruction selected from storage. To a large extent, of course, this depends on the design of a particular computer (parallel, serial, etc.) and the instruction format chosen. However, certain similarities are present in all computers, and it may be illustrative to consider a simplified control sequence for a single-address, serial-storage computer. A functional block diagram for this type of control is shown in Fig. 146.

Program Register. Essentially, the following must be done: (1) an instruction must be located in the memory and transferred to control; (2) the operand (or operands) must be located and extracted from storage; (3) the operation specified in the instruction must be interpreted and performed on the operand; and (4) the next instruction must be selected from storage and transferred to control, whereupon the entire sequence is repeated. The sequence of steps suggests that the entire operating cycle should be carried out under the supervision of a 'control tower', which initiates the steps in proper sequence, counts, and times them. This is indeed the case, and the 'control tower' is variously known as program register, control register, control counter, or 'major cycle counter'; we prefer the term program register, since it stores the current part of the program. When the machine is first turned on the operator places the address of the first instruction into the program register by setting the counter to this address. The operator then presses the 'start' button, whereupon the required sequence of steps for the first instruction is carried out. While the instruction is being executed, the counter of the program register advances itself by 1, which becomes the address of the next instruction. After the operation is completed this new address locates the next instruction, and so on

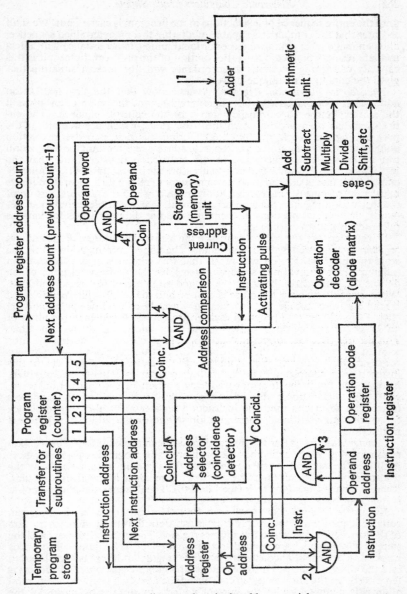

Fig. 146. Simplified control diagram for single-address, serial-storage computer.

until the entire sequence of instructions in the program is carried out. We shall see later that the computer is capable of altering this predetermined sequence, and can make what is known as conditional jumps, in accordance with intermediate results obtained during the solution of the problem. Indeed, it is this capacity of making decisions in accordance with the present situation that gives the computer its unique power.

Transfer to Instruction Register. Assume now that the first instruction address has been placed into the program register and the operator has pressed the starter button to commence step 1 of the control sequence. (The entire cycle has been broken down into four steps, which are marked at the program-register block, in Fig. 146.) During this first step the instruction address is temporarily stored in a separate address register until the instruction can be located and transferred to the instruction register. Note that the instruction register, which will eventually store the instruction word, is initially empty. This fact is utilized in some computers to eliminate a separate address register, by placing the instruction address directly into the address portion of the instruction register. Such an arrangement, where complete computer elements, or identical portions of them, are used at different times for various purposes, is known as time sharing.

During step 2 of the control sequence the instruction address in the address register is compared with a continually changing 'current address' in the store (memory unit). You will recall that this comparison is performed in different ways for various storage systems by the address selector, which detects the coincidence between the desired and current instruction address. When coincidence occurs, a pulse is emitted and the selected instruction word is read out from storage. The instruction word, the coincidence pulse, and the 'step 2' signal from the program register are applied together to the input of an AND gate, which opens only when all three are present at the same time. This assures that the correct instruction word is transferred during the time allotted for step 2. When the gate opens, the instruction word is inserted into the instruction register, which stores it until the entire operating cycle for this instruction is completed. The instruction register is actually two registers in one: one holds the (operand) address portion of the instruction word, while the other stores the operation code. If a two- or three-address instruction format is used the instruction register must also provide space for storing an additional operand address and/or the address of the next instruction.

Counter Advance. After the instruction has been placed into the instruction register and is being carried out the counter of the program register advances itself by 1 to the next instruction address. This is done automatically by the transmission of the current address count to the adder in the arithmetic unit, where a binary 1 is added to it (see Fig. 146). The incremented address count is then sent back to the program register.

Transfer of Operand to Arithmetic Unit. During step 3 of the program counter sequence the operand address is transferred from the address portion of the instruction register to the address register. The operand address and the step 3 counter signal are applied to the input of an AND gate, which opens when both are simultaneously present. The operand address is then inserted into the address register. (If there is no address register the operand address is fed directly to the address selector.) During the next step (step 4) the operand address is compared with the current address in storage by means of the address selector. Again—as in the instruction address selection—coincidence between the desired and currently stored address causes the emission of a coincidence pulse from the address selector and the read-out of the operand word from storage. The operand word, coincidence pulse, and the step 4 counter signal are applied to the input of another AND gate, which is enabled only when all three are simultaneously present. The operand word (number) then passes to the appropriate register in the arithmetic unit.

Operation Decoding. While the operand is being transferred to the arithmetic unit, the operation to be performed must be interpreted and the arithmetic unit must be prepared to carry it out. The operation code of the instruction is decoded by the operation decoder, shown in the lower right-hand portion of Fig. 146. The decoder consists essentially of a diode matrix, which can activate any one of a number of output gates in accordance with the binary operation code. These gates determine whether an addition, subtraction, multiplication, division, or other operation will be performed by the arithmetic unit. We shall study encoders and decoders in greater detail during the discussion of the computer input–output section.

Operation. The operation code is transferred from the operation part of the instruction register to the operation decoder, where the information is decoded. The selected decoder output gate is not activated, however, until the arrival of an activating pulse from another AND gate. This AND gate receives the coincidence pulse from the address selector at the completion of the operand selection and, also, the step 4 counter signal. The gate is enabled only when both signals are present, assuring that the decoder output gate is not activated before the proper operand has been selected and that the operation is actually carried out during the interval allotted to step 4 of the cycle. When the AND gate opens, the selected decoder output gate directs the coincidence (activating) pulse to the arithmetic unit. Here the control sequence is subdivided into smaller substeps necessary to carry out the arithmetic portion of the problem. The arrival of the pulse from the decoder output gate activates the subcontrols needed to transfer the operands into the proper arithmetic registers and carry out the operation. These control subdivisions are frequently numerous, and a separate 'minor cycle counter' is generally provided to keep track of the sequence of steps and whether enough time (clock cycles) has elapsed for the sub-portions of the problem to have been accomplished.

Selection of Next Instruction. Step 5 (shown separately in the diagram) is identical to step 1; it consists of the transfer of the next instruction address to the address register. The entire cycle (steps 1 to 4) then repeats itself.

Other Instruction Formats

Although the control diagram (Fig. 146) has been drawn for a single-address system, the general operation of the computer would not be very different for other instruction formats. In a two-address system, for example, the next instruction address would not be derived from the separate counter in the program register, but it would be contained within the instruction stored in the instruction register (see Fig. 145). The last step of the control sequence would consist of transferring the next instruction address from the instruction register to the address register (instead of taking it out of the program register).

The chief difference in three-address or four-address machines compared with the single-address type is the greater length of the instruction register needed to store the additional addresses and the extra control steps required to carry out a single instruction. Thus, in the three-address format first one operand would have to be selected, then the second operand, then the operation would have to be performed, and finally the next instruction address would have to be obtained from the instruction register. In the four-address structure an additional step is required to transfer the address of the result and store it before the next instruction address is taken out of the instruction register. In partial-address [1 + (1)] systems the sequence is essentially the same as described for the single-address format, except that the next instruction address is generated in the arithmetic element, as the sum of the operand address and the partial address of the next instruction.

Branch Instructions

The power of the present-day digital computer resides in its capacity to deal effectively both with the drudgery of endlessly repeated calculations, and with new conditions, arising during the solution of a problem, which require a modification of its normal operating sequence. A computer chews up instructions at a tremendous rate. If it were not able to instruct itself to repeat certain calculations over and over again, the programmer would have to prepare all the detail steps of repetitious computations himself—an extremely time-consuming process. Moreover, the programmer cannot foresee all the possible intermediate results occurring during the solution of a problem, upon which the further course of the problem solution may depend. If the computer could not make certain decisions at crucial points of the solution and then follow the required course of action, it would have to stop at these points and the programmer would have to decide what to do next. Actually, both types of situations described above are resolved automatically by the device of branch instructions. The first type, involving repeated calculations, does not depend on the conditions within the problem and, hence, is known as an unconditional branch instruction, or a program loop. The second type, which depends on intermediate results, is called a conditional branch instruction, transfer, decision, or jump.

Program Loops. In making out a payroll or computing an electricity bill, for example, essentially the same calculations are repeated over and over again, a situation which is handled by program loops. A loop is simply a series of instructions that are repeated a number of times. For example, a complete division operation may, perhaps, require seven consecutive instructions: (1 and 2) to obtain the operands (requiring two instructions); (3) test-subtract the divisor; (4) restore the remainder; (5) shift left; (6) store the quotient; and (7) test for next quotient digit. If a number of divisions are to be carried out the same seven instructions will suffice for each, and only the new operands for each operation need be specified. These instructions are therefore looped together by means of a branch instruction, which tells the computer to go back and repeat the first seven steps. You will recall that the program register controls the consecutive order of instructions by advancing the instruction address by one each time. Instruction (8), in this case, would simply state (in binary terms) 'branch to instruction (1)'. The program register would then repeat the address sequence of the first seven instructions over and over again, until told to 'break the loop'.

Since the instruction addresses contained in the program register are identical for each loop, but the operands change, the addresses of the new operands must be specified in some way. This can be done by the method of partial addressing we have described before. The operand addresses for each set of instructions are simply incremented by a specified number, which gives the new operand address. The arithmetic unit then adds the specified digit to the operand and sends the new operand address to storage to obtain the operand for the next loop sequence.

Iterations. Program loops are used not only for brief sequences of instructions, such as a division operation or other subdivision of the problem, but sometimes a program loop includes the entire problem. In a complicated problem some parameters may not be initially known and certain assumptions may have to be made to obtain an approximate solution. This approximate solution does not completely bear out the original assumptions, but can be used to obtain a more accurate answer on the second try. Again the answer can be put back into the problem for a still more accurate solution, and by repeating this process over and over again, as required, a solution to any desired degree of accuracy can be obtained. This is known as the method of successive approximations, and it requires that the computer go back to the beginning after each solution and repeat the entire process again, using the new answer. Each successively more accurate cycle of solving the problem is called an iteration.

Breaking the Loop. Although a computer can be programmed in advance to make a loop over a certain number of instructions at some point, a condition within the problem determines when the repeated operations have been completed and the loop must be broken. In a division operation, for example, the loop must be broken when all possible quotient digits have been generated (and the remainder becomes negative); in a payroll computation the loop must be broken after the take-home pay of each individual has been calculated, and so on. In each case the computer must look for some condition in deciding when to break the loop. Breaking a loop, therefore, requires a conditional branch instruction, or a jump to a new instruction.

Conditional Branch Instructions (Decisions or Jumps). The conditional jump is a special instruction that tells the computer to sample some condition, such as the sign of a number in the accumulator register or the relative magnitudes of two numbers. Depending upon the outcome, the machine then either continues its original instruction sequence or branches to the address of a new instruction. To be able to do this, the address of the new instruction must, of course, be included in the present address. This is true even for a single-address system, which normally does not contain the next instruction address.

As an example, a conditional branch instruction may specify: 'Check sign of number in accumulator; if number is negative, branch to address 465 for next instruction; if it is positive, obtain next instruction address from program register, as usual.' (Although here written out in full, such an instruction requires only a few coded digits plus the branch address.) To carry out this instruction, the operation decoder is activated immediately to decode the instruction after it has been transferred to the instruction register, instead of first selecting the operand address. After checking the sign with a sign comparator, the control either branches to instruction address 465 if the sign is negative or obtains the next instruction address from the program register (as usual) if the sign is positive. To save space, the conditional address may be a simple modification of the normal address of the next instruction, such as the addition of 1 (partial address).

Conditional branch instructions are used in many situations, not only to break out of program loops. For example, in computer-prepared bank account statements, the machine must decide if cheque withdrawals result in a negative balance, resulting in an overdrawn account. Thus, whenever a withdrawal from the old balance results in a negative balance, the computer automatically restores the old balance and, perhaps, prints out a notice of 'account overdrawn'. As another illustration, electricity bills are made out by comparison with a standard rate table, which is full of conditions. A certain minimum fee is charged (and stored in the memory) regardless of the amount of electricity used. With some tariff systems, above certain fixed amounts the rates decrease. The computer subtracts the meter reading from the fixed amounts stored in the memory, and selects the rate by conditional branch instructions. If the balance is positive, less than the fixed amount has been used, and the higher rate is charged; if the balance is negative, more has been used, and the bill is computed on the basis of the lower rate.

Types of Instructions

There are at least four basic types of instructions, of which we have thus far considered three. Let us briefly review these:

1. *Arithmetic Instructions.* These are instructions that tell the computer to add, subtract, multiply, divide, compare, etc.
2. *Transfer Instructions.* These concern the transfer of data (operands) and instruction words within the computer proper. Thus, they may specify transfer of an instruction from storage to the instruction register, transfer of an operand from the auxiliary register to the accumulator, transfer of a result back to storage, etc.
3. *Branch Instructions.* As we have seen, branch instructions may be either unconditional, telling the computer to start a new instruction sequence at a certain point or make a loop; or conditional, presenting a choice between two subsequent instructions depending upon a condition to be sampled.
4. *Input-Output Instructions.* There is another type of instruction, which concerns the transfer of instructions and data from the input section to the computer, the transfer of results to the output section, the type of print-out to be used, and other input–output operations. We shall consider these later in this chapter.

TECHNIQUES OF PROGRAMMING

Programming a computer is more an art than a science. Given the same problem, there are a multiplicity of ways of instructing a computer to solve it—some good, some bad—and no two programmers are likely to come up with the same program. However, any programmer must essentially accomplish two major tasks: he must put the problem into a mathematical form that the computer can solve, and he must translate this mathematical scheme into a language that the computer can understand. In a complex scientific or business problem the job is usually broken down into at least four major steps, each generally carried out by separate personnel. These may be classified as follows:

1. *Numerical Analysis.* A scientist, mathematician, or department supervisor may hand a computer analyst the original problem, either completely described in mathematical form or, perhaps, as an idea of 'something the computer should be able to do'. In either case the analyst must define the problem and analyse it in all its aspects. He must look over the present procedures used for solving the problem and determine whether or not the computer can make a significant contribution in time saved, efficiency, and cost. He must also analyse the mathematical formulae and equations of the problem and simplify them to the common arithmetic operations the computer is capable of performing. He may need more information and may possibly consult with a number of persons to properly define the problem and find a mathematical format for solving it.

2. *Programming Proper.* The programmers (usually several in a large installation) examine the mathematical procedures worked out by the analyst in the light of the computer's capabilities, suggest possible changes and simplifications, and then proceed to prepare the actual procedures and instructions that the computer will carry out in solving the problem. At this stage the problem is represented in the form of flow charts, which graphically describe the procedures to be performed in arriving at a solution (see Fig. 147). The flow charts are an English-language description of the steps the computer must follow, showing all the loops, re-runs, and iterations needed

to solve the complete problem. It is here that the greatest ingenuity is required to reduce the problem—and hence the computer's task—to its simplest possible form.

3. *Coding.* Since the computer does not understand English, the detailed instructions in the flow charts must be translated into computer language; that is, into the coded instructions with which we have become familiar. This job is usually done by hand, either by the programmers or by specialists known as coders. The coder may be assisted in this time-consuming job by various automatic coding techniques, libraries of computer sub-routines, and computer self-programming, as we shall see in more detail later.

4. *Running the Computer.* When all the instructions have been coded and placed into the input of the machine the operators take over and proceed to activate the computer. Early runs of the problem with simulated data may reveal errors in programming or in the machine itself (or both), requiring an inevitable 'debugging' process. The programmed runs are checked and rechecked, changes are made in the program and, perhaps, in the computer itself until a smooth sample run of the problem is obtained. The actual problem is then fed to the machine and regular 'production runs' can begin.

As an example of what the computer must do to solve an actual problem, a flow chart of payroll processing by means of a general-purpose computer is shown in Fig. 147. Note the numerous program loops that are used to reduce the job to its simplest repetitive terms. Before the computer can run through this chart to print out pay slips, it must run through an even more complicated table to figure out the various deductions to be made for each employee.

Random-access Programming. In the example of a typical control sequence (see Fig. 146) we have assumed serial storage, where all information must be selected by coincidence detection of the required and current addresses (through an address selector). If the storage system is of the random-access type, coincidence detection is not required and the stored information can be read out immediately, as soon as the address of the information arrives in the address register. With the time of access to the information being fixed, it is no longer necessary to wait for the completion of address selection before beginning the arithmetic operation. The problem can therefore be programmed without regard to the time of access for the information in the registers. This is known as random-access programming. Even when the storage system is not of the random-access type, it is always possible to program a computer in a manner that wastes a minimum amount of time in obtaining information from storage; this is called minimum-access or minimum-latency programming.

Microprogramming. We have mentioned that an arithmetic operation consists of many small subdivisions that require control sub-sequences under the supervision of a 'minor-cycle' or subdivider counter. This is also true for other operations, such as 'write' instructions, transfer of data, etc. Computers are sometimes designed to make all these small steps and subdivisions a part of the regular instruction sequence under the control of the programmer. Each small control step—called micro-operation—is then governed by the program sequence, a technique known as microprogramming. Although microprogramming necessitates extremely detailed preparation of the instruction sequence, with attendant considerable programming effort, it has the advantages of requiring a minimum of control logic devices within the computer and providing a great deal of flexibility in programming.

Automatic Programming. You may have gathered from the descriptions of the last few pages that coding, and programming in general, are major bottlenecks that appear to cancel some of the time saved by the amazingly rapid operation of electronic computers. As a result, much effort and ingenuity has

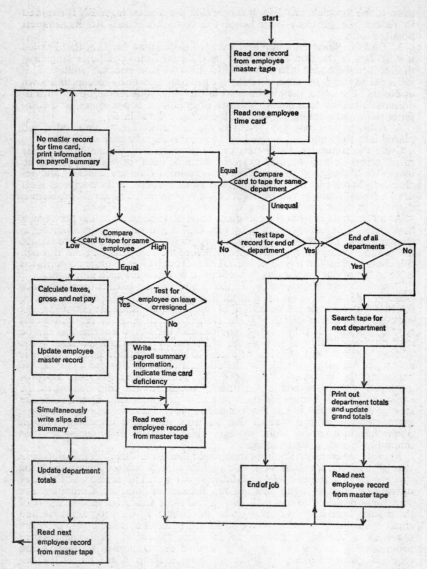

Fig. 147. Flow chart showing payroll processing with Datatron computer. (Courtesy Burroughs Corporation.)

been expended in the last few years to 'educate' computers to take over a part of the programming burden and, in essence, program themselves to a considerable extent. Ever more sophisticated automatic coding machines are being devised, which accept problems in almost the ordinary language of mathematics and 'programming English' and then convert the equations and instructions into the chain of coded commands that is digestible by a particular computer.

Among the various synthetic languages that can code instructions automatically, the following three are the most widely used: ALGOL, COBOL, and FORTRAN. ALGOL (algebraic-oriented language) is a universal algebraic language, developed by an international committee of computer specialists, that permits a programmer to write instructions for solving mathematical and scientific problems almost in the terms and symbols of the problem itself. COBOL (common business-oriented language), developed by the US Defence Department, allows the writing of business data-processing instructions in simplified business English. FORTRAN (formula translator), developed by International Business Machines Corporation, is another algebraic language 'compiler', which has been adopted by a number of computer manufacturers.

Sub-routines. A part of the 'education of a computer' is directed towards the preparation of frequently needed routine problems and program 'packages' in advance—as a 'library of routines'. The computer is then instructed to select the required programs from the library and assemble them in a special program for the solution of the particular problem. Thus, a typical program may consist of a main, or 'executive' routine that ties together a number of sub-programs and sub-routines, which the computer obtains or copies from the indexed library of sub-routines and assembles in the required sequence. Typical sub-routines may include the sorting of data, making reports, loading information into storage, finding square roots, trigonometric and logarithmic functions, series for rapid approximations, iteration formulae for various types of problems, and so on. By the use of such sub-routines compiled by the computer, the total programming time may be cut by a factor of 100 or greater, in comparison with all-human programmings.

In addition to 'libraries of routines', there are now available complete application programs, which can run a computer through a previously encountered business or scientific problem, such as information retrieval, hospital accounting, budget control, sales forecasting, etc. These complete programs can be procured from a number of computer manufacturers for use in their machines.

While a computer is performing an automatic sub-routine its regular (main) program is interrupted, of course; provisions must be made, therefore, to resume the main program when the sub-routine is completed. To effect the required transfer, the instruction addresses of the main program usually held in the program register must be temporarily stored elsewhere during the sub-routine and then reinserted into the program register after completion. This is indicated in the sample control diagram (Fig. 146) by the block entitled 'temporary program storage'. When a sub-routine is to be initiated an appropriate transfer instruction transfers the contents of the program register to the temporary program storage register, where it remains during the course of the sub-routine sequence. A new address is inserted into the program register which specifies the branch instruction that actually transfers computer control to the sub-routine program. A conditional branch instruction at the completion of the sub-routine then returns the contents of the temporary program register to the program register, thus transferring control back to the main program.

COMMUNICATION WITH THE COMPUTER THROUGH THE INPUT-OUTPUT UNITS

Every computer is linked to the outside world through its input and output equipment. To obtain the data of the problem and the instructions for solving it, the computer requires appropriate input equipment. To transmit the results of the calculations back to the human operator—in a form in which he can understand it—the machine needs output devices. The form that the input-output equipment takes depends on the type of information the computer must handle—that is, the problem to be solved—and on the design of the machine itself. Thus, the input-output devices for computers designed to process business data—compute payrolls and keep track of inventories—differ markedly from those needed to observe physical variables for computer process control. The former must process large masses of numerical data, already in digital form, though not as yet converted into machine language, while the latter must continuously measure the process variables with some sort of analogue device and then convert this analogue information into digital form, a process known as 'digitizing'. We shall take a glimpse at computer process control and analogue-digital converters in the next chapter and shall restrict ourselves to digital information here.

The type of input device also depends upon whether the information is to be inserted into the machine by a human operator using some sort of keyboard device—a slow process at best—or whether information previously stored in computer language is to be machine-inserted into the computer, which can be done as fast as the computer can use the information. Small or fixed-purpose computers generally use one or two types of input-output devices, and therefore all information must be prepared in the proper form for these devices, either directly or by conversion from some other format. Large general-purpose installations can frequently handle a variety of input-output equipment, as well as devices for converting one type of information into another, and 'files' for permanent storage. However, regardless of type, all input devices must accept the problem information and convert it into (coded) language suitable for the computer, while the output devices must accept the results computed by the machine and convert it into language that can be understood by human beings.

Mechanical input-output equipment, especially that operated by human operators, is necessarily considerably slower than the electronic devices that process the information inside the computer. Since no system can be any better than its weakest link, the actual effectiveness of a computing system depends upon the balance between its input-output equipment and the computer proper (usually called 'central processor'). Even a million calculations per second are of little use if the input-output devices can accept and print out only a few thousand binary digits each second. In order to restore the necessary balance of information flow through the system, great strides are being made at present to bring the processing speeds of input-output equipment to the ever-rising level of computing speeds. All-electronic character recognition and display devices are well on the way to solving the input-output bottleneck.

Input Devices

Some of the commonly used devices for communicating with the computer proper (central processor) are illustrated in Fig. 148. In general, these devices can be divided into 'on-line' equipment, which communicates coded information directly into and out of the computer and is under its control, and 'off-line' or peripheral equipment, which is used for preparing, storing away, or converting information and is not under the direct control of the computer.

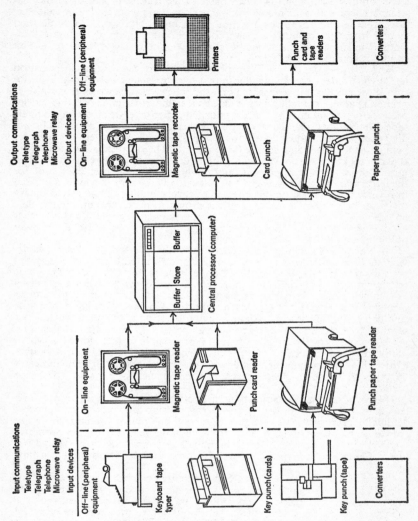

Fig. 148. Communication with computer through input-output devices.

(The 'on-line' equipment is shown next to the central processor in the diagram.) There is also a class of communication equipment, including teletype, telegraph, telephone, microwave radio links, radar, etc., which transmits data to and from the computer location for use in other areas or computer locations. The latter equipment has little to do with computer operation, and we need not consider it further.

Keyboard Devices. Most frequently used for preparing input information are keyboard devices, which fall into two categories. One is an 'on-line' device (not shown in Fig. 148) consisting of a special electric typewriter that converts printed characters directly into code signals; the other is an 'off-line' device used for preparing punched cards, punched paper tape, or recorded magnetic tape. The 'on-line' manual keyboard is very slow, processing about 15 binary digits per second, and hence is used only in small computers or in larger computers for manual checking ('debugging'). The depression of each key in the manual keyboard activates a combination of six or more relays that transmit an equal number of current pulses, which represent the binary code for the particular character (see Fig. 149). Essentially the same device can be used in reverse to print the output characters corresponding to the coded binary digits of the computed results. (This is explained in the section on encoders and decoders in this chapter.) The use of the manual keyboard for both read-in and print-out represents the simplest, most direct, and also the slowest possible input–output system.

The key punch is perhaps the most frequently used 'off-line' keyboard device. Its basic function is to receive character information on a keyboard and produce punched cards or punched paper tape perforated in a binary code corresponding to the keyboard impressions. Various types of card and paper-tape punches exist. They may be separate keyboard devices or simple attachments on cash registers or typewriters with the punches operating in response to a depression of the keys. All are slow, processing about 10 characters per second, which is equivalent to approximately 60–120 binary digits. (A character is composed of 6–12 binary digits.) Processing is further slowed down by the need for the detection of errors, a process called verification. A punch-card verifier manually compares separate recordings of the cards and indicates by appropriate signals when holes have been punched in the wrong places or not at all.

The keyboard magnetic-tape typer is another 'off-line' device for preparing coded information for later high-speed transmission to the computer memory. Manual depression of the keys produces combinations of voltages, which are converted by magnetic heads into corresponding magnetic fields to produce a coded record on magnetic tape. Hundreds of characters can be stored per inch (of length) of the tape. The tape is stored for later insertion into a tape reader.

Readers. A variety of devices are available that can 'read' a stored form of computer language and regenerate the original coded signals. Like most input–output devices, the readers enjoy a sort of double existence: they are used 'on line' as input devices to give information to the computer from the stored medium (cards, tape, etc.); they are also used 'off line' at the computer output to read the previously stored results and transmit them to an outside communication link, to a printing device, or to a 'converter' for conversions into another storage medium. The most widely used reading devices are punch-card readers, punch-tape readers, and magnetic-tape readers.

Punched-card systems were pioneered by IBM for data-processing long before the arrival of the automatic computer. The cards can easily be sorted into any desired sequence and can be 'read' through an electrical or mechanical hole-sensing device at a speed varying between 150 and 1000 cards per minute, depending on the equipment. A typical IBM card consists of some 80 columns with 12 spaces per column, into which holes may be punched. The holes in each column or row represent a particular character in a coded pattern that depends on the hole and column position. Ten of the 12 spaces represent decimal digits and the others—called zones—are reserved for sign and additional information. Letters and alphanumeric characters are represented by combinations of zone and numerical bits according to the code used.

A card reader is a mechanism for the high-speed transport of punched cards from an input hopper past the sensing (reading) device to an output stacker. As each card moves past the reading station the holes are sensed by flexible wire brushes or 'mechanical fingers' that ride against the card and make electrical contact with a metal plate behind the card whenever a hole goes by. Holes can also be sensed photoelectrically by passing a beam of light through them to photoelectric cells behind the card, and in a variety of other ways. A punch tape reader operates essentially in the same way, except that the reading process is continuous and the pattern of holes is usually directly interpreted as binary data. The storage capacity of punched paper tape is higher than that of cards, but the tape cannot be sorted, which creates checking and editing difficulties; moreover, it must be rewound at the end of a run.

Magnetic-tape readers sense the recorded information by passing the tape underneath magnetic heads, which pick up the small variations in the magnetic field and re-create the original coded signals in electrical form (see Chapter 11). Tape readers are usually part of complete magnetic-tape units, which can both record (write) and read back data immediately, similar to the popular home tape recorders. In contrast to audio tape recorders, however, computer tape transports are operated at high speeds of over 100 inches per second (as against 3·75 or 7·5 inches per second for audio recorders) and the tape forms long buffer loops at each of the reels, to take up the slack during rapid start and stop operations. Tape units have a high recording density of several hundred characters per inch of tape, and many are equipped for immediate 'read-during-write' operation, thus permitting very fast processing and immediate error detection.

Converters. It is frequently necessary to convert one kind of computer language into another. Small computers with only one type of input–output system must be able to process information from other computers with a different type of input–output device. Large computers usually transcribe the data stored on punched tape or punch cards on to magnetic tape for high-speed input processing and use the reverse process at the output. These requirements are fulfilled by an 'off-line' device known as a converter, which can translate one kind of machine language into another. Almost any type of converter is commercially available; among them punched card to paper tape, punched card to magnetic tape, paper tape to magnetic tape, and the reverse operations for each of these. There are also tape-to-tape converters that translate one type of (magnetic) tape code into another.

Increasingly sophisticated universal converters are being developed. For example, one commercially available unit translates teletype, magnetic tape, or punch cards into a common computer language. Information can be taken from teletype circuits at the rate of 3000 words per minute and is fed directly to the computer, or vice versa. Such converters can be used, for example, to relay computer-assembled 'buy' orders by teletype to warehouses for shipping. Another recently manufactured converter can take information from punched cards or tape at various locations and transmit it by telephone to a central computer at the rate of better than 1500 words per minute.

Buffers. The 'on-line' input–output devices rarely communicate directly with the internal memory (storage) of the computer, 'buffers' being generally necessary to compensate for the inevitable differences in processing speed between the central processor and the input–output equipment. These buffers, or data synchronizers, have other functions as well. They not only must change the timing from the relatively slow rate of the input–output units to the rapid processing rate of the internal memory but also, when used in converters, must convert one type of code used by the input or output device to another type used within the computer. Occasionally they must change over from serial to parallel operation, or vice versa. Temporary storage registers, known as buffer memories, are provided at both the input and output to transfer data at high speed to internal storage from the input devices, or to transfer data from storage to the output devices at a lower speed. Buffer memories generally consist of shift registers, which can accept data at one rate and form (serial or parallel), and shift it out at another rate while simultaneously changing its form. (See also Chapter 12 on 'Memory'.)

Output Devices

The magnetic tape, punch card, and punch tape units also serve as 'on-line' equipment at the output of the computer. Card and paper tape punches at the output do not use keyboards, however, but are high-speed units that directly accept the computer signals and convert them into the corresponding coded patterns of perforations. 'Off-line' card and tape readers are available for subsequent read-out of the stored information to communication links or printers. Again, converters may be needed to convert one type of storage medium into another. High-speed magnetic tape units are frequently used for direct recording of the output data, with later conversion into punched cards or tape for permanent storage.

Print-out. Some sort of 'plain language' print-out of the computed results is usually required. Printing devices may be either 'on line' or 'off line' and they range all the way from electric typewriters typing one character at a time (about 20 each second) to high-speed 'on the fly' printers, which can print out 1000 lines per minute of 120 characters each, or approximately 2000 characters per second. Which device is used depends very much on the size and speed requirements of the particular computer installation.

Electric Typewriters. The simplest type of 'on-line' print-out device is an electric typewriter with the same type of keyboard device as we have described for the input. Each combination of a six-digit computer output signal actuates a particular key through a series of relays. The typewriter may be of the ordinary kind or it may be 'Varityper' style. Some typewriters also have attachments for punching paper tape during print-out ('Flexowriters'). In general, however, electric-typewriter print-out suffices only for the smallest desk-type computers.

High-speed Printers. Much progress has been made in developing high-speed mechanical printers. Earlier types, which could print out only one line of characters at a time, were frequently operated 'off line' from magnetic tape or card punch units to avoid delaying the processing of data. Present high-speed computers print out a 'surface-at-a-time' consisting of several rows of characters, with a speed of a 1000 lines per minute being common. Various versions of these machines exist, depending on the manufacturer, but two have become particularly popular. One, called an 'on-the-fly' or Shepard printer, uses multiple revolving print wheels with a complete row of symbols on each wheel. A comparator checks the characters in printing position against those required by the computer output and actuates corresponding printing hammers when coincidence occurs. The other type, called a matrix printer, forms several rows of characters simultaneously by varying combinations of 5 by 7 arrays of wire points (per character). The coded output signal from the computer determines which of the 35 (5 by 7) wires are pushed forward to make a dot impression against paper.

Other Printers. Various types of computers have been developed in which the electrical computer signals are used more directly to effect print-out. Electrochemical printers rely on chemical changes occurring in specially prepared paper when an electrical current is passed through it, causing a dark spot to appear at specified points. Thermal printers use the electrical computer signals to heat a stylus, which then blackens heat-sensitive paper in accordance with the signals. (This process is not suitable for high-speed printing, however.) Magnetic printers make use of magnetic tape bonded to paper to attract a mixture of ink and iron powder to the magnetized areas, thus rendering the pattern visible. By passing the paper through a heating chamber for a short time, the ink can be permanently fixed to the paper. In this context, it may be of interest to note that magnetic ink is being used by banks for marking account numbers and amounts. The visible ink is deposited in the form of slightly modified conventional letters and numerals, which can be easily spotted by a magnetic detector. A magnetic field magnetizes the inked pattern shortly before read-out.

Electronic Printers. Photographic print-out in conjunction with a cathode-ray character display tube permits print-out speeds comparable to computer processing rates. Various electronic character display tubes have been developed, some of which use exterior masks of characters, while others place the mask right inside the cathode-ray-tube envelope. Signals from the computer, or magnetic tape, deflect the beam of electrons and 'write' a page of information on the screen of the tube. When a page is completed it is either photographed for later reproduction, or the image on the tube face can be printed out directly by a photoconductive (Xerographic) process. The original master sheets can be used for later duplication by conventional methods. Printing rates beyond 25,000 alphanumeric characters each second are made possible in this way.

Encoders and Decoders

Since computers work best with binary data, while human beings prefer to use decimals, various codes have been evolved for translating decimal digits into combinations of four or more binary digits, one for each decimal. The most frequently used binary-coded decimal notations were explained in Chapter 8, and you may want to refresh your memory on these before going on. We must now take a brief look at the devices which encode the decimal or other data at the input of the computer into the required binary form, and also at the decoding devices that translate the coded data from the output of the computer back into conventional form. The various keyboard and other input devices that accept decimals and conventional letters must, thus, be equipped with an encoder for generating the particular code the computer is designed for, while the output devices must have a corresponding decoder. You would also expect to find, in the control unit of the computer, decoders for decoding the various instructions, the memory addresses, and states of counters and flip-flops, etc. In general, a decoder must produce a single, meaningful output describing the equivalent of a particular input combination.

Encoders. The function of a binary encoder is to convert discrete inputs, consisting of decimal characters 0 to 9 or letters of the alphabet, into binary-coded output combinations. Thus, a single input signal representing a decimal digit (0 ... 9) must result in a multiple output signal consisting of the pulse or voltage level combination that corresponds to the input in the selected binary-decimal code. Four binary digits permit 2^4 or 16 combinations, and hence are sufficient to represent the decimal digits 0 to 9. For example, the decimal digit 9 is represented by binary digits 1001 in straight binary (8421) code, by 1100 in excess-three code, by 1010 in 7421 code, by 1100 in 5421 code, and by 1111 in 2-421 code. If alphanumerical data is to be encoded, the 26 letters of the alphabet in addition to the 10 decimal digits, a total of 36 characters, must be represented by binary digits. At least six binary digits, having 2^6 or 64 possible combinations, are then needed to represent each alphanumerical character. For this reason most modern codes use at least six bits for letters and numbers. Since the principles are the same in either case, however, we shall confine ourselves to four-digit encoders.

Keyboard Encoder. Encoders are used in the keyboard devices at the input of a computer. Consider a simple keyboard where each of the decimal digits 0 to 9 are selected by pressing a single key, as illustrated in Fig. 149. Let us

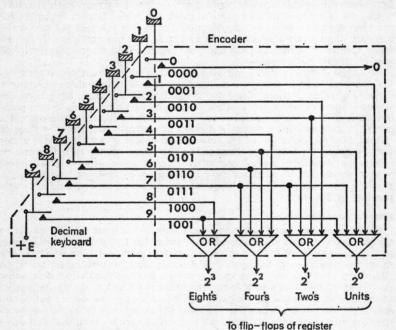

Fig. 149. Decimal keyboard with straight binary encoder.

assume that each decimal digit is to be represented by a straight four-digit binary (8421) code combination, i.e. by the ordinary binary number system. Pressing a particular key must close a circuit that activates the corresponding binary output combination; for example, pressing the 7 key must establish connexions resulting in the output pulse combination 0111. Since binary 0 is represented by the absence of a pulse, the encoder need provide output pulses only for the 'one' bits. In the example (0111), output pulses must be supplied for the unit (2^0) column, the twos (2^1) column, and the fours (2^2) column, but none for the eights (2^4) column. Checking in Fig. 149, you will see that pressing the 7 key results in sending a d.c. pulse ($+E$) to each of the OR gates representing the units, twos, and fours columns. Similarly, pressing each of the other keys results in passing a d.c. pulse to the OR gates that represent the particular binary code combination for the decimal digit. The OR gates are used to pass signals from the alternate input sources (keys 1 to 9) without tying the input together. (If the inputs were not kept separate, or buffered, confusion would result.) The output pulses from the OR gates may be used to set the flip-flops of a register, which has initially been cleared to 0000. Pressing the 7 key, for example, would result in setting the units, twos, and fours flip-flops to one, while the eights flip-flop would remain at zero.

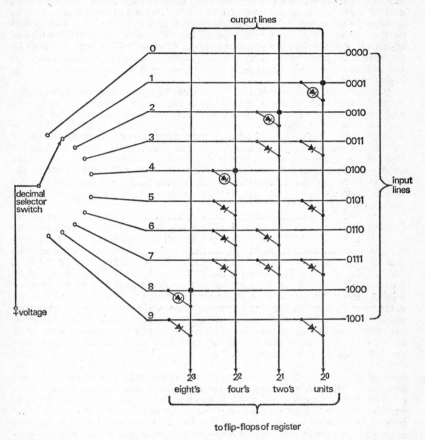

Fig. 150. Diode matrix for encoding decimals into straight binary (8421) code. (Encircled diodes can be eliminated by joining intersections directly, as shown by dots.)

Implementation. Each input leg of the OR gates in Fig. 149 represents some unilateral device, such as a diode or magnetic core, which conducts the input (decimal) signal to the proper output channel, while at the same time keeping the inputs buffered apart. The encoding matrix that results from implementing the OR gates of the keyboard encoder (Fig. 149) by means of diodes is illustrated in Fig. 150. For simplicity, the keyboard is replaced by a selector switch that can connect a positive voltage pulse to any one of the ten input lines representing the decimal digits 0 to 9. Fifteen diodes pass the discrete input pulses to one or more of the four output lines representing binary ones, twos, fours, and eights. Normally, a diode is needed for each leg of an OR gate; thus, diodes must be placed at all input–output line intersections that must conduct pulses to a particular output column. For example, all odd decimals (1, 3, 5, 7, 9) require a binary one in the units column; hence, diodes must be placed at the intersections of the units (2^0) output line and the '1', '3', '5', '7', and '9' input lines, as is shown in Fig. 150. The eights output column, in contrast, is activated only by decimal digits 8 and 9; hence, only two diodes need be provided at the intersection of the '8' and '9' input lines with the eights (2^3) output line. As you can see, the diode encoding matrix—with its horizontal decimal inputs, vertical binary outputs, and gating diodes placed at the intersections—has a certain symmetry about it which is well suited for logical design.

Diode Saving. The function of the diodes is to pass an input signal to the proper output channel without permitting interaction between the inputs, as we have mentioned. However, where an input is connected to only a single output, no interaction can take place and, hence, no diode is needed. For example, the '1', '2', '4', and '8' decimal input lines in Fig. 150 connect directly to the corresponding single outputs in the units, twos, fours, and eights columns, respectively, and hence a direct wire connexion can replace the diodes at these intersections. This is shown in Fig. 150 by the dots at the junctions (indicating a connexion), which replace the encircled diodes directly beneath them. Four diodes can be eliminated in this fashion, resulting in a matrix with 11 diodes.

Decoders. In general, decoders must perform the reverse function of encoders: they must recognize a multiple (binary) input signal combination and convert it into a single (decimal) output signal. Each element of a binary-to-decimal decoder must respond only to the unique combination of binary bits that represents the decimal character in the particular code used. For instance, the decoder element that senses a decimal 5 must respond only if the combination of binary digits 0101 is simultaneously present, and must reject, or inhibit, all other combinations. It is apparent from this description that AND gates must be used as decoder elements, since AND gates give a true output pulse for only unique combination of inputs and inhibit all others. Moreover, the number of input legs used for each AND gate must equal the number of binary digits (columns) used to represent the decimal character. Thus, a four-bit code requires four-legged AND gates, a six-bit code requires six-legged AND gates, and so on. An AND gate output must be provided for decoding each binary input combination. Since a four-bit code has 2^4 or 16 unique combinations, 16 AND gates must be provided to decode all combinations; a six-bit code requires 2^6 or 64 AND gates at the output, and so on. Finally, each leg of an AND gate requires a diode or similar element, which can result in a large number of components. For example, decoding a four-bit code normally requires 16 four-legged AND gates, resulting in a total of $16 \times 4 = 64$ diodes; decoding a six-bit code requires 64 six-legged AND gates, which amounts to a total of $64 \times 6 = 384$ separate diodes. With diodes costing several shillings each, decoding can become an expensive business.

Decoding a Three-stage Binary Counter. As we have seen earlier, counters are frequently used to initiate control sequences at a predetermined count of clock pulses. To do this, the binary combination corresponding to a given (decimal) count must be decoded by a binary decoder, whose output is then routed to the control circuit to be activated. Fig. 151 illustrates the principles involved in decoding successive counts from a three-stage binary counter, consisting of three flip-flops and sensing elements. The step count pulses are applied to the input of the unit (2^0) counter, which is reset every second count. Whenever the unit counter is set to one, it applies an output pulse to the twos (2^1) counter, which thus is reset every four counts. Whenever the twos

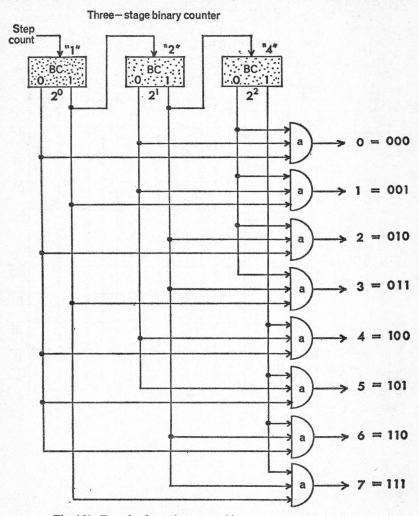

Fig. 151. Decoder for a three-stage binary counter with 8 counts.

counter is set to one (every fourth input count), it applies an output pulse to the fours (2^2) counter, which, therefore, alternates its state every four counts. After eight counts (0 to 7), the entire arrangement is reset to zero, and the count starts over again.

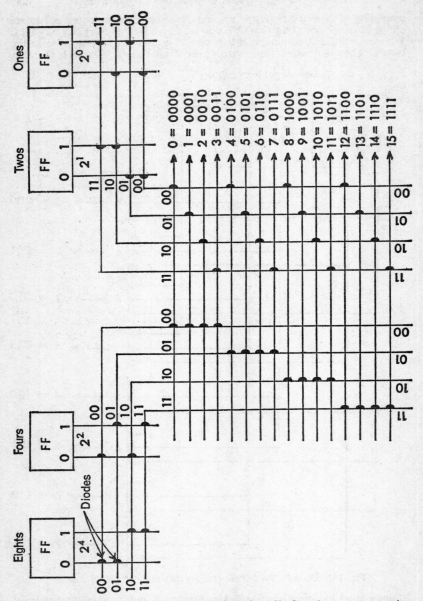

Fig. 152. Decoding the 16 unique outputs from four flip-flops in two stages, using 48 diodes.

To decode the eight possible three-bit output pulse combinations (or states) of such a three-stage counter, eight three-legged AND gates are required, as shown in Fig. 151. (A total of $8 \times 3 = 24$ diodes are needed to implement this circuit.) Note that each AND gate responds only to one unique binary output combination of the counter. Thus, the '1' output gate responds only to binary digits 001; that is, when the units (2^0) counter is set to one the twos counter is reset to zero, and the fours (2^2) counter is reset to zero. Similarly, the '7' output gate responds only when all three counter stages are set to one, representing binary 111. The 'zero' gate responds only when all counter stages are in the zero state, representing 000.

Decoding in Two Steps. When the outputs of more than three binary counters or flip-flops are to be decoded, a large number of AND gates must be driven by the flip-flops, which usually necessitates separate driving amplifiers in each flip-flop output line. Moreover, the number of AND gate diodes becomes excessive, as we have seen. Recall that the 16 possible four-bit outputs of four flip-flops, for example, require 16 AND gates and $4 \times 16 = 64$ diodes. A considerable saving in diodes can be attained by decoding the flip-flop output in two steps, as shown in the schematic diagram (Fig. 152). The D-shaped symbols at the intersections of the matrix rows and columns represent gating diodes that steer the flip-flop pulses to the proper output line.

The four flip-flops are divided into two groups, which are first separately decoded. Each of the two groups of horizontal output lines (at the top of Fig. 152) contains four two-input diode AND gates, resulting in a total of eight first-stage output lines. The second-stage decoder then combines the outputs from the left and right groups by four more two-input diode AND gates for each of the four output line combinations of the first-stage decoder. This results in decoding the 16 possible states of the four flip-flops by means of 24 two-legged AND gates, or a total of 48 diodes, as against the 64 diodes originally required.

CHAPTER 14

PRESENT TRENDS AND FUTURE PROSPECTS

We have come almost to the end of our journey through the realm of automatic computers. To clarify the technical complexities, we have had to classify and subdivide computers into various types and component parts, in addition to the major analogue and digital categories. Let us now put the pieces back together again and look at the computer as a whole. What role are computers destined to play in industry and in our lives? What new tricks will we be able to teach computers and what shall we learn from them in return? Shall we be able some day to construct computers that approximate the capacity and complexity of the human brain without becoming monstrously large or, for that matter, prohibitively expensive? If and when such capacious robots have been sired and have learned all we can teach them, will they then dominate us intellectually, make us their physical slaves, or at least put us all out of work? If, on the other hand, we should learn to control this race of 'giant brains', will the millennium have arrived? Will we all become poets, painters, composers, computer designers, or technicians?

To find out just what the realistic prospects are for the next few years, let us take a glimpse at some of the areas where computers are beginning to play an increasingly important part. One is the control of industrial processes by computers—a development which is already with us and which will have revolutionary consequences for our economy before it has run its course. The multiple relations between computers and human beings is another area of increasing interest to science and education. Finally, dramatic new computer techniques and improvements in component parts will give us additional clues to the computer's eventual role in this era, which is frequently characterized as 'the Second Industrial Revolution'.

COMPUTER PROCESS CONTROL

Computers have been used for a long time as part of larger systems and operations, such as radar fire control, automatic navigation, flight simulation, air-traffic control, the automatic programming of machine tools, and many others. Analogue computers have found their place in automatic control of many industrial processes, nuclear reactors, oil refineries, etc. There has been considerable hesitation, however, about using the digital computer for industrial process control, although it potentially offers great advantages over the analogue type because of its high accuracy and flexibility of programming. The tendency in the past has been to build a special-purpose analogue machine for one particular industrial or military application, and ignore the digital computer, though its general-purpose design could fit it for various changing applications and purposes.

There are several reasons for this. All actual processes take place in the real time of the world and not in the abstract time of mathematics. The analogue computer, being a model of a real system, can and does solve problems in the actual (real) time of the system it represents, and hence it has the answer (controlled operation) ready when it is needed. The digital computer, on the other hand, converts all system variables into mathematical operations and computes the answers in its own electronic time, which is unrelated to the time of the real occurrence. Thus, one difficulty in using a digital computer for process control is the synchronization of the real time of the process with the computer's own time. The proper phasing in of the computer into the physical time of a process is known as real-time process control.

Analogue–Digital Converters

The major difficulty, however, is how to feed a mathematical machine—a digital computer—the elements physical happenings are made of. The digital computer fits in naturally with the numerical data of business and mathematical problems. As we have seen earlier, these are discontinuous and discrete, and hence can be fed into the input of a digital computer one at a time. Most physical occurrences and quantities, in contrast, are continuous and non-numerical. They can be measured and sensed by various analogue devices, but they cannot be fed as numerical data into a digital computer without major modification. Similarly, the output of a digital computer consists of discrete numerical data, which are not suitable for controlling a physical quantity or process. Obviously, some sort of converter is necessary to match the input and output of a digital computer to the physical (analogue) quantities involved in industrial processes. Devices that match a physical variable to the input of a digital computer are known as analogue-to-digital (A–D) converters, and those that match a physical variable to the output of the computer are called digital-to-analogue (D–A) converters.

Sampling. The analogue quantities that must be 'digitized' when fed to the input of a digital computer typically may be a time interval, a frequency, a voltage or current, velocity, acceleration, pressure, temperature, and the like. To convert these continuous functions of time into discrete numerical values, sample values of the function must be taken at frequent intervals to show its numerical value at particular instants of time. The function is then represented as a number or digit during the sampling instants, but nothing is known about it during the intervals when no sample is taken. This is not too serious if the function varies smoothly with time (a sine wave for example), but it may lead to serious errors if the function undergoes abrupt changes at times, such as a square or triangular wave. Thus, the number of samples taken of the analogue function must be sufficient to define the function sufficiently well. If the function varies slowly with time, only a few sample values need be taken; if it varies rapidly, many samples must be taken.

As an example of a slowly varying function, consider a sine wave that is sampled, say, every 30° of its cycle. Assume the sine wave represents an a.c. voltage of 100V peak amplitude; if its frequency is 60 c/s, sampling every 30° corresponds to successive time intervals of

$$\frac{30°}{360°} \times \frac{1}{60} = \frac{1}{720} \text{ s.}$$

We can then make a table of values of the sine-wave voltage for successive samples every 30°, or $\frac{1}{720}$ s, as follows:

Degrees	30	60	90	120	150	180	210	240	270	300	330	360
or Time	$\frac{1}{720}$	$\frac{2}{720}$	$\frac{3}{720}$	$\frac{4}{720}$	$\frac{5}{720}$	$\frac{6}{720}$	$\frac{7}{720}$	$\frac{8}{720}$	$\frac{9}{720}$	$\frac{10}{720}$	$\frac{11}{720}$	$\frac{1}{60}$
Voltage:	50·0	86·6	100	86·6	50·0	0	−50·0	−86·6	−100	−86·6	−50	0

If we plot these sample values of the sine-wave voltage against time or degrees we obtain a series of vertical line segments that represent the value of the voltage at 30° intervals, as shown in Fig. 153. The vertical lines indicate the numerical values that would be fed to the input of a digital computer by an analogue-to-digital converter. By connecting the tips of the vertical segments with a smooth line, as shown by the dotted line in Fig. 153, it becomes apparent that the 30° samples represent a fairly good approximation of the actual sine wave. For greater accuracy samples should be taken, perhaps, every 10° or $\frac{1}{2160}$ s.

Sampling techniques are not new; they have been used in radar equipment and data transmission systems for more than twenty years. Things become somewhat complicated, however, if a number of physical variables must be sampled at the same time and their digital equivalents must all be fed to the

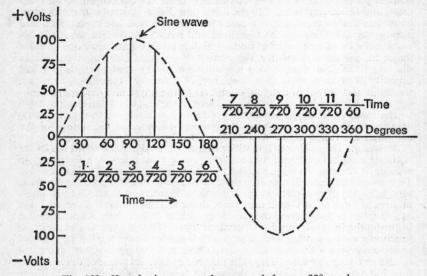

Fig. 153. 60-cycle sine-wave voltage sampled every 30° or $\frac{1}{720}$ s.

input of the same digital computers. This can be done only by sampling each of the input variables in turn by some sort of scanning device, and feeding the combined data sequentially to the input of the digital computer, a technique known as multiplexing. The computer must, of course, have some sort of discriminating (decoding) device for sorting out the input data according to the source. Moreover, the control of the process by the output of the computer must be shared in time between the various input variables.

Methods of Conversion

How are samples of analogue quantities obtained and how are they converted into digital form? There are essentially two methods of analogue-to-digital conversion, or 'digitizing': counting and comparing. We can always count the number of units in a quantity, whether it be seconds in a time interval, cycles in a frequency, volts, amperes, or whatever. By representing the units of the quantity in the form of pulses, we can use for digitizing the high-speed electronic counters with which we have become familiar. We have also become acquainted with the principle of comparison, which involves feedback. In the comparison method the unknown analogue quantity and a series of known trial values are fed to the input of an equality comparator, such as the 'exclusive-OR' circuit we have studied. The comparator rejects all trial values until one that is equal to the analogue quantity is applied; it then provides a digital output equal to the analogue input. To get better acquainted with these two methods let us look at a few examples of analogue-to-digital conversion.

'Digitizing' Time and Frequency. Figs. 154 and 155 illustrate, respectively,

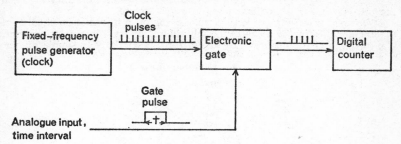

Fig. 154. Time-to-digital conversion with electronic gate and counter.

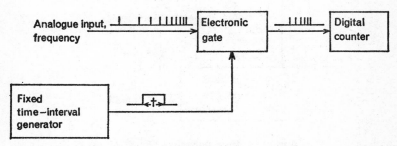

Fig. 155. Frequency-to-digital conversion with electronic gate and counter.

the analogue-to-digital conversion of time and frequency by means of the counting method. Conversion in each case is accomplished by an electronic (AND) gate in conjunction with a digital counter. In the case of time-to-digital conversion the analogue input consists of a rectangular gating pulse, which is equal in duration (width) to the desired time interval. This pulse is applied to the input of an electronic gate, thus turning it on. Also applied to the gate is the pulse output of a fixed-frequency pulse generator, which can be one of the clock oscillators we have described. During the time the gate is kept open by the enabling pulse (analogue input) the clock pulses are allowed to pass through and are counted by a digital (binary) counter. At the conclusion of the time interval the negative-going gating pulse turns off the electronic gate and no more clock pulses can pass through. The number of pulses counted during the time (t) the gate is open represents the digital equivalent of the analogue time interval.

The frequency-to-digital conversion uses essentially the same method. Again, a rectangular pulse is used to enable an electronic (AND) gate. This pulse represents a known time interval (t) obtained from a fixed-time-interval generator. Such a fixed time interval can be produced by passing the output of a clock oscillator through a frequency divider and then using the lower-frequency output to synchronize a multivibrator. The multivibrator thus generates precisely spaced rectangular pulses, whose duration is an exact submultiple of the clock oscillator repetition rate. During the fixed time the gate is kept open by the rectangular pulse a series of pulses at the analogue frequency are allowed to pass through, which are counted by the digital counter. The number of pulses counted during the known time interval (when the gate is open) is the frequency in pulses per second.

Digitizing Shaft Position. Very often the analogue information consists of the angular position of a shaft, possibly that of a servo or a potentiometer. Converting a shaft position into digital form is a relatively simple matter.

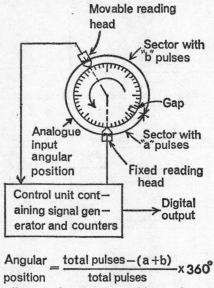

$$\frac{\text{Angular}}{\text{position}} = \frac{\text{total pulses} - (a+b)}{\text{total pulses}} \times 360°$$

Fig. 156. Schematic diagram of an angular position analogue-to-digital converter.

Essentially, all we need do is make a number of engraved marks around the periphery of the shaft (or a disc attached to it), which divide the complete circle into convenient subdivisions, 360°, for example. The marks themselves can then be made to generate some sort of pulsed signal as they are scanned by a pick-up device, which may be a fixed brush in contact with the periphery of the disc, a photoelectric cell and light source, or possibly a magnetic reading head. To determine the shaft position or the number of revolutions made by it we simply add up all the marker pulses by means of an electronic counter. The scheme of one such angular-position converter, which uses a relatively complicated set-up of two engraved discs driven by a 3600-rev/min synchronous motor and two reading heads, is illustrated in Fig. 156. The formula shows how the angular position in degrees is determined from the pulses counted.

Shaft-position Encoders. By attaching a disc to the shaft whose position is to be measured, we can obtain many outputs in addition to straightforward counting of degrees by marker pulses. The marks on the disc can be arranged to conform with any desired output code, such as a pure binary output, one of the binary-coded decimal notations we have studied, or even some function of shaft rotation, such as the sine or cosine of the shaft angle.

Fig. 157 illustrates a typical binary-coded disc, which is part of an analogue-to-digital converter. As you can see, the disc is laid out in concentric rings, starting at the periphery and working inward. The number of rings determines the number of digits (columns) of the binary read-out. The disc is further divided into equal radial segments, whose number depends upon the desired precision. The areas formed by the intersections between the rings and the radial segments represent the digits of a binary code, which are read radially along a segment corresponding to the shaft position. Printed-circuit techniques are used to prepare the marked areas. For optical read-out by means of photocells, the dark areas in Fig. 157 are made opaque and the white

Fig. 157. Librascope binary-coded disc for brush read-out.

```
 0 = 0000
 1 = 0001
 2 = 0010
 3 = 0011
 4 = 0100
 5 = 0101
 6 = 0110
 7 = 0111
 8 = 1000
 9 = 1001
10 = 1010
11 = 1011
12 = 1100
13 = 1101
14 = 1110
15 = 1111
```

Fig. 158. Coded segment representation of binary progression (0–15).

areas are left clear, while for electrical read-out by means of brushes the dark areas are made of insulating material and the white areas of conducting material. A photocell or brush is associated with each concentric ring for separate read-out of the binary digits.

To clarify the principle of disc encoding, we have 'straightened out' a portion of the disc, to show the printed-circuit representation of a sample four-column binary progression (see Fig. 158). The solid dark segments, made of insulating material, represent the zeros of the binary progression at left, while white (blank) segments, made of conducting material, represent the ones. The four white circles next to binary 0111 represent brushes riding along the four digit tracks. Whenever a brush rides over a white (conducting) segment it picks up a current pulse that represents a binary one, while the absence of current pulses along the dark (non-conducting) segments represents binary zeros.

You can see that with the brushes positioned as shown in Fig. 158 the extreme left (most significant) brush rides on a dark (insulated) segment, and hence provides no current pulse, which corresponds to a binary zero. The remaining three brushes ride on white (conducting) segments and provide a current pulse corresponding to a one. The reading of the brushes, thus, is

0111 (decimal 7), as shown in the binary progression at left. You can check each of the other brush positions in the same way.

You can see that this coding system works very well as long as the brushes are positioned at the centre of the segments, but whenever the brushes straddle the dividing line between insulated and conducting segments a positional ambiguity results. To avoid the large errors caused by this ambiguity in brush position, various methods have been devised which either displace the brushes with respect to each other in a V-shaped pattern, change the code to a cyclic form in which only one digit changes at a time, or mutually displace the segments themselves to prevent the brushes from simultaneously contacting the dividing line between conducting and insulated segments. The principle of displaced segments and a typical 'V-disc' encoder (with displaced segments) is shown in Fig. 159. This particular disc yields a binary-encoded decimal

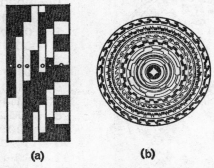

Fig. 159. Principle of V-disc encoder. (*a*) Displaced segments for binary 0100 to 1100; (*b*) Typical V-disc.

count in degrees of angular shaft position. Other disc-type A–D converters are available that produce coded sine–cosine outputs of shaft position, or arbitrary functions.

'Digitizing' Voltage. In analogue computers special transducers change most physical variables into corresponding input voltages. To apply a voltage representing some physical variable to the input of a digital computer, it must be 'digitized', or converted to numerical form. A number of methods exist for doing this, a relatively simple one being illustrated in the block diagram opposite (Fig. 160).

The analogue input voltage is passed through an amplifier to one input of a comparator, which is pulsed by a 200-kc/s clock oscillator. This means that the analogue input voltage is being sampled 200,000 times each second, at the frequency of the clock oscillator. The second input to the comparator is the feedback (trial) voltage, which is generated in the following manner. An electronic forward–backward counter (i.e. one that can either increase or decrease its count) generates a series of trial numbers, which are converted into proportional voltages with the aid of a digital-to-voltage conversion network. (We shall see later how this is done.) Each of these trial voltages is compared at the clock rate to the unknown analogue voltage in the comparator. If the unknown analogue input is greater than the trial voltage the clock pulse from the comparator switches a logical decision unit (gate) to provide an output that drives the forward–backward counter one count in the forward direction. The increased count, in turn, increases the feedback voltage by one unit and the comparison is made again. If the unknown input voltage is still greater the count and feedback voltage is increased again during the next clock pulse, and so on, until equality is reached.

Similarly, if the unknown voltage is less than the trial value applied to the comparator, the clock pulse passed by the comparator switches the decision unit to drive the counter backward by one count, which decreases the feedback voltage by one unit. Comparison is made again during this and the succeeding

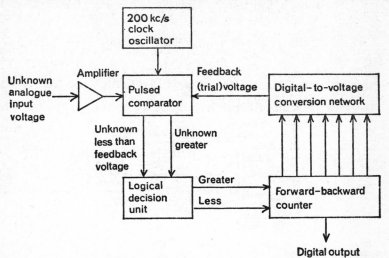

Fig. 160. Analogue-to-digital voltage conversion scheme used in Model A converter of Epsco Company.

clock pulses until equality is reached. The forward–backward counter, which is coupled to a mechanical counter, then displays the numerical value of the analogue input voltage. If the input voltage should change, the counter follows the variation at the 200 kc/s rate (one pulse every 5 microseconds) and continues to display the true voltage as long as the input voltage does not vary faster than the counter rate. Such a scheme permits direct visual read-out of a voltage in numerical form by a digital voltmeter. Various digital voltmeters

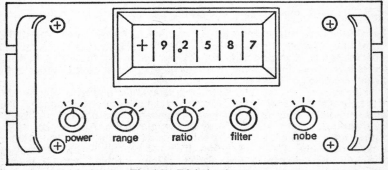

Fig. 161. Digital voltmeter.

employing different conversion methods exist. The front panel of a digital voltmeter is illustrated in Fig. 161.

Digital-to-Analogue Converters

Since the output of a digital computer is numerical in nature, it cannot be used directly for process control but must first be reconverted to analogue form. A whole class of digital-to-analogue (D–A) converters is available to do this job. To illustrate the principles involved, let us look at a few of the more important ones.

Digital-to-voltage Conversion. An electrical voltage can energize relays, operate switches, and actuate almost any controller you can think of. Generating a voltage proportional to the numerical output of a digital computer is, thus, one of the most important digital-to-analogue conversions required. In the process of studying this type of conversion we shall also learn how to convert a number to a proportional time interval.

The block diagram of Fig. 162 illustrates a frequently used scheme for

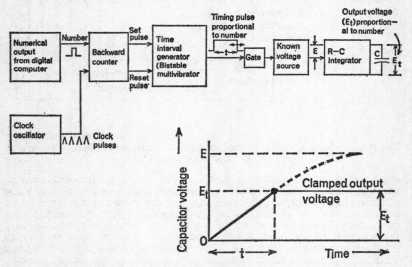

Fig. 162. Block diagram of digital-to-analogue voltage converter.

digital-to-voltage conversion. The conversion is performed in two steps. First a time interval proportional to the numerical (digital) output is generated; a capacitor is then charged during this time interval so that the voltage build-up across it is proportional to the time interval and, hence, to the output of the digital computer.

At the start of the time interval to be generated the numerical output pulse of the digial computer is set into a backward-counting electronic counter. This initial pulse also triggers a time-interval generator, which may be a bistable multivibrator, set to 'one' (high) by the counter pulse. Also applied to the counter are a series of clock pulses with known repetition rates, generated by a clock oscillator. The clock pulses drive the counter backward from the originally inserted number towards zero. Upon reaching zero the counter emits another pulse that resets the bistable multivibrator to 'zero', thus concluding the time interval. The result is a rectangular timing pulse, whose duration (width), 't', is proportional to the numerical output of the computer. If this number is supposed to represent a time interval (instead of a voltage) this first step is all that is needed to convert it to analogue form. If the number is to represent a voltage, however, a second step, described next, is required.

The timing pulse from the time-interval generator is applied to a gate that turns on a known voltage source. This voltage, E, is in turn applied to an R–C (or other) integrator (which we studied in the analogue section).

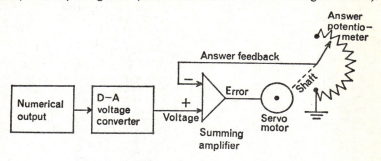

(a) By means of a shaft positioning servo

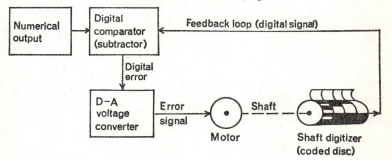

(b) By means of digital comparison (digital servo)

Fig. 163. Two methods of digital-to-shaft-position conversion.

Essentially, the voltage charges a capacitor linearly with time, so that the capacitor voltage is always a proportional function of the charging time. At the end of the given time interval, t, the capacitor is charged to a voltage, E_t, which is proportional to the charging time and therefore also to the numerical output of the computer. This voltage is 'clamped' by means of a diode so that it remains constant until the numerical output of the computer changes.

Digital-to-shaft-position Conversion. If the numerical output of the digital computer is to be represented by the angular position of a shaft some sort of positioning servo is required. This may be the standard type we discussed in the analogue section, or it may be a 'digital servo', which, as we shall see, has some advantages. First let us consider the standard servo arrangement, illustrated in Fig. 163 (*a*).

The numerical output of the computer is first converted into a voltage with a digital-to-analogue voltage converter, such as that illustrated in Fig. 162. The resulting voltage, which is proportional to the digital output, is applied to a summing amplifier, which forms part of a standard positioning servo. The amplifier 'error' output drives a servo motor, whose shaft in turn positions an 'answer' potentiometer. A voltage that is opposite in phase to the input voltage is picked off the wiper of the pot and fed back as 'answer' to the input of the summing amplifier. The amplifier sums the input and answer voltages, and when the two are equal the servo is 'nulled' with its shaft angular position proportional to the input voltage. Note that the feedback loop in this case [Fig. 163 (a)] includes only the positioning servo itself, but excludes the D-A voltage converter and the digital output.

For greater accuracy, some indication of the shaft position should be available in digital form, so that the 'error' can be compared directly with the digital computer output. A 'digital servo', which permits this digital comparison, is illustrated in Fig. 163 (b). Here a digital-to-analogue converter directly drives a motor whose output shaft is equipped with a 'shaft digitizer'. The shaft digitizer is a binary-coded disc, such as that shown in Fig. 157, which converts the shaft position to digital (numerical) form. The resulting digital signal is fed back to a digital comparator, which consists of a subtractor (complementing adder) in this case. Also fed to the comparator is the numerical output of the computer. The comparator computes the difference between these two inputs, and depending upon which is larger, emits a positive or negative digital error pulse, which is applied to the D-A voltage converter. The error signal from the converter, thus, is proportional to the digital error between the computer output and the shaft position. This error signal drives the motor and output shaft until the error becomes zero, at which time the angular position of the shaft is proportional to the numerical output of the computer. Note that the feedback loop in this case includes the digital portion of the system, thus assuring greater accuracy than is possible with that shown in Fig. 163 (a).

Complete Digital Process Control System

Although few real-time digital process control systems have been put into operation so far, the basic arrangement of such a system can be shown by a simple functional block diagram, as illustrated in Fig. 164.

The various analogue variables of the process to be controlled are sampled in sequence by automatically operated channel selectors under the control of a digital computer. The sample from the selected process variable is applied to a suitable analogue-to-digital converter, whose output is fed to the input of a digital computer. A number of converters are available for various process parameters. The digital computer is a standard general-purpose unit, the input-output sections being omitted in Fig. 164, for simplicity. The computer performs operations on the numerical outputs of the A-D converters, in accordance with a stored program designed to optimize the controlled process. The results are converted back to analogue form by digital-to-analogue converters at the output of the computer. Various control functions —throwing switches, turning shafts, etc.—are then performed by the analogue outputs of the D-A converters. The performance of the process itself in accordance with the programmed objectives provides error signals, which are fed back to the input of the system. The interrogation, or sampling, of the process variables by the channel selectors need not necessarily be sequential, but varies according to the process and the computer program.

An example of complete computer control of a chemical process is given in Fig. 165, which illustrates the flow diagram of an ammonia plant. Control of all important process variables is accomplished by a digital computer, in conjunction with input and output transducer cabinets, which house the process selectors, A–D and D–A converters, and associated equipment. (The computer control portion is enclosed by a broken line in Fig. 165.) Essentially, the computer stores a mathematical model of the ammonia-making process and continuously computes all important variables to assure optimum performance.

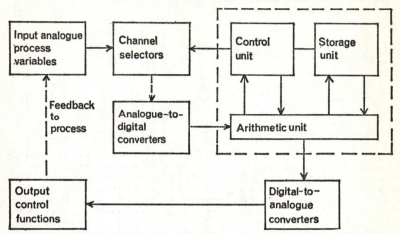

Fig. 164. Functional block diagram of digital process-control system.

Hybrid Computers

The foregoing illustrations of mixed analogue-digital techniques indicate that analogue and digital computers as completely separate breeds are probably on the way out. More and more analogue computers take on digital characteristics—such as storage of information in a 'memory'—while digital computers adapt themselves to real-time computations using analogue parameters. Computers of the future will undoubtedly borrow freely from both digital and analogue techniques, and may eventually become true 'hybrids', merging the characteristics of both types so completely that they are no longer separately identifiable.

COMPUTERS AND MAN

When Norbert Wiener defined the comparative study of automatic control, or 'cybernetics', in his now classic book of that name, he noted the many parallels between human behaviour and the response of automatic (feedback) control systems. We cannot perform the simplest physical action, such as raising our hands, without multiple feedback signals coming into play similar to those used to correct the performance of automatic machines. When we subject a person to conflicting stimuli he may suffer a nervous breakdown; similarly, an automatic computer with contradictory inputs may suffer a breakdown, with signals circulating endlessly in a schizoid pattern that results in no positive action.

Computers are the heart of most modern automatic control systems, and they are at least semi-intelligent mechanisms. As a result, there have come

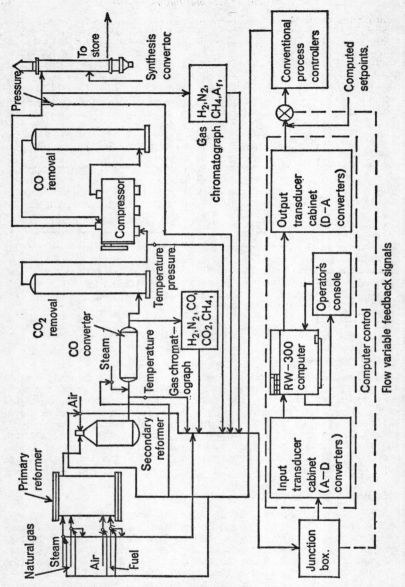

Fig. 165. Flow diagram of Monsanto Chemical Company's ammonia process using digital computer control. (Reprinted from *Control Engineering*, copyright by McGraw-Hill Publishing Co., Inc. All rights reserved.)

into being multiple relations between computers and human beings, and as computers become more intelligent, these relations take on a bilateral character. Not only do we teach the machine to imitate us in carrying out increasingly more complex tasks but we also begin to learn from the machine something about the organization and behaviour of the human organism. This statement may sound startling, but the fact is that computers play a key role in present-day biological and psychological investigations. Moreover, this role is not confined to data processing, but computers are actually used to imitate or duplicate the performance of 'self-organizing' biological systems such as 'artificial neurons' constructed of electronic parts. As increasingly larger aggregates of these complex units are built up, their behaviour will begin to resemble the patterns of the human brain, and we shall be able to gain an insight into its functioning not otherwise possible.

The relations between computers and man are, of course, much broader than those indicated above. A machine capable of taking over not only the menial tasks of man but also his most developed skills raises serious questions about man's future economic and social organization. If computing systems can perform all the tasks of production and free man from non-creative work, what economic organization will then be required to keep a growing population 'gainfully' employed? The spectre of technological unemployment brought about by this second industrial revolution looms larger than that envisaged by the economic theorists of our present economic system, and no ready answers are forthcoming. The corollary problem of the creative use of man's fast-growing leisure time will keep social scientists busy for some time to come. If these problems are successfully solved the technological revolution gives promise of a 'golden age' far exceeding any of the classic periods of antiquity.

We obviously cannot go into the economic and social implications of the computer-caused technological revolution in this study. However, we can trace briefly the scientific aspects of the computer-man system. Let us start with two relatively simple concepts, known as the adaptive principle and self-organization.

The Adaptive Principle

The adaptive principle applies to control systems. It simply means that a control system should be able to adapt itself to changing environmental conditions with the object of providing optimum performance for all conditions. To do this, an adaptive control system must continually measure all characteristics of the process being controlled and those of its own performance, then compare its performance with the desired optimum, and finally adjust its parameters to obtain this goal. Such a self-optimizing system would permit automatic control of processes whose parameters undergo wide fluctuations in value in an unpredictable manner. You might wonder how an adaptive system differs from a conventional closed-loop feedback control system, such as a servo, which also is provided with a self-correcting feature. The difference appears through the concept of optimum performance. A servomechanism simply carries out an order and makes sure that it does so correctly by 'zeroing out' the error between its actual performance and the given command. It makes no attempt to secure optimum performance. An adaptive control system, in contrast, continuously monitors the system's performance in relation to a desired optimum condition, or figure of merit, and automatically modifies the system parameters by means of closed-loop (feedback) action to approach this optimum.

Figure of Merit. As you might guess, the most crucial aspect of any adaptive control system is the figure of merit, or optimum performance index, against which the system compares and corrects its performance. The final response of an adaptive system can be no better than the criterion used as a figure of merit. This may be a measured, a calculated, or a theoretical quantity. The simplest situation occurs, of course, if the figure of merit is a directly measurable quantity, such as an engine pressure ratio, signal/noise ratio, etc. After a figure of merit has been chosen a method must be provided for measuring or calculating it, and finally a controller must be devised that will drive the system towards the desired optimum performance. All these steps involve large difficulties of mathematical and physical feasibility, system stability, complexity, and expense.

An Adaptive Controller. It is not surprising, in view of the difficulties outlined above, that few self-optimizing control systems have been built so far.

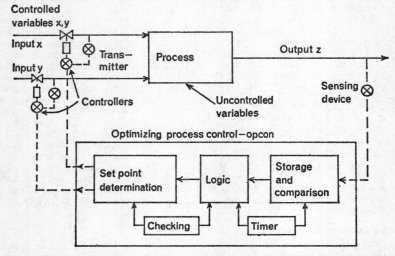

Fig. 166. Block diagram of process controlled by the OPCON (Optimizing Process Control) Controller of Westinghouse Electric Corporation. (Reprinted from *Control Engineering*, copyright by McGraw-Hill Publishing Co., Inc. All rights reserved.)

Most of the controls that do exist have been devised for a special purpose, such as flight control, and are not easily transferable to other processes. An example of an adaptive control system that is more general in nature and uses digital-computer techniques to control a number of variables, is illustrated in Fig. 166.

In the process illustrated only two input variables, x and y, are controlled and optimized by the OPCON controller. Other (uncontrolled) variables are also applied. A sensing device monitors the output, z, of the process and feeds an error signal to the input of the controller. The OPCON controller itself consists of transistorized digital-computer circuitry, including storage, comparison, logic, timing, and control circuits. The function of these circuits is to search in two dimensions for the optimum given by the stored figure of merit, and then adjust the controlled input variables automatically, in incremental steps, until optimum performance is attained.

Self-organizing (Biological) Systems

A step beyond adaptive (optimizing) control are self-organizing, or biological, systems which imitate almost-human behaviour. Dubbed 'bionics' by the Americans, such self-organizing systems (which really do not exist as yet) are modelled on biological organisms and are, in principle, capable of adjusting their own organization and structure in accordance with changes of the environment, based upon past experience and without advance information. The human eye and brain, for instance, have performed incredible feats of self-organization as a matter of course since the beginning of creation. If it were possible to duplicate such humanlike behaviour by a computing system a host of non-numerical problems could be tackled. Among these are the recognition of patterns of speech, the interpretation of pictures, faithful machine translation of languages, automatic reading and writing, and, perhaps, the free association of ideas to produce hitherto unknown thought combinations. As the English mathematician and computer pioneer Turing once expressed it, if a machine and a man in different rooms could ever carry on a conversation, and the man could not tell that he were conversing with a machine, then the machine could truly qualify as 'thinking'. While we have not as yet achieved this, we are well on the way to devising self-organizing mechanisms that can solve many non-numerical problems. Let us glimpse at a few of the intriguing ideas in this fascinating area.

The Artificial Neuron. At least a dozen research organizations are actively searching for a device that will perform the functions of a living nerve cell, or neuron, (which acts somewhat like an electronic gate). Since these functions are not completely known, this is not really possible as yet, but considerable progress has been made to justify the belief that an electronic model of a neuron can be devised. For example, Bell Telephone Laboratories has developed an electronic model of a neuron, consisting of conventional transistors, diodes, capacitors, and resistors. The model exhibits some of the characteristics of a living neuron; it delivers electrical impulses when stimulated. Like the neuron, the electronic model has an 'all-or-nothing' response and it exhibits fatigue. In one experiment, where a network of these artificial neurons was subjected to a light stimulus through a battery of photocells, the network distinguished specific patterns of bright and dark, similar to the eye's reaction to light. Similar experiments explore the nature of the hearing process.

The United States Air Force has an ambitious million-dollar programme in 'bionics', of which the search for an artificial neuron is an important part. An artificial neuron the size of a portable radio is being built, and will have 45 stimulating inputs and only one output, which increases or decreases with the stimulation. It is hoped that this unit can become part of a system for recognizing patterns and for discriminating between missiles and decoys during detection. Other uses are possible.

The Stanford Research Institute has proposed a 'neuristor' that should be able to do many of the things a neuron can do. Rather than being a single device, this is an extended circuit that will carry electrical stimulating pulses along a sort of transmission line to a branch point, along their respective branch lines until each is branched off again into two lines. As many branching points as needed can be connected together to simulate a 'sea of energy' in which pulses move about at constant speed. One important property of the proposed neuristor, similar to that of an actual neuron, is that any line along which an electrical impulse has passed becomes non-conducting and blocks the passage of another pulse along this line. Each line thus acts like a negative logic gate, which prevents pulses from nearby lines from passing through the gate to other lines. No one can tell at this point what theoretical understanding or practical use will come from the neuristor.

Interesting as it is, a single artificial neuron constructed of conventional components will tell us little about the behaviour of biological organisms or the human brain. What is needed is a tiny (molecule-sized), inexpensive neuron element, making it possible to connect together neural networks consisting of billions of these components. Molecular electronics may be able to supply such a component. Only when networks of billions of neurons are constructed will it become feasible to simulate and analyse biological organisms.

Miracle Property Filters. Not long ago the Massachusetts Institute of Technology (MIT) published a report, 'What the frog's eye tells the frog's brain', which opened up a new field of study. In essence, the report described that although the frog's eyes are very poor, it is able to pick out clearly the trajectory of a fly in which the frog is interested for obvious reasons. Thus, the frog may not be able to see many stationary things well, but it sees exactly

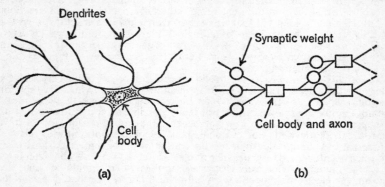

Fig. 167. (*a*) A nerve cell or neuron. (*b*) Network analogue.

what it wants to see: moving edible objects. Similarly, it was discovered that a frog cannot hear many of the sounds we are able to hear; but any frog can hear accurately another frog, which, after all, is the thing that matters most to him. Such miraculous properties are ascribed to special biological 'property filters', which screen out unnecessary information while allowing vital matters of existence to pass freely. It is believed that most living organisms have a large number of such special property filters, which permit them to make sense out of incoming information and avoid confusion. These filters are being studied with great interest by a number of researchers in the hope of being able eventually to duplicate them by electronic means. The use of such filters as a screen at the input of a computer is expected to simplify the job of data processing, especially in learning machines, to which we shall now turn briefly.

Learning Machines

The process of learning, one of man's self-organizing activities, has long intrigued psychologists. Recently, computer people have moved into the field to attempt the construction of learning devices that would incorporate some of the psychologists' findings, such as patterns of recognition, repetition, reinforcement, etc., involved in learning. The Cornell Aeronautical Laboratory constructed and demonstrated one successful, though limited, learning device known as the Perceptron. The Perceptron is essentially a pattern recognition (reading) and printing device that utilizes association of known patterns. The visual patterns or characters to be identified are converted by a matrix of photocells, known as stimulation, or S-units, into equivalent electrical signals in a retina (see Fig. 168). Several stimulating sources are randomly connected to each of a number of association, or A-units, which are activated if certain threshold values are exceeded. The association units, which were originally mechanized by servo-positioned potentiometers, such as the one shown for θ in Fig. 168, serve as the neurons of the system. An association unit that is activated by the summing of several stimulating signals triggers an appropriate response, or R-unit, which identifies the character.

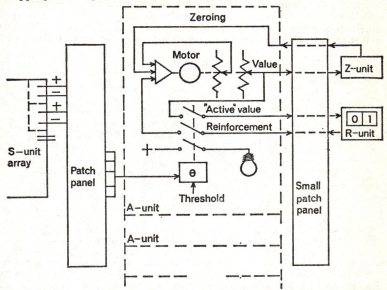

Fig. 168. Functional diagram of Mark I Perceptron, a learning machine built by the Cornell Aeronautical Laboratory. (Reprinted from *Control Engineering*, copyright by McGraw-Hill Publishing Co., Inc. All rights reserved.)

The Mark I Perceptron has only 512 motor-driven A-units, which is too small a number for recognition of sophisticated patterns. Presently under development is a magnetic association unit using a tiny ferrite core. Though this will permit the use of thousands of A-units, what is really needed is a machine with a 'billion neurons', as described earlier.

Data Processing of Operant Behaviour

Continuous automatic studies of the behaviour patterns of rats, monkeys, and human beings have yielded valuable results concerning the effects of drugs on animals and human beings. As part of the automatic behaviour studies, a special 'rat rotor' has been devised, in which eight white rats are cycled through a series of testing stations that provide access to a water dispenser and two or more bars. The test station also has an array of signal lights and other stimulators. In a typical experiment a rat may be conditioned to press bar A six times and bar B once, if it is to obtain a 'reward', consisting of a drop of water. The experiment is repeated after a drug to be tested has been administered. The rat's responses are recorded as relay openings and sometimes as the force exerted on the bars.

The relay outputs of thirteen such rat rotors are digitized, recorded on magnetic tape, and passed through a special-purpose computer, which analyses the data and converts it into punched-card format suitable for a general purpose computer (see Fig. 169). The experimental results are also displayed

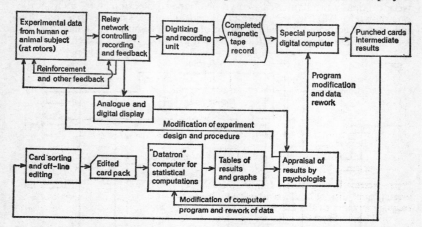

Fig. 169. Flow of data in Schering Drug Company's Automatic Operant Behaviour Laboratory. (Reprinted from *Control Engineering*, copyright by McGraw-Hill Publishing Co., Inc. All rights reserved.)

on analogue and digital computing equipment and, after an appraisal by the psychologists, either the experiments, the computer programming, or both may be modified. Results are statistically evaluated by the computer.

Teaching Machines

Though they are still in the learning stage, digital-computer techniques are rapidly branching out into teaching. Everyone may soon have his own electronic tutor to help him come up with the right answers and reinforce correct responses. Many companies have already brought out a variety of teaching machines ranging from simple mechanical and electromechanical projection devices to fully-fledged, though small, electronic digital computers. The educational principles and programming upon which these machines are based differ as widely as the controversy that rages on the right kind of teaching. All the machines agree in first exposing the student to some sort of written material or explanation and then quizzing him by requiring him to make an active response. There is no agreement, however, on the type of material and sequence (programming) the student is to be taught and what kind of response is needed to check on his understanding. As shown in the accompanying chart (Fig. 170) taken from *Control Engineering* magazine, the disagreements stem from opposing viewpoints of educational psychology, concerning the type of skill to be taught and the kind of programs and answers required. In some machines the student writes in the answers and checks them against the machine; in others he pushes a multiple-choice button, and if he is right the machine places a new page or new card before him; if he is wrong he may be told why, or he may be told nothing at all.

A new teaching machine, called 'Digiflex', relies on the principle of the conditioned response to train future keyboard operators (of key punches and

Present Trends and Future Prospects 307

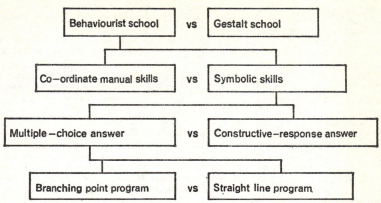

Fig. 170. The teaching machine controversy. (Reprinted from *Control Engineering*, copyright by McGraw-Hill Publishing Co., Inc. All rights reserved.)

typewriters) through their fingers instead of by conventional discussion and association. Fig. 171 illustrates the schematic of a Digiflex system built to teach operators of the US Post Office's semi-automatic mail-sorting machines.

The Digiflex system consists of a student's keyboard, which looks like the operator's portion of a mail sorter, an instructor's console that controls up to twenty students' stations, and a projector-programming unit. The student's keyboard has ten keys that exactly duplicate the movement of the mail sorter keys. The instructor's console has a lighted panel, through which the instructor can set the speed of presentation, watch for error lights that indicate the student's mistakes, and program the action of the keys in various modes. For

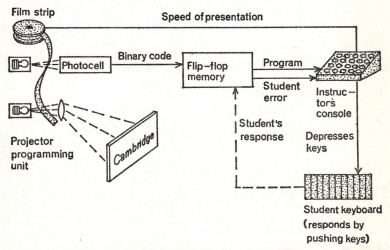

Fig. 171. A Digiflex Electronic Teaching Machine for training operators of mail-sorting machines. (Reprinted from *Control Engineering*, copyright by McGraw-Hill Publishing Co., Inc. All rights reserved.)

instance, after the student's interrogation the instructor may require the correct keys to be pushed up, the incorrect keys to drop away, he may perform both actions, or neither.

At the start of a training program the programmer-projector unit repeatedly flashes an alphanumeric representation, say 'Cambridge', on the screen before the student. At the same time a photoelectric cell reads a binary dot code for 'Cambridge' and transmits it to a solid-state flip-flop memory. The memory stores the information and transmits the signals associated with the proper sorting keys for 'Cambridge' to the instructor's console, where the mode of operation is selected. The instructor may decide to have the keys associated with the code for 'Cambridge' push up against the student's fingers while the rest of the keys drop away. The student then pushes the affected keys back by reflex action. The response of the student's fingers—pushing back the keys—actuates limit switches in the student station, which transmit signals back to the memory. There they are compared with the original code from the projected film and the appropriate action is taken. If the student has erred, a light flashes on the instructor's console. If his response was correct, the next representation is initiated.

Game-playing Computers

Automatic computers have been programmed to play many popular games of strategy, including billiards, draughts, and chess. Since computers have memories and can be taught the rules of any game, they can learn to play these games reasonably well. However, an expert can outclass any machine in highly skilled games, such as chess, which involve complex memory associations of varying aspects of the game, rather than simple logic. This will remain so until an associational computer with a 'billion neurons', resembling the human brain, is eventually built.

Language Translation Machines

Ever since 1954, general-purpose digital computers have been able to translate from one language to another, in a fashion. A computer can translate crudely if it is provided with an adequate dictionary of words (with multiple meanings) in its memory and also with the rules of syntax. The selection of the right meaning, correct prefix or suffix, and proper sentence structure, which is not always logical, are still the stumbling blocks which occasionally turn the translation of a piece—a poem, for example—into ludicrous hash.

NEW TECHNIQUES AND COMPONENTS

The computer world swirls with new concepts and terms—multiple-value logic, thin films, tunnel diodes, master programs, 'soft-wave' audio recognition, micro-miniaturization, packing density, integrated electronics, molecular electronics, and so forth. These new ideas, which are in various stages of being translated into 'hardware', indicate that we are at the very beginning of the computer era, with the best things still to come. Many of the new developments are concerned with increasing the speed and memory capacity of the computer; improving programming ease, input–output devices, and general computer organization; and drastically reducing computer size, weight, and power consumption. Before concluding this book, let us take a brief look at some of these *avant-garde* activities.

Computer Organization

The step-by-step computer organization we described in Chapters 12 and 13 was conceived in the 1940s, primarily by the mathematician John von

Neumann. Since the logic circuitry of that period could operate many times faster than either memory circuits or input and output devices, von Neumann devised the step-by-step method of computer operation, where each bit of information was processed in a stepwise fashion to permit the slow computer portions to keep up with the fast ones. To add two numbers, for example, the address of an operand has to be taken from storage, moved to a special register, the number to be added has to be fetched from memory and stored in another register, the addition is performed, the sum is returned to memory for storage, and so forth, as we have seen. Since input–output and memory speeds have recently almost caught up with logic speeds, the stepwise computer organization is fast becoming obsolete. The trend is towards purely parallel operation, which will allow a computer to work on different parts of a problem at the same time or on more than one problem at a time. Typically, the most recent computers are organized around two information channels, which carry information in and out. Between these channels as many computer elements (input–output, arithmetic, storage, control, etc.) as are needed are connected in parallel, so that a number of operations can be carried on simultaneously.

An even more radical approach is polymorphic design, in which all elements of a computer remain unconnected until the program is inserted. The program itself then determines the number of arithmetic modules, logic elements, etc., which are to be connected together to solve the problem.

Computer Logic

Binary logic came into computer usage partially because of the ease of mechanizing two-value (true-false) logic by bistable elements and partially because logic systems have been two-valued since Aristotle's days. Since the decimal system is used almost exclusively in all non-computer calculations, this has led to the binary coding of decimal numerals, which, you will recall, requires at least four bits for each decimal digit. The resulting waste is carried all through the computer, requiring many additional logic circuits and greatly increased storage capacity. However, there is nothing inherent in 'the laws of thought' that prevents a logical system from being multiple-valued, as long as it is used consistently. Thus, a logic could be three-valued, for example, with truth values of 0, 1, and 2, or -1, 0, and $+1$, corresponding to 'false', 'probable' and 'true', or whatever interpretation one wishes to place on the extra truth value. A number of three-state elements, such as magnetic cores, exist for mechanizing a three-value logic. Even a ten-value logic is feasible, provided reliable and inexpensive ten-state elements could be devised. Such ten-state elements would have considerable advantages over binary devices in the direct storage of decimal numerals. Interest is also increasing in what is called 'majority logic', where a logic gate produces a 'true' output only when two out of three input variables are in the same state. Such a system is valuable for ambiguous decisions that cannot be represented by a single YES or NO, and also to improve the reliability of computers in general.

Improved Storage (Memory)

Present-day memory storage systems are expensive and relatively slow. A great deal of attention is being given to developing inexpensive bistable devices with high switching speeds. We have already mentioned the superconducting cryotrons, which can switch extremely rapidly from a superconducting to a normal state by the application of a small magnetic field. In addition to these, various solid-state, photosensitive, and chemical devices, as well as improved wiring schemes, are under investigation to improve switching speed and storage capacity.

Tunnel Diodes. A new semiconductor (crystal) diode consisting of a single P–N junction has been invented by the Japanese scientist Esaki, which by far outperforms transistors in respect to high-frequency (gigacycle) operation; it has switching speeds of nanoseconds (10^{-9} s), extremely small size, low operating power, and high reliability. The increase in high-frequency and switching

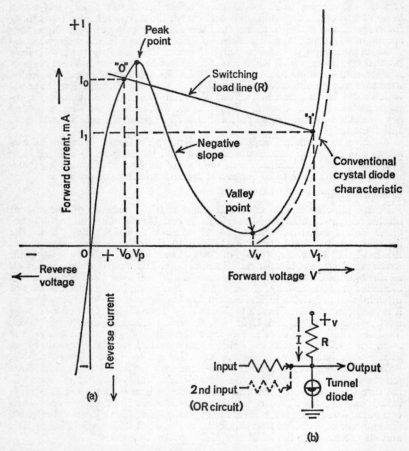

Fig. 172. (*a*) Tunnel diode characteristic with switching load line. (*b*) Schematic diagram of (OR) logic circuit.

capability comes about through a highly conductive, extremely narrow junction of P- and N-type germanium. Because of this extremely narrow junction, electrons can rapidly 'tunnel through' to the other side of the junction, though they do not possess sufficient energy to surmount the 'potential barrier', or wall, always present at such a junction. You can conceive of this effect in terms of a billiard ball rolling over a considerable hump in the table, although it has hardly been pushed at all and does not (or should not) have the energy to do it. Neither common sense nor classical physics can explain this surprising situation, but it is explained by quantum mechanics as 'quantum-mechanical tunnelling', a matter into which we cannot go here.

What occurs in a practical way is apparent from the characteristic of a typical tunnel diode, illustrated in Fig. 172 (a), which also gives the characteristic of a conventional diode for comparison. When a reverse (negative) bias voltage is applied to the anode of a conventional crystal diode, it will not conduct, while a tunnel diode will conduct. (Not that this bilateral characteristic is an advantage; it causes a great deal of trouble.) At low values of an applied forward (positive) anode voltage, V, the tunnel diode passes a considerable current, which reaches a peak at a low voltage, V_p, when the conventional diode has not even begun to conduct as yet. As the applied forward voltage is increased, the tunnel diode current starts to decrease again and reaches a minimum at the valley point for a voltage V_v. This decrease in tunnel current with increasing voltage causes a negative slope or negative conductance characteristic, which is typical for all tunnel diodes. It is this negative conductance characteristic that permits the tunnel diode to be used either as an amplifier, an oscillator, or a bistable switching device (gate or flip-flop). As the forward voltage (V) is increased further, beyond the valley point, the tunnelling effect ceases and the current increases in a manner identical to that of a conventional diode.

To use the tunnel diode as a bistable switching device, it is inserted in series with a voltage source (V) and a resistor (R), as is illustrated in Fig. 172 (b). The value of the resistor is so chosen that the currents flowing through it for the two switching conditions occur within the stable, positive-conductance portions of the tunnel diode characteristic. Graphically this is ascertained by drawing a 'load line' whose slope equals $1/R$ so that it intersects the characteristic [Fig. 172 (a)] at two points in the positive regions of the characteristic, before the peak and after the valley. The currents (I_0, I_1) and voltages (V_0, V_1) at these two switching points define the truth values 0 (false) and 1 (true), respectively. The diode can be switched from one stable state to the other by applying appropriate positive or negative signals through an input resistor. By applying several inputs in parallel, logical gating functions can be obtained. If any one of the input currents is sufficient to switch the diode an OR gate results; if simultaneous currents from all inputs are required to switch the diode an AND gate is obtained. Similar simple circuitry permits mechanizing a tunnel diode flip-flop or a free-running (astable) multivibrator. Tunnel diodes are particularly suitable for 'majority-rule' logic.

Thin Films. As the tunnel diode demonstrates, switching time and energy depend to a considerable extent on the thickness of the active material used in the switching device. For this reason, thin films are being developed by various groups for performing high-speed memory and logic functions. We have already mentioned the cryotrons, which combine the switching speed advantage of thin films with the principle of superconductivity. Thin magnetic films, consisting of small dots of a metal alloy (such as nickel–cobalt) evaporated on flat plates, are another promising material. The film dots act as tiny magnets, similar to ferrite cores, while providing a vast reduction in size and weight and a hundredfold increase in switching speed. Remington Rand Univac has developed a storage system constructed of such thin magnetic film memory planes, which can be switched in a few billionths of a second. IBM, preferring cryotrons, has developed a thin-film cryogenic memory frame the size of a postage stamp. Consisting of 135 cryotrons built up in 19 layers of material, the IBM memory plane combines storage and logic functions for 40 bits of information.

Ferroelectrics. Certain crystalline materials deform when a voltage is applied by displacing charged particles within the molecules. Applying a reverse voltage causes the charged particles to shift to another position. This behaviour, which is similar to the alignment of domains in a ferro-magnetic material, is known as ferroelectricity. When a ferroelectric crystal is placed into a circuit and a voltage is applied, the shift of charge results in a brief displacement current pulse. For storage purposes the crystals are placed in electric fields that deform them in two different directions, representing binary '0' and '1' respectively. By applying a field in the '0' direction for read-out, the crystals storing a '0' are unaffected, while those storing a '1' supply a current surge, or read pulse. Ferroelectric materials can be fabricated into highly compact printed-circuit memory matrices in the form of thin sheets.

Luminescent Phosphors and Photoelectric Materials. The search for better and faster switching circuits extends to many little known electro-chemical effects. Bistable chemicals that change colour when subjected to certain wavelengths of light, known as photochromic dyes, are being investigated for storage and high-speed printing applications. Also of great interest is a possible combination of the action of electroluminescent phosphors, which glow when a voltage is applied to them, with photoconducting materials (which allow a current to pass when illuminated). A combination of the two effects may be able to provide storage or logic functions.

Improved Input–Output Equipment

Present-day input–output equipment is in an elementary stage of development compared with potential capacities and operating speeds. Magnetic tapes are continually being improved in character packing density, number of channels per inch of width, and speed of character transfer. Tapes are likely to be superseded, however, by character-recognition devices at the input and display tubes coupled with photographic or electrostatic printers at the output. Optical or magnetic character recognition devices will read characters printed in ordinary writing or in special magnetic ink and convert them into machine language as fast as the computer can accept the information. Fluorescent or phosphorescent screens will read out data at electronic speeds and print it equally fast by use of photochemical processes, such as described above. Another distinct possibility, primarily investigated by the Japanese, is audio (phonetic) recognition devices, which will convert the human voice directly into a binary code suitable for the computer.

Data Communications with Central Processor. As computers become larger, faster, and more expensive, few business organizations can afford to buy or even rent them, especially if there is a considerable amount of idle or 'down' time. The answer for departments or branches of the same organization, or possibly even for separate organizations, is to buy a single large central processor, capable of handling the computing problems of all. Co-axial cables, microwave relays, or high-speed telephone circuits link the inputs of different departments or from the various offices with the central processor and supply them with data at rates exceeding several thousand words per minute. A master program permits all companies to process their data almost simultaneously. Actually, the central computer handles brief segments of each company's problem in turn, but at such speeds that processing appears to be continuous.

Even the new 'generation' of coming computers, however, is not quite big and fast enough to process certain types of problems (such as weather forecasting) in the brief time permissible. For maximum efficiency in these problems, computers are being linked together into networks that will permit each computer to solve portions of the problem at the same time.

Improved Programming Methods

Various schemes are being developed to utilize the computer to do much of the time-consuming work of developing instructions for each problem under the guidance of a master program. Such a master program for a given problem would permit the computer to write its own optimum program, employing libraries of sub-routines, prepackaged application programs, and pseudo-instructions. A pseudo-instruction is a single command instructing the computer to carry out a complicated operation, such as 'compute π to 200 decimal places'. The command simply refers the computer to its prepared sub-routine for this particular operation.

Programs themselves are already being written in synthetic programming language, such as ALGOL and COBOL, which resemble our ordinary language (see Chapter 13). A translating mechanism swiftly converts the synthetic instructions into detailed machine instructions in the binary code used by the machine. Micro-programming, mentioned earlier, may also assist in the conversion of simple pseudo-instructions into detailed computer steps. Future computers may no longer be equipped with complicated logic networks, such as adders or coders, but may have simple logic (switching) devices with small 'read-only' memories, which would be 'micro-programmed' to carry out any required complex function. Finally, as computer processing speeds increase, more emphasis will be placed on simultaneous operations, such as 'read-while-process-while-write' and on the multiple processing of several programs.

Micro-miniaturization (Integrated Electronics)

Micro-miniaturization—the concept of making electronic devices many times smaller and lighter than ever before possible or envisaged—originated with missile and space research, where rigid space and weight requirements made it an absolute necessity. Man's reach into space has become an incidental boon to the entire electronics industry, which not only profits by the vast reduction in size of electronic components, but by the attendant reduction in power requirements and increases in reliability, the latter being one of the reasons why the miniaturization programme was undertaken in the first place. Present-day electronic systems have become tremendously complex, consisting frequently of hundreds of thousands of separate parts and components, which leads to increasing system unreliability and the possibility of breakdowns. By reducing the number of components and connexions to fewer, highly reliable parts and compact 'integrated' circuits, micro-miniaturization promises to achieve marked increases in system reliability. Depending on the manufacturer and techniques used, the concept of micro-miniaturization goes under many names, such as integrated electronics, molecular electronics, micro-modules, micro-circuits, and others. Regardless of the name, however, you can get an idea of what is involved and may be achieved in the not-too-distant future from the fact that work is in progress on the preliminary design of computers composed of some hundred thousand million (10^{11}) active components, which are to be mounted on 100 one-inch square frames, with the entire assembly to be packed into a space of no more than one cubic inch.

Finally, there can be no doubt that research is going on at this moment which will render obsolete many of the advanced techniques described in this book. The computer era has hardly begun.

APPENDIX:
GLOSSARY OF COMPUTER TERMS

Italicized terms within individual definitions are defined elsewhere (in alphabetical order) in the Glossary.

A

ACCESS TIME—(1) The time interval required to communicate with the *memory* or *storage unit* of a *digital computer*. (2) The time interval between the instant at which the *arithmetic unit* calls for information from the *memory* and the instant at which the information is delivered. Also, the time interval between the instant at which the *arithmetic unit* starts to transmit information to the *memory* and the instant at which the storage of information in the *memory* is completed.

ACCUMULATOR—(1) The unit in a *digital computer* where numbers are totalled. (2) A *register* in the *arithmetic unit* of a *digital computer* where the results of an arithmetic or logical operation are first stored.

ADDER—A device that can form the sum, and if necessary the 'carry', of two or more quantities delivered to it, such as an *accumulator*, a differential gear assembly, etc.

ADDRESS—A name or number that designates the location of information in a *storage* or *memory* device; a source or destination of information.

ADDRESSED MEMORY—The sections of the *memory* in a *digital computer* where each individual register bears an address.

ADDRESSABLE REGISTER—A *register* that can be specified by means of an address.

ALPHABETIC OR ALPHANUMERIC CODING—A system of abbreviations, used in counting or preparing information for the input of a *digital computer*, in which numbers as well as letters are used (for example 0 to 9 and A to Z).

ANALOGUE—A physical quantity or measurement used to represent and correspond with a numerical variable occurring in a computation; a symbolic model. Usually contrasted with '*digital*'.

ANALOGUE COMPUTER—A computer which receives non-numerical, physical analogues as inputs and uses these to perform calculations by a process *analogous* to the one about which information is desired. A one-to-one correspondence usually exists between each numerical variable of the problem and the analogous physical variable used in the analogue computer.

AND CIRCUIT OR AND GATE—(1) A circuit or *gate* that performs the function of the logical AND. (2) A circuit with a multiplicity of inputs and a single output, which emits a binary *one* output signal only when all its input signals are *one*.

AND LOGIC OPERATOR—A logical (*Boolean*) operator which has the property that the statement '*A* AND *B* is true' (written *A . B*, or *AB*) holds only when both *A* and *B* are separately true; defined by *truth table*.

ARITHMETIC UNIT—The part of a *digital computer* which performs arithmetic and *logical operations* upon information.

ASYNCHRONOUS COMPUTER—An *automatic computer* in which any operation is started as the result of a signal that the previous operation has been completed (i.e. the elements are not synchronized on the same time base).

Automatic Coding—The automatic preparation in *code* or *pseudo-code* of a list of successive computer operations required to solve a specific problem.

Automatic Computer—A computer that automatically performs sequences of arithmetical and *logical operations* upon information.

B

Base or Radix—The total number of different digits used to form numbers in a *numbering system*.

Basic Cycle—The time required to complete a set of operations for the execution of each *instruction*.

Binary—Involving the integer two. The *binary number system* uses 2 as its base of notation.

Binary Cell—An element that has two stable states for storing a unit of information.

Binary Code—A system employing *binary digits* (*ones* and *zeros*) to represent a letter, digit, or other character in a computer.

Binary-coded-decimal Notation—A system of writing numbers in which each decimal digit of the number is represented by a *binary code*.

Binary Digit or Bit—A digit in the *binary scale of notation*. This digit is either zero (0) or one (1), representing 'false' or 'true', 'off' or 'on', 'no' or 'yes', respectively.

Binary Notation—The writing of numbers in which the positions of the digits designate increasing powers of two (from right to left). Thus, the first ten decimal numbers (0 to 9) in the binary scale of notation are 0, 1, 10, 11, 100, 101, 110, 111, 1000, and 1001.

Binary Point—The counterpart, in *binary notation*, of the decimal point.

Bit—A *binary digit*; the smallest unit of information; a single pulse in a group of pulses; a unit of *storage capacity*.

Block—A set of computer *words* arranged sequentially on magnetic tape.

Boolean Algebra—An algebra, named after George Boole (1815–64), which deals with logical propositions, classes, on–off circuit elements, etc., and employs logical operators, such as AND, OR, NOT, EXCEPT, NOR, etc.

Branch—One of two alternative paths of computation that a *digital computer* can follow at a given point of its program.

Buffer—(1) An isolating circuit used to avoid any reaction of a driven circuit (load) on the corresponding driving circuit. (2) An *OR-circuit*.

Bus—A path over which information is transferred in a *digital computer* from any of several sources to any of several destinations; also, a *channel*, line, highway, or trunk.

C

Capacity—(1) The number of digits or *characters* which may be regularly processed in a *digital computer*. (2) The upper and lower limits of the numbers that may regularly be handled in the computer. (3) The capacity, in *bits*, of a *store*; it is equal to the logarithm to the base two of the number of possible states of the device. (See also *Storage Capacity*.)

Card—(1) A *punch card*. (2) An assembly of *logic elements*.

Card Punch—A machine that punches cards according to a *program*.

Card Reader—A mechanism in a punch card machine that causes the information in *punched cards* to be read.

Cell—A memory location in a *digital computer* that stores one unit of information, usually either one *character* or one machine *word*.

Central Processor—The cabinet of a computer containing *arithmetic* and *control* circuits.

CHANNEL—(1) A path along which information, particularly a series of digits, or *characters*, or units of information, may flow or be stored. (2) That portion of a *storage* medium which is accessible to a given reading station, such as in magnetic tape or in magnetic drums.

CHARACTER—(1) One of a set of elementary marks that may be combined to express information. (2) A decimal digit, 0 to 9, or a letter, A to Z, or any other symbol (such as the keys of a typewriter) that a machine may take in, store, or put out.

CHECK DIGIT—One or more digits (*parity digits*) carried along with a *machine word*, which reports information about the other digits to permit checking for errors. (See also *Parity Check*.)

CIRCULATING MEMORY—A device in a *digital computer* using a *delay line*, which stores information in a circulating train of pulses or waves.

CLEAR—To replace *information* in a *register* by 'zero' in the number system used. In general, to reset any bistable device, such as a *flip-flop*, to its initial state.

CLOCK RATE OR FREQUENCY—The speed or frequency at which the *master clock* oscillator produces control pulses, which schedule the operation of the computer.

CODE—A system of symbols for representing *information* in a computer and the rules for associating them; to express information in a code.

CODED PROGRAM—A *program* that has been expressed in a specific computer code.

COLUMN—(1) The place or position of a *character* or digit in a unit of *information* (*word*). (2) A position or place in a number corresponding to a given power of the radix (base) in the scale of the number system.

COINCIDENT CURRENT—A current resulting from the simultaneous application of two or more current pulses. Used to select a desired *register* in a coincident-current *memory*.

COMMAND—A pulse, signal, or set of signals that occur in a computer as a result of an *instruction* and which initiate one step in the process of executing the instruction.

COMPARATOR—(1) A circuit that compares two signals and provides an output to indicate agreement or disagreement. (2) A device for verifying agreement of two different transcriptions of the same information.

COMPLEMENT—To reverse the state of a storage device or of a control level (e.g. changing a *flip-flop* from the *one* to the *zero* state, or vice versa). Also, the *complement* of a number (base or base-minus-one).

CONDITIONAL JUMP OR TRANSFER INSTRUCTION—A *digital computer instruction* which, when reached in the course of the *program*, will cause the computer either to continue with the next instruction in the original sequence or to transfer control to another *instruction* or *routine*, depending on a condition regarding some property of a number or numbers.

CONTROL CIRCUITS—The circuits in a *digital computer* that affect the carrying out of the instructions in the proper sequence.

CONTROL REGISTER—The *register* or *counter* that stores the current *instruction*. Same as *Program Register*.

CONTROL UNIT—The portion of a *digital computer* that directs the sequence of operations, interprets the coded instructions, and initiates the proper signals to the computer circuits to execute the instructions.

CONVERTER—A device that changes information in one kind of machine language into corresponding information in another kind of machine language (for example, conversion from punch cards to magnetic tape).

COUNTER—A mechanism, or *register*, that can be reset to zero, which either totals digital numbers or adds binary 1 to a *column* of digital numbers.

CYBERNETICS—The comparative study of control, communication, and complex-information-handling machinery in higher animals and in the machine.

CYCLE—(1) The smallest period of time or complete action in a computer that is repeated in order. Some computers distinguish between *major* and *minor cycles*. (2) A shift of the digits of a number (or the *characters* of a *word*) in which digits removed from one end of the word are inserted in the same sequence at the other end of the word, in circular fashion.

D

DATA—(1) Meaningful combinations of symbols. (2) *Information*. (3) The *program*.

DATA PROCESSING—The handling, storage, and analysis of information in a sequence of systematic and logical operations by a computer (*data processor*).

DATA WORD—An ordered set of *characters* that has meaning and is stored and transferred by the computer circuits as a unit of information. Ordinarily, a data word has a fixed number of characters or digits.

DECIMAL NOTATION—The writing of quantities in the scale of ten.

DECISION ELEMENTS OR CIRCUITS—An element that performs *logical operations* (AND, OR, NOT, etc.) on binary digits representing 'true' or 'false'.

DECODER—A circuit network in which a combination of inputs is excited at one time to produce a single output, of, for example, binary code.

DELAY LINE—A device that stores information in a train of pulses or waves in a transmission medium with reflecting walls, such as mercury in a pipe, or in another acoustic or electrical medium.

DIFFERENTIAL ANALYSER—Usually an *analogue computer* (sometimes digital) designed particularly for solving many types of differential equations.

DIFFERENTIATOR—An *analogue* device, such as a resistance–capacitance network, whose output signal is proportional to the derivative of the input.

DIGIT—A symbol expressing an integral value ranging from 0 to $n-1$ inclusive in a scale of numbering to the base n. For example, in the scale of ten ($n = 10$) the digits range from 0 to 9; in the scale of two from 0 to 1.

DIGITAL—Using numbers expressed in *digits* in a scale of notation to represent the variables of a problem.

DIGITAL COMPUTER—A computer that operates with *information*, numerical or otherwise, with all variables represented in *digital* form, usually as ones and zeros.

DYNAMIC STORAGE—Storage of information in acoustic *delay lines*, magnetic drums, or in similar devices where the information is not always available instantly.

E

ELECTROSTATIC STORAGE—Storage of information in the form of the presence or absence of spots bearing electrostatic charges, usually on the screen of a cathode-ray tube.

ENABLE—To activate a circuit, gate, or similar device by means of a logic level, pulse, etc.

ERASABLE STORAGE—Storage media, such as magnetic tape, which can be *erased* and re-used.

ERASE—(1) To remove information from *storage* and leave the space available for recording new information. (2) To replace the binary digits in a storage device by binary zeros; equivalent to *clearing*.

ERROR—Loss of precision in a quantity; the difference between an accurate quantity and its calculated approximation.

EXECUTIVE ROUTINE—A *digital computer programming routine* designed to process and control other subordinate routines.

F

FAULT TIME (DOWNTIME)—Time when a computer is not operating correctly.

FEEDBACK—The return of a fraction of the output of a system or process to its input, either by addition or subtraction. Subtracting the returned fraction (negative feedback) results in self-correction or control of the process, and is used in amplifiers and *servomechanisms*. Adding the returned fraction (positive feedback) results in runaway or out-of-control process; used in oscillators.

FIXED-CYCLE OPERATION—See *Synchronous Computer*.

FIXED-POINT REPRESENTATION—All numerical quantities are expressed by the same specified number of digits, with the decimal or binary point located at the same specified position with respect to one end of the number. Contrasts with *Floating-point Representation*.

FLEXOWRITER—An electric typewriter that serves either as an *on-line* output device for printing out copy directly from the computer or as an *off-line* device for producing punched paper tape.

FLIP-FLOP—An electronic circuit having two stable states (bistable multivibrator), which can store one binary digit of information. Flip-flops may have *one input* and *one output*, such that each successive input pulse changes the output voltage from low to high, or high to low. Alternatively, a computer flip-flop may have *two inputs* and *two* corresponding *outputs* (*zero* and *one*) such that a signal is produced on either of the output lines only if the last input pulse received is on the corresponding input line. A flip-flop is set to *one* if its *one* output generates a *one* level and its *zero* output generates a *zero* level. The flip-flop is cleared (re-set to *zero*) if its *one* output generates a *zero* level and its *zero* output generates a *one* level.

FLOATING-POINT REPRESENTATION—System taking into account varying location of the decimal or binary point; consists of writing each number by specifying separately its sign (+ or −), its coefficient or significant digits, and the power (exponent) of the base.

FLOW CHART OR FLOW DIAGRAM—Graphical representation of sequence of *digital computer* programming, using symbols to represent operations.

FOUR-ADDRESS INSTRUCTION—Each complete *instruction* specifies the operation and the addresses of four *registers*, usually those of three *operands* and that of the next instruction.

G

GATE, ELECTRONIC—(1) An electronic circuit having a multiplicity of inputs and one output, so designed that the output is activated only when certain input conditions are met. (2) A circuit having one output and two inputs so designed that a pulse appears on the output only if some specified combination of pulses occurs on the two input lines. See also *AND-gate* and *OR-gate*.

H

HALF-ADDER—A circuit that performs part of the function of binary addition by obtaining the sum of two numbers and producing a 'carry', but which cannot take account of the 'carries' from lower-order columns. Full binary addition can be accomplished with two half-adders.

HOLD—To retain the information contained in one *storage* device after copying it into a second storage device; opposite of *clear*.

I

INFORMATION—A set of marks that has meaning or that designates one out of a finite number of alternatives.

INHIBIT—A signal which prevents a circuit, gate, or other device from being triggered or activated.

INPUT UNIT OR EQUIPMENT—The equipment used for taking *information* into a computer.

INSTRUCTION—A *machine word* or a set of characters in *machine* (artificial) *language* that directs the computer to take a certain action. Part of the *instruction word* specifies the *operation* to be performed and another part specifies one or more *addresses* that identify particular locations in *storage*. *Note:* The term 'instruction' is preferred to the terms 'command' and 'order'.

INSTRUCTION CODE—An artificial language consisting of symbols, names, and definitions for describing *instructions* that are directly intelligible to, and can be carried out in sequence by, an *automatic computer*.

INTEGRATOR—A device in an *analogue computer* whose varying output is proportional to the integral of a varying input quantity.

INTERNAL MEMORY OR STORE—The total *memory* or *store* that forms an integral physical part of the computer and is directly controlled by it.

ITERATION—The process of taking an approximate computer result, representing a first guess, and using it to obtain a better approximation, which is used, in turn, to obtain a still better approximation, and so forth, until the desired accuracy is obtained.

ITERATIVE LOOP—A means by which a desired computer operation is performed repeatedly.

J

JUMP—An *instruction* or signal in a *digital computer*, which conditionally or unconditionally specifies and directs the computer to the next instruction. A jump is used to alter the normal sequence of instructions in a *digital computer*. See also *Conditional Jump*.

K

KEY—A set of *characters* used to identify and select information or a computer operation.

L

LATENCY—The delay in a *digital computer* while waiting for information called for from the *memory* to be delivered to the *arithmetic unit*.

LEVEL—A voltage with a time duration equal to or greater than the interval between two clock pulses. Used as input to logical *gates*.

LIBRARY—A collection of standard *routines* and *sub-routines* through which many types of problems and parts of problems can be solved.

LINE PRINTER—A machine that prints a whole line of characters at one time under computer control, usually with one type bar for each character space in the line.

LOCATION—A storage position, or *register*, in the main *internal memory* that permits storing one *computer word*.

LOGICAL DESIGN—(1) The planning of a computer for handling logical and mathematical interrelationships. (2) The design (synthesizing) of a network of *logical elements* to perform a specified function; frequently called the *logic* of the system, machine, or network.

LOGICAL ELEMENT—The smallest building blocks in a *digital computer* that can be represented by operators in a system of *symbolic logic*; for example, *AND-gates*, *OR-gates*, and *flip-flops*.

LOGICAL FUNCTION—A basic expression of *Boolean algebra*, such as NOT, AND, OR, etc.

LOGICAL OPERATION—An operation on a number, or on two or more numbers having the same number of digits, the operation being such as to result in a single number the *r*th digit of which depends only on the *r*th digit of each operand.

LOOP—Repetition of a group of *instructions* in a *routine*. See also *iterative loop*.

M

MACHINE LANGUAGE—*Information* in a physical form that a computer can handle, such as punched paper tape.

MACHINE WORD—A unit of *information* of a standard number of *characters*, which a *digital computer* regularly handles in each *transfer*. See also *instruction* and *word*.

MAGNETIC CORE—A form of *storage* in which *information* is represented by the direction of polarization (magnetization) of a magnetically permeable, wire-wound core.

MAGNETIC DISCS—A memory element consisting of discs made of magnetic material.

MAGNETIC DRUM—A rotating cylinder whose surface is coated with magnetic material on which *information* may be stored as polarized spots.

MAGNETIC HEAD—A small electromagnet used for reading, recording, or erasing small magnetized (polarized) spots on a magnetic surface.

MAGNETIC MEMORY—Any portion of the *memory* which makes use of the magnetic properties of a material to store *information*.

MAGNETIC TAPE—Tape made of paper, metal, or plastic, coated with magnetic material, on which *information* can be stored magnetically.

MAGNETIC WIRE—Wire made of magnetic material on which *information* can be stored in the form of polarized (magnetized) spots.

MAJOR CYCLE—The time interval between successive appearances of the same storage position in a memory device that provides sequential (*serial*) access to stored information (e.g. one rotation of magnetic drum).

MARGINAL CHECKING—A preventive maintenance procedure in which certain operating conditions are varied about their normal values, to detect incipient defective components.

MASTER CLOCK—The primary source of electronic timing signals that govern the sequence of computer operations.

MATHEMATICAL LOGIC—See *symbolic logic*.

MATRIX—A regular assembly of circuit elements (diodes, transistors, *magnetic cores*, etc.) designed to perform a specific logical function.

MEMORY—See *store*.

MEMORY CAPACITY—See *storage capacity*.

MEMORY CORE—A bistable magnetic device for storing *information*. See also *magnetic core*.

MERCURY MEMORY—*Delay lines* using mercury as the medium for storage of a circulating train of waves or pulses.

MICROSECOND—A millionth of a second.

MILLISECOND—A thousandth of a second.

MINIMUM-ACCESS PROGRAMMING—*Programming* in such a way that minimum waiting time is required to obtain information from the *memory*. Also called minimum-latency programming.

Appendix 321

MINOR CYCLE—In a *digital computer* using *serial transmission*, the time required for the transmission of one *machine word*.

N

NAND (NOT AND) GATE—A logical gate whose output is *zero* only when all its inputs are *ones*.

NON-ERASABLE STORAGE—Storage media, such as punched cards, that cannot be *erased* and re-used.

NON-VOLATILE STORE—Store that retains information in the absence of power, such as magnetic drums, cores, or tapes.

NOR GATE—A logical gate whose output is *one* only when all its inputs are *zero*.

NOTATION—A manner of representing numbers.

NUMERIC CODING—A system of *coding* in the preparation of *machine language* such that all *information* is represented by numbers.

NUMERICAL ANALYSIS—The conversion of a complex problem into a series of simple arithmetical steps, suitable for processing by a *digital computer*.

O

OCTAL DIGIT—One of the symbols 0 to 7, when used as a digit in numbering in the scale of eight.

OCTAL NOTATION—Notation of numbers in the scale of eight. Used in binary machines because octal numbers are easier to read than binaries, but nevertheless can be converted directly into binary digits.

ODD–EVEN CHECK—See *Parity Check*.

ONES COMPLEMENT—The complement of a *binary number* or *word* formed by changing all *one* bits to *zero*, and all *zero* bits to *one*.

OPERAND—Any one of the quantities entering into or arising from an operation. An operand may be an argument, a result, a parameter, or an indication of the location of the next *instruction*.

OPERATION CODE—(1) That part of an *instruction* that designates the kind of operation of arithmetic, *logic*, or *transfer* to be performed, but not the location of the *operands*. (2) The list of operations occurring in an *instruction code*.

OR-GATE OR -CIRCUIT—A *logical element* or device with a multiplicity of inputs and a single output, so that the output is activated ('true' or *one*) whenever one or more of its inputs are in the state prescribed for the logical 'true' or *one*; performs the function of the logical 'inclusive-OR'. Alternatively, a device that produces an output pulse whenever a pulse is present on one of its inputs.

ORDER—(1) Sequence. (2) Synonym for instruction.

OUTPUT UNIT OR EQUIPMENT—The equipment used for transferring information out of a computer, in an acceptable language.

OVERFLOW—In a *counter* or *register*, the production of a number that is beyond the *storage capacity* of the counter or register. The extra number may be held in an 'overflow element'.

P

PACKING DENSITY—The relative number of *information* units (bits) contained in a certain dimension of a storage medium; for example, the number of bits of polarized spots stored on magnetic tape per linear inch of tape.

PARALLEL OPERATION—The flow of information through a *digital computer*, or part of it, using two or more lines of channels simultaneously.

PARALLEL STORAGE—Storage in which all *bits*, *characters*, or *words* are equally accessible in time; contrasts with *serial shortage*.

PARITY CHECK—Use of a digit, called the 'parity digit', carried along as a check which is 1 if the total number of *ones* in the *machine word* is odd, and 0 if the total number of *ones* in the machine word is even (odd parity). *Even parity* uses the reverse conditions.

PERMANENT MEMORY—*Storage* of *information* that remains intact when the power is turned off; for example, magnetic drums.

PLOTTING BOARD—An *output unit* which plots curves of variables.

PLUGBOARD—A removable board holding terminals into which connecting cords may be plugged in a desired program pattern.

POINT—In a scale of notation the position designated with a dot that separates the integral part of a number from its fractional part. Called 'decimal point' in the scale of ten and 'binary point' in the scale of two. See also *Fixed-point System* and *Floating-point System*.

POSITIONAL NOTATION—A scheme for representing numbers by the arrangement of the digits, so that successive digits are to be interpreted as coefficients of ascending powers of the base (radix) of the number system.

PROGRAM—(1) A plan for the solution of a problem. A complete program includes plans for the transcription of data, *coding* for the computer, and plans for the effective use of the results. (2) A precise sequence of coded instructions, or *routine*, for solving a problem with a computer.

PROGRAM REGISTER—The *register* in the *control unit of a digital computer* that stores the current *instruction* of the *program* and controls the operation of the computer during the execution of that instruction. Also called '*control register*', 'control counter', or 'program counter'.

PROGRAM STEP—A step in a program, usually one *instruction*.

PROGRAM TAPE—The tape that contains the sequence of *instructions* to a *digital computer* for solving a problem.

PSEUDO-CODE—An arbitrary *code*, not related to the circuitry of a computer, which must be first translated into a computer code if it is to direct the computer.

PULSE—A sharp voltage change, or generally any sharp difference between the average height of a wave, representing its normal level, and its crest or trough, representing a high or a low level, respectively.

PULSE CODE—A set of pulses that carries a particular meaning.

PUNCH CARD—A card of uniform size and shape, suitable for carring a pattern of holes that has meaning and can be sensed mechanically by metal fingers, electrically by wire brushes, photoelectrically, and in other ways.

PUNCHED-CARD MACHINERY—Devices that operate with *punch cards*.

PUNCHED TAPE—Paper tape punched in a pattern of holes that convey *information*.

Q

QUANTITY—Numeric data. Mathematically, a positive or negative real number.

R

RANDOM ACCESS—Access to the memory or *store* under conditions where the next position (*register*) from which *information* is obtained is in no way dependent on the previous one, i.e. chosen at random. For example, access to the names in a telephone book is 'random access'.

RANDOM-ACCESS PROGRAMMING—*Programming* a problem for a computer without regard to the time of *access* to the *registers* holding the *information*.

Appendix 323

RANDOM-ACCESS TIME—The maximum time required for *random access* to a piece of *information*.
READ—To acquire *information*, usually from some form of store.
REAL TIME—In solving a problem, a speed sufficient to give an answer in the actual time during which the problem must be solved.
REAL-TIME OPERATIONS—Processing data in the time scale of a physical process so that the results are useful in guiding the physical process. Also, solving problems in *real time*.
REGISTER—A device capable of retaining information, often that contained in one *machine word*. See also *store*.
RELATIVE ADDRESS—The position of a memory location in a *routine* or *sub-routine*. Reference must be made to some specific address, such as the first word of a routine, so that the actual (absolute) address can be computed.
REPETITION RATE—The fastest rate of electronic pulses used in the circuits of an electronic computer.
RERUN—To repeat a *run* of a *program*, or a portion of it.
RESET—To return a *register* to zero; for example, to reset a flip-flop to zero.
RESOLVER—A device in an *analogue computer* for resolving a vector into two mutually perpendicular (sine and cosine) components.
ROUTINE—A set of coded *instructions* arranged in proper sequence to cause a computer to perform a desired operation, or to solve a problem. See also *program* and *sub-routine*.
RUN—One performance on a computer of a *program* consisting of one or more *routines*.

S

SCALE FACTOR—One or more factors used to multiply or divide quantities occurring in a problem and to convert them into a range suitable for a computer (such as the range from plus one to minus one).
SEQUENCE—A punch-card mechanism that will put items of information in sequence.
SERIAL OPERATION—The flow of information through a computer, or any part of it, using only one *line* or *channel* at a time. Contrasts with *parallel operation*.
SERIAL STORE—*Store* in which *words*, *characters*, or *bits* appear one after another in time sequence and in which the *access time*, therefore, includes a variable waiting (*latency*) time from zero to many word (character, bit) times. For example, magnetic drums are serial by word, but may be serial or parallel by bit, or serial by character and parallel by bit, etc.
SERIAL TRANSFER—A system of data transfer in which elements of *information* are transferred in time sequence over a single path.
SERVO FUNCTION GENERATOR—A servomechanism whose output shaft drives a usually non-linear potentiometer, whose wiper voltage is a function of the servo input voltage.
SERVOMECHANISM—A self-correcting, closed-loop *feedback* control mechanism that carries out a desired command, usually by some sort of power amplification. In analogue computers a *positioning servomechanism* positions an output shaft in accordance with a single, or the sum of several, input signals, while an *integrating servomechanism* rotates at a rate (velocity) proportional to an input signal, so that the total number of its shaft revolutions represents the time *integral* of the input function.
SERVOMULTIPLIER—A *servomechanism* whose output shaft drives a potentiometer, so that the wiper voltage is the product of the potentiometer input voltage and the potentiometer setting, or servo input voltage.

SET—To switch a bistable device from the *zero* state to the *one* state.
SHIFT—To move the characters of a unit of *information* from one column to another, right or left. In the case of a number, this is equivalent to multiplying or dividing by a power of the base of the number system.
SIGN DIGIT—A 1 or a 0 used to designate the algebraic sign (+ or −).
SIMULATION—The representation of a physical system by a computer, or model and associated equipment.
SINGLE ADDRESS (One Address)—A system of *digital-computer programming* in which each complete *instruction* includes an operation and specifies the location of only one *register* in the *memory*. This register may contain either the destination of a previously prepared result or the location of the next instruction.
SONIC DELAY LINE—A *delay line* which uses pulses moving in an acoustic medium (sound pulses), in contrast to an *electrical delay line*, which utilizes electrical pulses in a wire or assembly of coils and capacitors.
STATIC STORAGE—*Storage* of *information* fixed in space and available any time the power is on; for example, flip-flops, magnetic cores.
STORAGE CAPACITY—The amount of information that can be retained in a *store* (memory) unit, expressed either as the number of standard *words*, *characters*, the number of decimal digits, or the number of binary digits (bits). Also called 'memory capacity'.
STORAGE REGISTER—A *register* in the *store* (memory) unit of the computer, in contrast with a register in one of the other units.
STORE—(1) Any device into which *information* can be inserted and held, and then extracted at a later time. (2) In a computer the unit which holds or retains items of *information*. Also called 'memory' or 'storage'.
SUB-ROUTINE—(1) A part of a *routine*; a short or repeated sequence of *instructions* for a computer to solve a part of a problem. (2) The sequence of *instructions* necessary to direct the computer to carry out a well-defined mathematical or logical operation.
SYMBOLIC LOGIC—Exact reasoning about non-numerical relations using symbols that can be used for calculation. Also called 'mathematical logic'. *Boolean algebra* is a branch of this subject.
SYNCHRONOUS COMPUTER—An automatic *digital computer* in which the performance of all ordinary operations starts with equally spaced signals from a *master clock*.

T

TAPE—*Magnetic tape* or *punched paper tape*.
TAPE FEED OR TAPE TRANSPORT—A mechanism that will feed tape to be read or sensed.
TEMPORARY STORAGE—*Internal storage* (*memory*) locations reserved for intermediate or partial results.
THREE-ADDRESS—A type of computer *programming* in which each complete *instruction* includes an operation and specifies the location of three *registers*.
TRACK—A single path containing a set of pulses in a *magnetic drum* or *magnetic tape*.
TRANSFER—(1) To transfer information from one register to another without modifying it; i.e. to copy, exchange, read, record, store, transmit, transport, or write data. (2) To transfer control of a computer from one instruction to another.
TRANSFER INSTRUCTION—An *instruction* or signal that conditionally or unconditionally specifies the location of the next instruction and directs the computer to it. See also *Jump* and *Conditional Transfer*.

TRUTH TABLE—A table for representing a logical expression and determining its truth value (1 or 0) for any combination of truth values of the basic elements of the expression.

TWO-ADDRESS—A type of computer *programming* in which each complete *instruction* includes an operation and specifies the location of two *registers*, usually one containing an *operand* and the other containing the result of the operation.

U

UNCONDITIONAL TRANSFER—An *instruction* that causes the following instruction to be taken from an out-of-sequence *address*; applies to a *digital computer* which ordinarily obtains its instructions *serially* from an ordered sequence. See also *Transfer Instruction*.

V

VARIABLE CYCLE OPERATION—Operation of an *asynchronous computer*.

VERIFIER—(1) A manually operated *punch-card machine*, which reports whether punched holes have been inserted in the wrong places in a *punch card* or have not been inserted at all. (2) An auxiliary device on which a previous manual transcription of data can be verified by comparing a current manual transcription of it.

VOLATILE MEMORY OR STORE—*Memory* or *store* in which the information vanishes when the power is turned off; for example, delay-line store.

W

WILLIAMS TUBE—A cathode-ray tube for electrostatic *storage* of *information*.

WORD—An ordered set of *characters* which has at least one meaning and is stored, transferred, or operated upon by the computer circuits as a unit. Also called '*machine word*' or 'information word'. A *word* is treated as an *instruction* by the *control unit*, and as a numerical quantity by the *arithmetic unit*.

WORD TIME—In *serially* stored words, the time required to *transfer* a *machine word* from one *store* to another.

WRITE—(1) To record *information* in a *register*, location, or other *store*. (2) To transfer *information* to an *output* medium. (3) To copy information from *internal* to external *store*.

Z

ZERO—The computer's conception of 'zero'. A computer may recognize two zeros: *positive* binary zero, represented by the absence of pulses or digits, or *negative* binary zero, applying to a computer operating with *ones complements*.

ZONE—(1) Any of the three top positions 12, 11, and 0, in *punch cards*. A punch in a zone position, in combination with a punch in one of positions 1 to 9, represents an alphabetic or special code character. (2) A portion of an *internal store* (memory) in a *digital computer* allocated for a particular purpose.

Index

Accumulator register, 246
Adaptive control, 301–2
Adders, binary, 199–204
Addition, algebraic, 251–53
   digital, 247–49
   mechanical, 36–39
   of voltages, 47–48, 77–79
Address selection, 241–45
Analogies, electrical–mechanical, 34
Analogue characteristics, 13
Analogue computers, 16–116
Analogue–digital converters, 115, 289–96
AND, logical, 150, 160
   gates, 174–78
   magnetic AND gates, 219–20
Arithmetic unit, 235, 245–51
   accumulator, 246
   adder, 246
   block diagram, 246
   complementer, 250
   fixed and floating point, 259
   multiplier, 253
   sign comparator, 251–52
Astable multivibrator, 192–95

Binary adders, 199–204
   full adder, 204
   half-adder, 200–4
Binary addition, 126–27, 203
Binary-coded disc, 292–94
Binary codes, 137–38
Binary counters, 197–99
Binary counting, 124
Binary division, 131
Binary multiplication, 130
Binary number system, 123–32
Binary subtraction, 127–30
Binary values, 124
Binary variables, combinations, 143–44
   functions of, 144–47
Bistable components, 124
Bistable multivibrator, 189–91
Boolean algebra, 140–72
Branch instructions, 270
   conditional, 271
Buffers, 279
Burroughs, William Seward, 8

Calculators, 6–9
Cam and follower, 44–45
Clock generator, 261
Closed-loop control system, 72, 100
Coding, 273

Coincident currents, 221
Complementer, 250
Computer block diagram, 106–9
Computer languages, 275
Computer logic, 140–72
Computer process control, 288–99
Computer word, 238–41
Conditional branch instruction, 271
Control transformer, 98
Converters, 279
   analogue–digital, 115, 289–96
   digital–analogue, 296–99
Cryotrons, 228–30

Decimal number system, 122–23
Decoders, 284–87
Delay lines, 225–28
De Morgan's rules, 162–64
Derivative, 25
Destructive readout, 217–18
Differential analyser, 16
Differential equation, 102–3
   of motion, 26
   second order, 107
   solution of, 109–10
Differential gear assembly, 38
Differential operators, 102–3
Differentiation
   definition, 25
   with operational amplifier, 67
Differentiator, operational, 67
   resistance–capacitance, 51–52
Digiflex, 306–7
Digital–analogue converters, 296–98
Digital characteristics, 12–13
Digital computers, 117–287
Digital differential analyser, 14
Digital voltmeter, 295
Diode AND circuits, 174–78
Diode OR gate, 182–83
Diode-transistor NAND gate, 187
Diode-transistor NOR gate, 185–86
Disc-and-wheel integrator, 40
Divider, servo, 86–87
Division, binary, 131
   by a constant, 39, 46–47
   digital, 256–59
   generalized, 40
   servo divider, 86–87

Electrical computing devices, 45–52
Electrical problem, 30–34
Electronic devices, 173–210
Electrostatic storage, 224

Index

Encoders, 281–83
Exclusive-OR, 157

Feedback, 55–57
  negative, 56
  positive, 56
Ferroelectrics, 312
Field intensity, 212
Fixed-point arithmetic, 259
Flight simulation, 110–14
Flip-flops, 189–91. *See also* Multivibrators
Floating-point arithmetic, 259
Flux density, 212
Full adder, 204
Function generation, diode, 92–96
  mechanical, 42–45
  photoelectric mask, 96–97
  servo, 87–89
  sine wave, 89

Half-adder, 173, 200–4
Harmonic synthesizer, 10
Hexadecimal notation, 136–7
Hunting, 82
Hybrid computers, 14, 299
Hysteresis loop, 213–15

Implicit function technique, 106–7
Information bus, 233
Initial conditions, 108
Input devices, 276–80
Instruction address, 264
Instruction formats, 264, 269
Instruction register, 264
Instructions, 263–72
  branch, 270–71
Integration, 26
  electrical, 49
  of functions, 41, 63–67
Integrator, ball-and-disc, 10
  bootstrap, 65
  disc-and-wheel, 40
  operational, 66
  perfect,
  resistance–capacitance, 49–51
  servo integrator, 83–84
Iterations, 270
Inverters, 180–81

Jumps, 271

Keyboard devices, 278
Key punch, 278
Kirchhoff's laws, 54

Laplace transforms, 104–5
Laws of rearrangement, 164–68
Logic, 140–72. *See also* Boolean Algebra
Logic chains, 197–208

Machine equations, 100–2
Magnetic cores, 215–18
  determining state of, 217
Magnetic-core logic, 218–20
Magnetic-core store, 220–22
Magnetic devices, 211–24
Magnetic drums, 222–24, 243–45
Magnetic heads, 222–23
Magnetic hysteresis, 213–15
Magnetic memory, 220–22
Magnetic storage, 220–22
Magnetic tapes, 224
Magnetic theory, 211–16
Magnetization, 212–16
Magnetostrictive delay lines, 227
Marchant calculator, 9
Mathematical logic, 140
Matrix, 220–21
Mechanical computing devices, 36–45
Mechanics problem, 27–30
Memory, 220–25, 236–45
  address, 237, 241–45
  locations, 237
  registers, 237
  storage capacity, 237–38
Mercury delay lines, 225
Microprogramming, 273
Micro-miniaturization, 313
Monroe calculator, 9
Multiplication, binary, 130–31
  by a constant, 39, 46, 60
  digital, 253–56
  generalized, 40
  servo, 84–86
Multiplier, servo, 84–86
Multivibrators, 187–97
  astable (free-running), 192–95
  basic, 187–89
  bistable (flip-flops), 189–91
  delay, 204–6
  monostable (one-shot), 195–97

NAND, logical, 156
NAND gates, 186–87
Negation, logical, 145–46, 181
Networks, passive, 45–52
  resistance–capacitance, 49
NOR gates, 184–86
NOR, logical, 152–53
Number systems, survey of, 122–39
  binary, 123–32
  binary-coded decimal, 137–38
  biquinary, 132–33
  decimal, 122–23
  hexadecimal, 136–37
  octal, 133
  unitary, 122
Numerical analysis, 272

Octal addition, 135
Octal multiplication, 135
Octal number system, 133

Ohm's law, 45
One-shot multivibrator, 195–97
ONE state, definition, 192, 217, 251
Operand address, 263–66
Operation code, 263–66
Operational amplifiers, 16, 54–71
   basic characteristics, 59–60
   feedback amplifier, 57–59
   integrator, 66
   solutions of equations, 109–10
   summing amplifier, 61–63
OR, logical, 153–54
OR gates, 181–84
   magnetic, 218
Oscillation, damped, 29
   frequency, 28, 30
   undamped, 29
Output devices, 280–81

Parallel addition, 249
Parallel division, 256–58
Parallel multiplication, 254–55
Parallel subtraction, 251
Parity bit, 240
Partial address, 266
Perceptron, 304–5
Permeability, 212–13
Photoelectric-mask generator, 96–97
Photoformer, 96–97
Potentiometer card, 84–85
Printers, 280–81
Program loops, 270
Program selection, 260
Programming, 260–75
   automatic, 273–75
   languages, 275
   micro-programming, 273
   random access, 273
   sub-routines, 275
   techniques, 272–75
Pulley and chain, 36
Pulse steering, 198–99
Punch-card reader, 278–79

Quartz delay lines, 227

Readers, 278–79
Readout, 217–18, 224
Register, 206–8, 237
Relays, 174
Repetitive computer, 144
Resolver, 89–90
Routines, 263–72

Sampling, 289–90
Scale changing, 60
Scale factors, 100–2
Serial addition, 247–49

Serial division, 258–59
Serial multiplication, 255–56
Serial operation, 233
Serial subtraction, 251
Servo. *See* Servomechanism
Servo divider, 86–87
Servo multiplier, 84–85
Servomechanism
   definition, 72
   direct-current, 81
   integrating, 83–84
   positioning, 77, 81
   velocity servo, 83–84
Shift register, 206–8, 219–20
Sign comparator, 251–52
Sign inverter, 253
Sine–cosine potentiometer, 90
Sine wave, 29
Single-address instruction, 256–66
Soroban, 5
Square hysteresis loop, 215
Store, delay line, 225–28
   electrostatic, 224–25
   magnetic, 220–24
   See also Memory
Sub-routines, 275
Subtraction, algebraic, 251–53
   binary, 127–130
   digital, 249–51
   mechanical, 36–39
   of voltages, 77–79
Summing amplifier, 61–63
Summing network, 47
Switching logic, 140–43
Synchro control transformer, 98
Synchro receiver, 98
Synchro transmitter, 98

Teaching machines, 306
Thin films, 311
Transformers, 47
Transistor multivibrator, 187
Transistor NAND gate, 186–87
Transistor NOR gate, 184–86
Transistor NOT gate, 181
Transistor OR gate, 183–84
Transistors, 178–80
Triode inverter, 180–81
Truth tables, 142–60, 168–72
Tunnel diodes, 310–11
Two-address instruction, 265
Two-phase a.c. motor, 79

Venn diagrams, 147–60

Write-in, 220–22

ZERO state, definition, 192, 217, 251